SPACE-TIME WIRELESS CHANNELS

ISBN 0-13-065647-X
90000
9 780130 656476

Prentice Hall Communications Engineering and Emerging Technologies Series

Theodore S. Rappaport, *Series Editor*

DOSTERT *Powerline Communications*

DURGIN *Space–Time Wireless Channels*

GARG *Wireless Network Evolution: 2G to 3G*

GARG *IS-95 CDMA and cdma2000: Cellular/PCS Systems Implementation*

GARG & WILKES *Principles and Applications of GSM*

HAĆ *Multimedia Applications Support for Wireless ATM Networks*

KIM *Handbook of CDMA System Design, Engineering, and Optimization*

LIBERTI & RAPPAPORT *Smart Antennas for Wireless Communications: IS-95 and Third Generation CDMA Applications*

PAHLAVAN & KRISHNAMURTHY *Principles of Wireless Networks: A Unified Approach*

RAPPAPORT *Wireless Communications: Principles and Practice, Second Edition*

RAZAVI *RF Microelectronics*

REED *Software Radio: A Modern Approach to Radio Engineering*

STARR, CIOFFI & SILVERMAN *Understanding Digital Subscriber Line Technology*

STARR, SORBARA, CIOFFI & SILVERMAN *DSL Advances*

TRANTER, SHANMUGAN, RAPPAPORT & KOSBAR *Principles of Communication Systems Simulation with Wireless Applications*

SPACE-TIME WIRELESS CHANNELS

Gregory D. Durgin

PRENTICE HALL PTR
UPPER SADDLE RIVER, NJ 07458
WWW.PHPTR.COM

Library of Congress Cataloging-in-Publication Data

A catalog record for this book can be obtained from the Library of Congress

Editorial/Production Supervision: *Nick Radhuber*
Acquisitions Editor: *Bernard Goodwin*
Editorial Assistant: *Michelle Vincente*
Marketing Manager: *Dan DePasquale*
Manufacturing Buyer: *Alexis Heydt-Long*
Cover Design: *Talar Boorujy*
Cover Design Director: *Jerry Votta*

PRENTICE HALL PTR

Publishing as Prentice Hall PTR
One Lake Street
Upper Saddle River, NJ 07458

Prentice Hall books are widely used by corporations and government agencies for training, marketing, and resale.

The publisher offers discounts on this book when ordered in bulk quantities. For more information, contact Corporate Sales Department, phone: 800-382-3419; fax: 201-236-7141; email: corpsales@prenhall.com
Or write: Corporate Sales Department, Prentice Hall PTR, One Lake Street, Upper Saddle River, NJ 07458.

Printed in the United States of America

10 9 8 7 6 5 4 3 2 1

ISBN 0-13-065647-X

Pearson Education LTD.
Pearson Education Australia PTY, Limited
Pearson Education Singapore, Pte. Ltd
Pearson Education North Asia Ltd
Pearson Education Canada, Ltd.
Pearson Educación de Mexico, S.A. de C.V.
Pearson Education—Japan
Pearson Education Malaysia, Pte. Ltd

To my parents,
Dave and Lisa Durgin

CONTENTS

PREFACE

Let me begin by saying that without my friends David A. de Wolf, Gary S. Brown, and Theodore S. Rappaport, this book would have never happened. Professor de Wolf, besides being the man who introduced me to scholarly research, has proofed much of the mathematical content in my work and has been a great collaborator during my time at Virginia Tech. Professor Brown taught me most of what I know about electromagnetics; I borrow much of his notation from well-crafted lectures on rough surface scattering and analytical propagation analysis. Professor Rappaport – my principle graduate advisor – has been a true friend by encouraging this project and giving me a first-rate graduate student experience at Virginia Tech's Mobile & Portable Radio Research Group.

Back in 1998 I was sitting through a presentation made by an elder statesman of radio, a very accomplished and respected engineering professor. The presentation included many wireless channel measurements. About halfway through the talk, an intense academic discussion (i.e., argument) broke out between the professor and his colleagues in the audience. An endless volley ensued about the nature of the fading observed in the measurements. As a lowly graduate student, I just took notes quietly in the back of the room. I observed that the argument – which was left unresolved – was not a problem in understanding, but in semantics. The arguing researchers were trying to describe a space–time wireless channel using archaic conventions.

These researchers – experts in narrowband analog communications – were desperately trying and failing to describe the radio channel experienced by mobile, broadband digital radios with antenna arrays. I got the impression that the field of channel modeling needed to be reworked to accommodate all these new, sophisticated space–time concepts in wireless. At the end of the presentation, I wrote down the following analogy: "Frequency is to delay, as time is to Doppler, as space is to wavenumber." I left that presentation with a great topic for a Ph.D. dissertation.

I began writing my dissertation as if it were a textbook in space–time channel modeling, not really believing that it would actually become that one day (a good lesson for other graduate students). Of course, that was a little too ambitious at the time, but there was enough content after my defense to justify pursuing a book after my graduate work. I took a one-year trip to the Land of the Rising Sun to complete what is now *Space–Time Wireless Channels.*

The goal of this book is the same as my Ph.D. work: to provide simple, cohesive concepts for understanding radio channels that fade randomly with respect to time, frequency, and space. And I wanted it to be a book that even *I* could read. This meant adding lots of pictures, gutting gratuitous mathematics, and inserting other understanding aides. In the process, I found that space–time wireless channels were not so difficult to understand, provided a few basic principles in other disciplines (communications, random process theory, and electromagnetics) are known.

My hope is that *Space–Time Wireless Channels* offers a great deal to both the novice radio engineer and the veteran wireless researcher. The text focuses on first principles in radio channel modeling; it does not provide the deepest treatment of all the signal-processing algorithms for space–time radios, since that type of discussion tends to multiply acronyms instead of genuine understanding. The book contains plenty of original material as well as new ways of looking at old problems. The seasoned researcher will notice the inclusion of many new concepts in channel modeling and characterization – and will also notice the intentional omission of others. I have avoided the temptation of turning this book into a "cut-and-paste" job, which so often constitutes engineering texts nowadays.

Since it contains problem sets and a pedagogic presentation of material, this book may be used in graduate or even undergraduate engineering courses. The book is also intended to be used by graduate students or industry engineers as a research aid or a self-study course. This book is written with wireless engineers in mind. Many colleagues have pointed out that space-time channel modeling theory applies to problems in optics, radar, acoustics, and imaging – to name just a few fields of study. I believe this text is useful to other engineers, physicists, or applied mathematicians, although I apologize to them in advance for all the references to wireless devices.

Combining disparate fields to synthesize a theoretical foundation creates all sorts of conflicts in notation. In fact, attempts to be consistent with the multiple conventions that exist in the research literature proved to be the most difficult part of writing *Space–Time Wireless Channels.* Although no desirable notation could be found, this book takes a "lesser-of-evils" approach to naming variables and functions in analysis.

(To underscore the notation difficulty, consider the convention of using R to describe the autocorrelation function of random processes. This notation conflicts with the convention for signal envelopes, so instead this book uses C to denote the autocorrelation function. But to describe the probability density function (PDF) of envelopes, we need a lowercase value of R to be the index of the PDF. However, r is commonly used to describe position in radial coordinate systems, so we defer to the Greek ρ for the PDF index. This move, however, conflicts with standard practice of using ρ to denote unit autocovariance of a random process, which becomes ϱ in this text. Without these precautions, there would have been ridiculous-looking functions such as $R_R(\Delta r)$. Do not get me started about ϕ.)

Much of the original research contained in this book was funded by a Bradley

Fellowship in Virginia Tech's Department of Electrical and Computer Engineering, ITT Defense & Electronics, and the MPRG Industrial Affiliates program. The completion of this manuscript was supported by the Japanese Society for the Promotion of Science (JSPS) in the form of a long-term fellowship for visiting researchers. And I cannot give enough thanks to my Japanese host professors, Dr. Norihiko Morinaga and Dr. Seiichi Sampei, and all of my great friends at Morinaga Laboratory in Osaka University.

大阪大学森永研究室の皆さん、どうも有り難うございます。
素晴しい一年でした。

I am also truly indebted to my long-time friends and officemates Neal Patwari and Hao Xu and to my friends David Wenzel and Jiun Siew for proofreading parts of the manuscript. I am also grateful toward the many colleagues at MPRG who assisted this effort with encouragement, support services, and feedback: Jason Aron, Chris Anderson, Keith Blankenship, Rich Ertl, Ran Gozali, Ben Henty, Kevin and Donna Krizman, Vikas Kukshya, Bror Peterson, Bruce Puckett, Cindy Reifsnider, Hilda Reynolds, Aurelia Scharnhorst, and Christopher Steger. Additional thanks to Dr. James Isaacs of ITT Defense & Electronics and to Dr. David Auckland of Etenna Corp.

Chapter 1

INTRODUCTION

There are few things in nature more unwieldy than the power-limited, space-varying, time-varying, frequency-varying *wireless channel.* Yet there is great reward for engineers who can overcome these limitations and transmit data through such harsh environments. The explosive worldwide growth of personal communications services through the 1990s is a testament to the business opportunities that result from conquering the wireless channel. However, given the emergence of newer wireless systems that require more and more bandwidth, the task of conquering the wireless channel is becoming more difficult. This task requires a thorough background in wireless channel modeling.

Furthermore, understanding and modeling the wireless channel provides beautiful insight into a number of other problems in the physical sciences. This book presents the basic principles for describing the random fading that plagues space–time wireless channels. Although most of the examples and discussions are in the context of commercial radio applications, it is possible to apply the theory to a wide range of problems in any field that involves dynamic wave propagation.

This chapter presents an overview of channel modeling history in the context of wireless communications and motivates the need for studying the full space–time wireless channel. This chapter includes

- Section 1.1: Historical context of wireless channel modeling.
- Section 1.2: Importance of the spatial channel interface.
- Section 1.3: Trends in wireless influencing channel modeling.
- Section 1.4: Content summary of this book.

Indeed, the current state of wireless communications points to a coming epoch when understanding the space–time channel is not a luxury, but an absolute necessity.

1.1 Perspectives in Propagation

To understand the importance of radio channel modeling, it helps to understand some history and background in the development of wireless communications. This section shows how the material presented in *Space–Time Wireless Channels* fits into the historical context of wireless technology development.

1.1.1 Early Years of Radio

The world's first digital radio system was actually the world's first radio system. Guglielmo Marconi's first wireless transmission in 1897 used Morse code (a digital representation of text) to communicate from ship to shore. He soon commercialized his technology by installing wireless systems in transatlantic ocean vessels. These Marconi wireless systems were first used to send distress calls to other nearby boats or shoreline stations, even in the famous luxury liner *Titanic*.

This first wireless system used a spark-gap transmitter, a glorified spark plug that sprayed electromagnetic waves in all directions at all frequencies. The spark-gap transmitter could be wired to send simple Morse code sequences, but the real challenge of the system was to receive the radio signal. For that, Marconi used a *coherer*, a device that could only detect the presence or absence of strong radio waves. This form of detection - coupled with the fact that only mechanical switching forms of signal amplification existed - meant that Marconi's wireless was only capable of digital transmission.

Note: What Is a Coherer?

A coherer is a glass tube that contains loose metal filings resting on the bottom, as illustrated in Figure 1.1. Two contact wires are placed on opposite ends of the tube, allowing an external apparatus to measure the overall resistance to electrical current through the tube. Normally, resistance across loose metal filings is large due to the loose, jagged contact points between the small shards of metal. If a strong electromagnetic wave (i.e., from a nearby radio transmitter) travels across the filings, they *cohere*; the overall resistance drops and the radio wave is detected. To repeat detection, a coherer must be shaken mechanically to return the filings to their uncohered state.

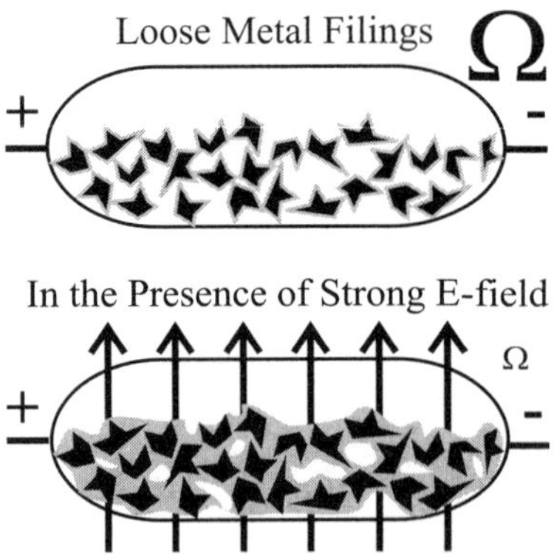

Figure 1.1 In the presence of a strong electric field, metal filings cohere and their overall electrical resistance drops.

The Marconi wireless was heavily limited in range and data speed by the power required to send and receive signals. However, radio communications - as well as every other electronic technology - changed in 1906 when Lee de Forest invented the first vacuum tube. The vacuum tube amplifies *analog* waveforms, so radio communication was liberated from its low-rate, on-and-off keying. It was now possible to transmit high-fidelity analog signals, such as voice and music, over amplitude modulation (AM). Commercial AM radio stations proliferated across the world in the 1920s, as marked on the timeline in Figure 1.2.

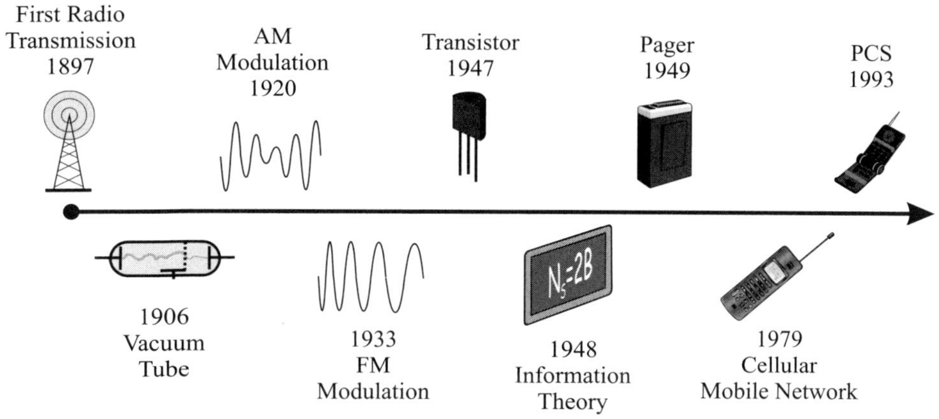

Figure 1.2 Some important milestones in radio communications.

The next great milestone in radio came in 1933, when E. H. Armstrong invented frequency modulation (FM). FM radio was the first example of signal-processing used to overcome the noisy, deleterious radio channel. In this case, the nonlinear modulation scheme of FM was capable of trading usable bandwidth for signal fidelity. For once, engineers could design radio links with a degree of freedom other than transmit power.

Many other wireless devices followed (television, military radios, radar, etc.), but perhaps the most important and sublime milestone occurred in 1948 with Claude E. Shannon's publication of his famous "A Mathematical Theory of Communications" [Sha48]. There are two extremely important principles outlined in this paper that revolutionized the design of communication links:

- All analog signals can be represented by sets of discrete digital symbols to a controllable degree of precision.

- The fundamental rate at which digital symbols may be sent through any channel is a function of bandwidth, signal power, and noise power.

In essence, Shannon's theory predicted that *digital* communications, rather than analog communications, was the best way to send data through any link. It was only a matter of time before most radio communications would use digital modulation. It turned out to be a *long* time, however.

1.1.2 Cellular on the Scene

Digital communications may have been *preferable* to analog communications, but with the technology of 1948 it was still not *possible.* Any type of digital communications requires discrete signal-processing operations that simply were not possible (commercially) with the vacuum tube technology of the day. Only with the advent of solid-state devices, such as the transistor in 1947, could engineers implement the signal processing required for digital communications. Even this was not possible overnight, as solid-state electronics had to undergo years of research and development before producing integrated circuits (IC) that were fast enough and cheap enough to implement signal processing. In the meantime, the wireless industry continued developing analog radios.

The modern cellular phone and paging industry was birthed in the post-Shannon period using analog radio technology. In 1949, Al Gross (the inventor of the walkie-talkie) introduced the first mobile pager, for use by hospital doctors. In 1979, the first commercial cellular phone market opened in Tokyo using a type of analog FM modulation to send voice signals to users. Similar systems in North America and Europe followed. By the late 1980s, analog cellular communications was a commercial success, and companies were pressing government regulatory agencies to open up new radio spectrum for more voice services.

Cellular telephony presented wireless engineers with a uniquely different design challenge. Previously, most radio systems operated in a *noise-limited* radio channel, where thermal noise was the sole source of signal degradation. As a mobile receiver moved in such a channel, fading would cause the signal power to fluctuate in space as the average noise power remained nearly constant; performance of the radio link depends on maintaining an adequate *signal-to-noise* ratio. Cellular networks, however, have *interference-limited* channels, because nearby cells reuse the same frequency spectrum. As illustrated in Figure 1.3, the interfering signal fluctuates in space along with the desired information signal. Careful spatial modeling becomes much more crucial for the noise-limited case.

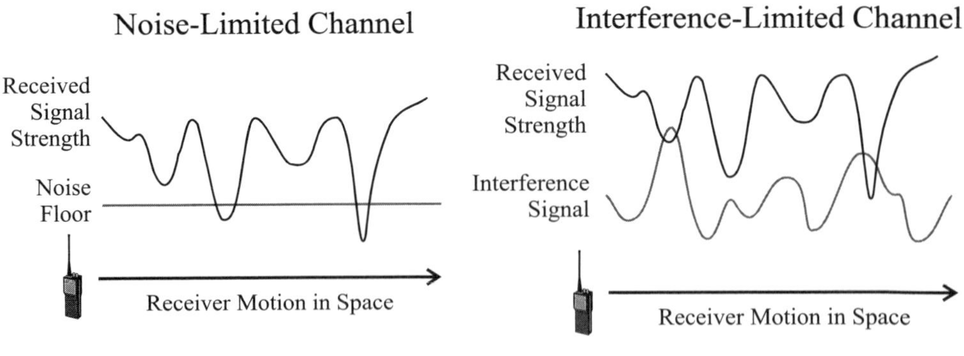

Figure 1.3 Fading for mobile communications causes sporadic moments of poor signal-to-interference+noise ratio (SINR) levels.

In 1993, the second generation of cellular telephone networks, called *Personal Communications Services* (PCS), were launched and rapidly spread throughout the world in just several years. Unlike their predecessors, these PCS networks were true digital communications systems, enabled by the cheap, fast, solid-state devices for signal-processing and radio-frequency (RF) electronics that became available at the end of the 1980s. Finally, the prophetic ideas of Shannon and digital information transmission had become a reality in the commercial wireless industry. In fact, the whole story of wireless communications is a story of great ideas followed by decades of incubation until the intellectual and industrial forces governing hardware development enabled implementation. Indeed, one must only look at recent vintages of proposed signaling techniques to understand the future of commercial wireless.

1.1.3 Origins of Channel Modeling

Most engineers are aware of the great inventions made at Bell Laboratories (the transistor, the laser, the communications satellite, to name just a few). Less appreciated, however, is the laboratory's enormous contribution to the theory of channel modeling and statistical communications analysis. The stochastic modeling work by Bell Laboratories researcher Stephen O. Rice stands out as one of the crucial achievements in describing radio communications. In 1944, Rice published a theory of random noise, which has since become a foundation of statistical analysis in communications [Ric44], [Ric45]. This work, originally used to characterize the noise in large-carrier AM and some FM signals, had extensive application for the description of fading signals, including level crossing rates, fade duration statistics, and the Rician envelope distribution that bears his name.

Still, the type of random processes described by Rice were signals with a single time dependency. In the early post-war period, there was little need for other types of analysis in radio communications. However, work began on the concept of mobile cellular telephony in the mid 1940s. As work progressed, it became obvious that newer channel descriptions were necessary - models that described the concept of *spatial multipath fading*. Unlike point-to-point microwave or satellite links, the cellular telephone user would operate receiver terminals buried within the clutter of a dense scattering environment, such as a cityscape or a neighborhood. In these types of propagation environments, objects such as terrain, buildings, and trees would block direct contact between a user terminal and a tower-top base station, but would instead provide a link between the two by scattering numerous low-powered radio waves from one to the other. These *multipath* waves arrive at the receiver from many different directions and, as illustrated in Figure 1.4, create pockets of constructive and destructive interference in space. As a receiver moves through space, it rapidly experiences the peaks and nulls of multipath fading, often losing a signal momentarily, even though there is a large amount of average signal power propagating through the immediate area.

Spatial fading is, perhaps, not a novel concept to most engineers. The first experience with spatial fading is usually encountered in a basic engineering or physics

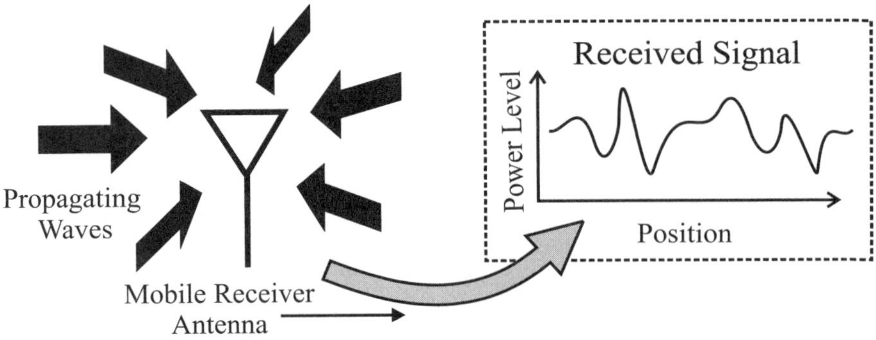

Figure 1.4 Small-scale fading for a mobile receiver in a multipath environment.

course on electromagnetism. The classic example is the transmission line, as shown in Figure 1.5. Recall that, unless the load is matched perfectly to the impedance of the transmission line, a wave sent down the line will partially reflect so that the net wave propagation is the superposition of two waves traveling in opposite directions. If we used a "mobile" power meter to measure the power along the transmission line, we would observe a standing wave interference pattern with peaks and nulls, each occurring at half-wavelength intervals.

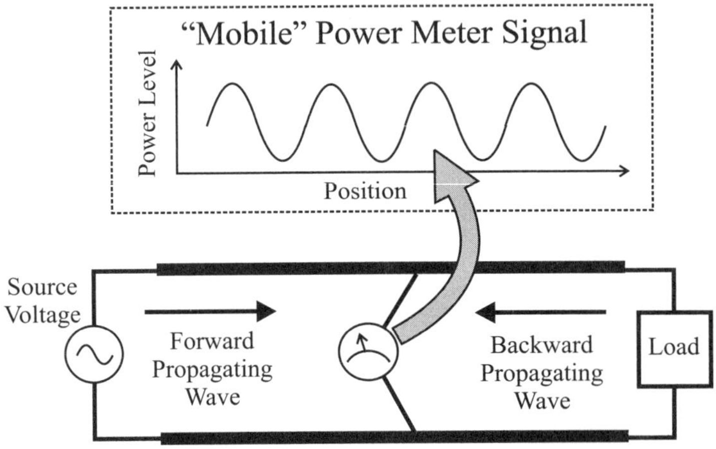

Figure 1.5 Signal fluctuation across a time-harmonic transmission line.

The transmission line example is a simple one. There are only two discrete waves. These waves, which are also scalar in the analysis, create a standing wave interference pattern that is regular and predictable. Contrast this to the case of multipath fading for a mobile wireless receiver. The multipath channel may have numerous waves of vastly differing magnitudes that obey *vector* wave propagation laws.

Yet the transmission line example is still a useful analogy. For example, as the frequency of the voltage source is increased, the wavelength of radiation decreases and the distance between peaks and nulls across the transmission line shrinks proportionally. The same effect is true of spatial multipath fading for a wireless mobile receiver. Modern cellular telephones operate in the upper UHF and microwave bands. In these bands, the wavelength of radiation is less than 1 m. At these wavelengths it is possible for a mobile receiver to receive a signal with a high-powered peak in one region of space and then, with just a few centimeters of movement, to receive virtually nothing.

As research on mobile telephony accelerated at Bell Laboratories during the 1960s, some important innovations in channel modeling were made. Researchers proposed the sum-of-waves model for spatial multipath [Oss64]. Clarke later extended this work to several basic scattering distributions, applying much of the Rician random process theory to the spatial fading for mobile receivers [Cla68]. Gans published a method for constructing a *Doppler spectrum* from the angles-of-arrival of multipath waves [Gan72]. Jakes published seminal work on the concept of *space diversity* - using multiple antennas to avoid deep signal fades [Jak71], [Jak74].

An important result in all of this research was the emergence of the omnidirectional Rayleigh fading model. This model assumes that radio waves arrive at the mobile receiver with equal power from all directions. This spatial channel models a fluctuating received signal strength with *Rayleigh statistics.* When a constant velocity is assumed by the mobile receiver, the model provides useful analytical expressions for channel coherence and fading statistics. This simple model is still a de facto standard in mobile radio system design.

1.1.4 Rayleigh Pessimism

The omnidirectional Rayleigh model is also unrealistic, but it is unrealistic in a way that endears it to engineers. It is a *pessimistic* channel model for conventional mobile receivers. There are two important characteristics of a mobile wireless receiver affected by small-scale channel fading: fade margin and update rate. The omnidirectional Rayleigh is useful for calculating both of these parameters.

Fade Margin

Fade margin is a critical parameter in the site design of a cellular radio system. The mobile handset requires a minimum signal-to-interference+noise ratio (SINR) to maintain the specified data rate and not drop calls. This minimum SINR is most difficult to achieve when the handset is operating on the fringe of a cell, where the distance between mobile unit and base station is the largest and received power is the weakest.

Because of small-scale fading, it is not enough to simply design a cellular network based on *average* power at the fringe of cells. If this were the case, a handset would drop a call at even the slightest signal fade, since the received signal power fluctuates over space. Instead, wireless engineers add anywhere from 12 dB to 18 dB of extra fade margin to the radio power link budget to ensure that small-scale fading does not drive the received signal below an acceptable level. On the business side of wireless, dB's translate into dollars, so the fade margin cannot be made arbitrarily high. The actual amount depends on the modulation scheme and the distribution of signal fades.

At this point, fading with Rayleigh statistics becomes a useful benchmark for a link design. Rayleigh statistics are considered by many to be a worst-case scenario of signal fading because the received signal strength experiences such deep fades (as we will see in Chapter 5). If a wireless receiver works in a Rayleigh fading channel, then it is likely to work in other types of channels. Thus, the fade margins of all systems are based on Rayleigh spatial fading statistics.

Note: Even More Pessimistic Statistics

Many have thought that worse-than-Rayleigh fading statistics were only possible in contrived mathematical analysis, but not in physical practice. In Chapter 5 we show a class of fading statistics that often leads to even deeper and more frequent signal fades than the Rayleigh fading channels. See Section 5.4 for a discussion of the *TWDP* fading distribution.

Update Rate

A narrowband receiver in a fading channel must apply some type of *automatic gain control* (AGC) to counter the unpredictable fading in received signal strength. An AGC unit in a handset is basically an amplifier with variable gain that increases when signal strength is low and decreases when signal strength is high. The key issue for an engineer designing AGC is update rate, the maximum speed that the receiver must change the gain in a realistic fading channel. To calculate this, we have to consider the worst-case scenario where fading is changing the fastest. The most pessimistic estimate would be the highest possible mobile user velocity in the propagation with the multipath arriving from all directions in space.

Once again, the omnidirectional Rayleigh model provides the pessimistic result. Since the angle-of-arrival is omnidirectional, it is impossible to generate fading with closer peaks and nulls in space (the reason for this becomes clearer in Chapter 6). A user talking in a car on the highway in an omnidirectional Rayleigh propagation environment will experience the fastest signal fading and will require the highest update rate. The update rate of most adaptive equalizers (of which the AGC is a simple example) are designed with this philosophy.

Changing Paradigm

Of course, Rayleigh pessimism is predicated on a purely passive design philosophy; the wireless engineer designs a communications system for worst-case fading scenarios and accepts whatever penalties are imposed upon a receiver by spatial fading. Emerging digital radios take more proactive efforts to overcome channel fading with dynamic modulation, diversity, and channel coding schemes. Furthermore, wireless systems of the future may employ *space–time processing* to further combat the fading channel. In fact, as we will see in Chapter 9, these wireless systems exploit the presence of multipath to *enhance* the transmission of data through a wireless link - a far cry from the design philosophy of early mobile radio. Ironically, Rayleigh fading becomes an optimistic, best-case scenario of operation for these future systems - and it is very bad engineering practice to design with optimism.

1.1.5 Channels with Multiple Dependencies

Despite the advances in spatial channel modeling in the 1960s, the treatment of the wireless channel was still usually restricted to a single scalar dependency. Since the analysis of fading channels requires some elegant random process theory, there was difficulty in characterizing channels with full space, time, and frequency dependencies. The seminal contribution to this part of the puzzle comes from work performed at MIT's Lincoln Laboratories during the 1950s. A research group performed groundbreaking research in the field of radio astronomy, where a theoretical framework for studying stochastic signals of multiple dependencies was first developed [Gre62].

Out of this work came a brilliant piece of scholarship by P. A. Bello in 1963, which was the first research to describe stochastic communication channels that were combinations of time-varying and frequency-varying random processes [Bel63]. The work itself may have been a little ahead of its time, however, since broadband communications were not in widespread use (thus, no need for the frequency-varying aspects of channel models). As time marched onward, bandwidths became larger and joint channel dependencies became more important. The work by H. Hashemi in the late 1970s was some of the first to study truly random wideband temporal channels in the context of wireless communications [Has79].

Extensions of the original Lincoln Lab theory to joint spatial modeling of wireless channels were absent as late as the year 2000. R. G. Vaughan presented joint-dependency Fourier analysis techniques for space in [Vau00], and B. H. Fleury presented the first description of joint space–time–frequency wireless channels in [Fle00]. Clearly, the field of space–time wireless channel modeling is still in its infancy, with a great deal of work and innovation left to be done.

1.2 The Case for Space

This book places particular emphasis on *spatial* channel modeling, because this is both the most complicated aspect of the wireless channel and, not surprisingly, the least understood. This section provides several arguments for why space is crucial in the future study of wireless communications.

1.2.1 Complexities of Wireless Channels

The wireless channel is often neglected in texts and courses on communications. Rarely does an engineer ever study anything more complicated than the *additive white Gaussian noise* (AWGN) channel. In AWGN channel modeling, the received signal is set equal to the transmitted signal with some proportion of Gaussian white noise added. The AWGN model, pictured in Figure 1.6, works well for very simple communications through band-limited channels corrupted mostly by thermal noise.

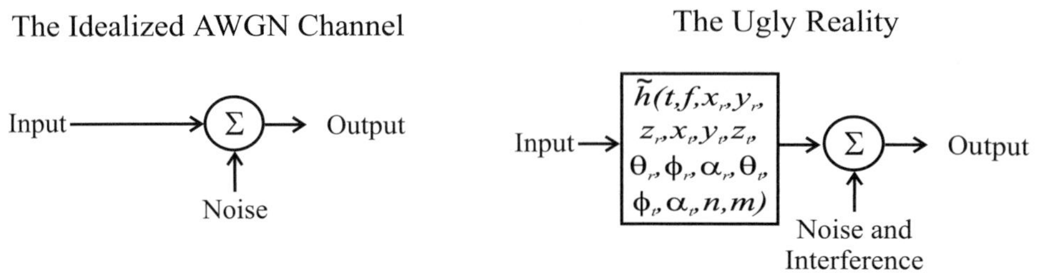

Figure 1.6 In contrast to the idealized additive white Gaussian noise (AWGN) channel, the true wireless radio channel has numerous dependencies.

For realistic wireless channels, however, there will be signal distortion that is described by a *channel transfer function*. Many engineers have working knowledge of "black-box" transfer functions that represent linear, time-invariant channel effects - those described by a single time dependency using the operation of convolution. The wireless channel, on the other hand, has multiple dependencies. In fact, some channels may have up to 16 different dependencies! Consider the following dependencies (we will discuss some of these in detail in the next chapter):

- **Frequency:** The wireless channel depends on the transmitted frequency.
- **Time:** The wireless channel is time-varying.
- **Receiver Translation:** The wireless channel depends on the position movement (called *translation*) of the receiver antenna. In three-dimensional space, translation is actually three different scalar dependencies in the wireless channel.
- **Transmitter Translation:** What is true for the receiver is also true for the transmitter. The freedom to change transmitter position in three-dimensional space adds another three dependencies to a wireless channel representation.
- **Receiver Orientation:** Translation is only one type of spatial dependency. A receiver antenna may be reoriented in any direction in space, which changes the polarization and gain pattern interaction of the receiver antenna with incoming radio waves. An antenna may be rotated in azimuth or elevation or be tilted laterally. Changes in orientation add up to three dependencies to a wireless channel representation.
- **Transmitter Orientation:** The radio channel also depends on the transmitter antenna orientation. This adds three more dependencies to the wireless channel.

- **Multi-Element Receiver:** The use of multiple antenna elements at the receiver adds one discrete dependency to a wireless channel.

- **Multi-Element Transmitter:** The use of multiple antenna elements at the transmitter adds another discrete dependency.

So, there are the 16 total radio channel dependencies. A characterization of the realistic wireless channel can be overwhelming.

Note that most of the dependencies can be categorized as spatial aspects of the wireless channel, summarized in Figure 1.7. Translation, orientation, and multi-element dependencies all influence the spatial filtering of the channel. They also happen to be the least-understood aspects of the wireless channel. Indeed, space really is the final frontier for channel modeling.

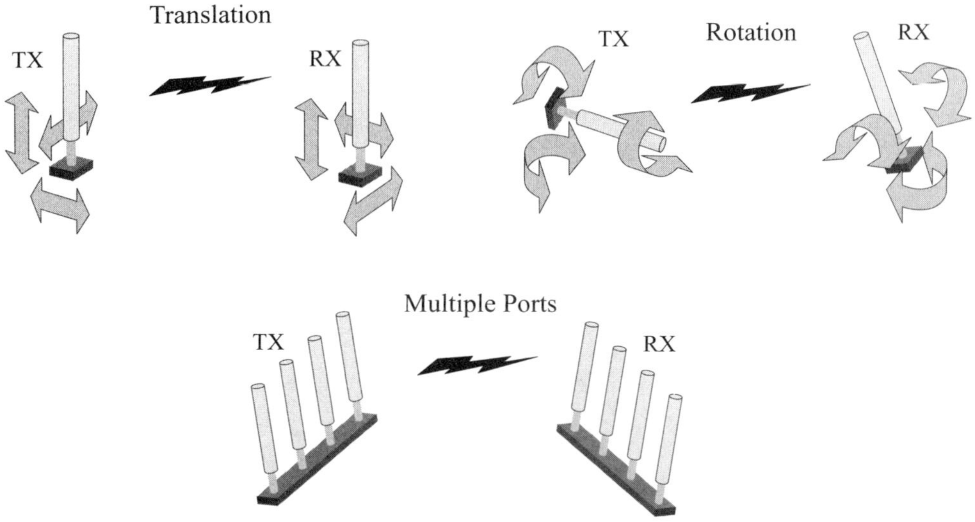

Figure 1.7 There are many types of spatial dependencies in a wireless channel: antenna translation, antenna rotation, and multiple ports for both transmitter (TX) and receiver (RX).

There is both a pessimistic and optimistic way of viewing all of these dependencies. Each dependency is another layer of complexity for understanding the transmission of wireless information. Each dependency is also a potential source of fading and unpredictability. On the other hand, each dependency may be viewed as an opportunity to increase channel capacity. Just as bandwidth or transmission time can be increased to send additional data, the spatial dimensions of a wireless channel may also be exploited similarly to increase capacity, so the 16 dependencies implies 16 potential opportunities to increase channel capacity. There are many fascinating design issues involving wireless channels with multiple dependencies.

1.2.2 Channel Primacy in Communications

When Shannon derived channel capacity equations, he showed that the fundamental limit on data rate in a communications system depended almost solely on the amount of power (in proportion to noise) delivered to a receiver. This limit is universal, independent of any of the signal-processing operations that may occur at either the transmitter or the receiver. The principle of the Shannon limit has key implications for wireless system design.

The modern approach to digital communication design requires wireless engineers to use a baseband block-diagram approach for building transmitters and receivers that span a radio link [Skl01]. In this scheme, the transmitter performs operations such as source coding, channel coding, multiplexing, modulation, and multiple access operations on a stream (or streams) of information. Each of these operations is represented by a functional block in Figure 1.8. The operation of the receiver is functionally the inverse of the transmitter, performing block operations in the reverse order to recover the original information.

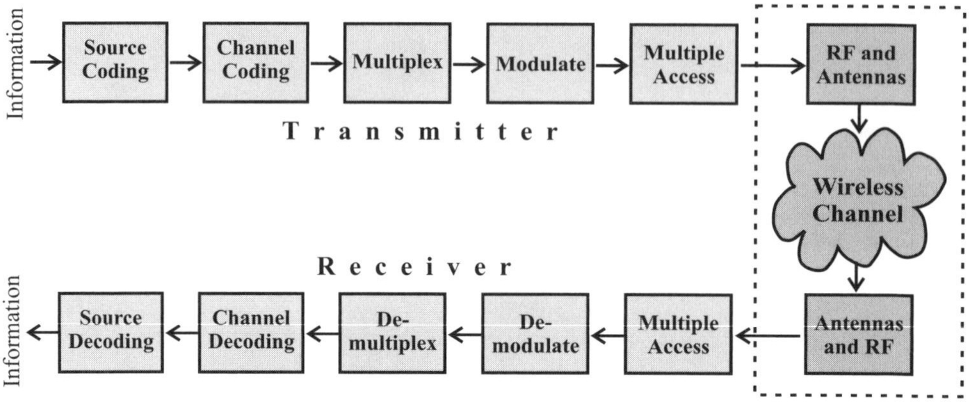

Figure 1.8 Some basic operations of a digital wireless communications system may be broken into functional blocks. (See B. Sklar's textbook [Skl01] for an outstanding, exhaustive block-diagram analysis of digital systems.)

In between the baseband block operations of transmitter and receiver, the signal passes through the wireless channel and the radio-frequency (RF)/antenna hardware that physically interacts with this channel. It is this part of the transmission, on the right side of Figure 1.8, that determines Shannon channel capacity. Thus, the communications engineer is at the mercy of whatever happens in this part of the link. The engineer selects appropriate blocks from an arsenal of signal-processing schemes and, upon success, has designed a system that approaches the available channel capacity. Engineering the baseband blocks of a digital communications link is a very close-ended problem: We know that an optimum design is achieved if the link performance approaches the Shannon limit.

Not so for the RF engineer, who works with the hardware and antennas that interface with the wireless channel. The physical and spatial interface with the radio channel is an *open*-ended problem, with no guarantee that even a successful solution is an optimal solution. From this viewpoint, studying the wireless channel interface has some of the richest design possibilities.

1.2.3 Wasted Space

Commercial wireless has mostly operated under the single-port paradigm: A single antenna element - usually a metal loop or whip - is attached to a user terminal. The first AM and FM radios, walkie-talkies, and pagers all used a single antenna. Despite all of the vaunted advances in signal processing and radio portability, a survey of digital handsets and wireless LAN terminals at the start of the 21st century would show the exact same single-port architecture. The single antenna element for radio terminals is a lot like the lead-acid battery for cars: A century of innovation has passed them by.

The single-port architecture has two Achilles' heels that make it terminally unattractive for radio design. First of all, a single-port receiver is always susceptible to *catastrophic fading.* In a spatial fading channel, there are receiver positions in space that cannot accommodate wireless communications. These "blind spots" occur unpredictably, even in environments with high average signal strength. Thus, all receivers operate in the fading equivalent of a mine field, encountering sporadic pockets of fades that threaten to sever the wireless link.

Space itself is not the only source of catastrophic fading. The immediate environment of the single antenna element can be problematic, as coupling with the human body, close-in objects, and even the casing and circuitry of the receiver itself can skew the pattern and radiation impedance of the antenna. There is no bullet-proof radio design for a single, static antenna element. When multiband operation is considered - using multiple, noncontiguous frequency bands - the antenna problem can seem hopeless.

Without an additional spatial port, the receiver is stuck with a single channel. The inability to overcome deep fades with a single-port receiver is the main reason cellular phones drop calls in mid-conversation. If there were additional spatial ports, the phone could at least employ some form of *selection diversity* (discussed in Chapter 10) and use an alternative signal if one port experiences a catastrophic fade.

The second critical problem with the single-port architecture is the wasted opportunities for power coupling into the receiver. Consider the case of the simple whip antenna fixed to the typical cellular handset. The effective electromagnetic aperture - roughly the area of space from which an antenna can sink radio power - is small compared to the handset itself. Most of the propagating radio power that impinges upon a handset is unused, reflected off into free space.

As Figure 1.9 shows, it would be much more desirable to design a handset that could somehow absorb radio power across its entire body. Furthermore, multipath radio waves have a polarization that cannot couple completely into any single antenna. From the electric-field point of view, there are three distinct polarizations (due to the three-dimensionality of space). Moreover, researchers have even shown benefits from separate sensing of the three magnetic-field components of propagating multipath waves *in addition to* the three electric-field components [And01].

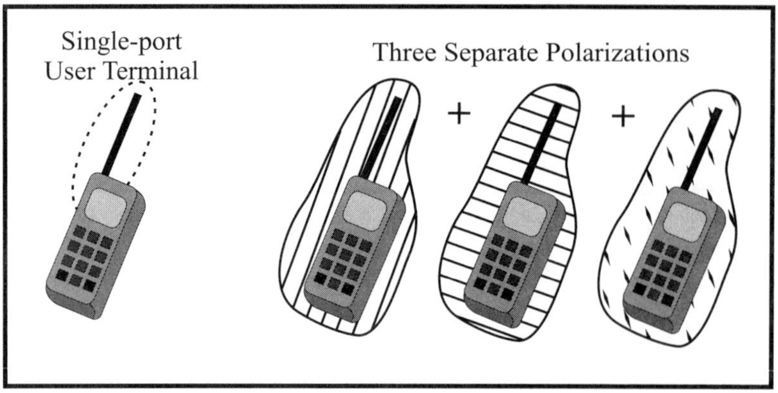

Figure 1.9 In a multipath channel, a single-port radio wastes much of the impinging signal power.

Clearly, a lot of potential radio power and opportunity is going to waste in a single-port radio. More power and multiple spatial ports can overcome the thermal noise and in-band interference that signal processing alone cannot remove.

Needless to say, space-wasting receivers have persisted for good reasons. There are two major challenges facing multiport receiver designs. First, multiplying spatial ports on a receiver also multiplies complexity in the radio-frequency hardware - a critical expense in the production of user terminals. Second, it is difficult and expensive to incorporate more than one low-profile antenna into a terminal. This problem is particularly acute for handsets, where aesthetics is important for user-acceptance. Nobody wants to carry around a pin cushion of antennas.

Still, multiport receivers are a certainty for wireless communication that desires to maximize data transmission. If the goal is to develop a receiver that can sustain a reliably high data rate, then the goal must be a power-stingy receiver that wastes no space.

1.3 Trends in Wireless Communications

The theory that engineers use to measure and model wireless communications has changed very little over the last 30 years. The main reason for this stagnation of development may be summed up as follows: The current theory still works for wireless systems that have been deployed to date. Do not expect this to hold much longer. There are six trends in wireless communications that emphasize the need for improved and expanded channel modeling theory.

1.3.1 Higher and Higher Data Rates

The capacity for data transmission of current wireless systems is still tiny when compared to wired forms of communications. But wireless data rates continue to increase. To understand the push for higher and higher data rates, it is useful to consider an analogy involving the trends of memory size and processor speed in the personal computer market. In the early 1980s, a typical personal computer had about 64 kilobytes of RAM and operated with a processor that clocked at speeds less than 1 MHz. In 2000, the typical personal computer had a processor that operated at a clock frequency close to 1 GHz and required as much as 100 MB of RAM. In short, as soon as computer hardware is enhanced, new commercial software applications are developed that exploit the newfound capacity for storing and manipulating data.

The computer hardware illustration provides a valuable lesson for the wireless industry. There is a basic rule that applies to all information technology: Technology that increases the capacity to store or manipulate data is eventually (and sometimes rapidly) followed by new applications that exhaust the resources. For wireless, this means that the current technology will continue to gravitate towards higher transmitted data rates [Rap02b].

Of course, higher data rates imply wireless systems that operate with wider bandwidths. Future wireless systems will operate with bandwidths that greatly exceed conventional channel models. New systems will require new channel models and measurements.

1.3.2 Ubiquity of Wireless Devices

Wireless personal communications has permeated nearly every environment on earth. It is now possible to use a wireless handset in a city, in a car, in the home, in an office building, on a boat - the list goes on. Future applications will involve wireless sensors and impersonal communications between engines, machinery, and appliances.

The wireless channel is heavily dependent on the environment in which it operates. Since future wireless applications will operate in nearly every imaginable environment, there will be an incredibly diverse variety of channels that require characterization. In fact, many of these new environments will defy characterization by the older paradigms of wireless channel modeling.

1.3.3 Smart Antennas

Adaptive arrays and other types of smart antenna techniques are emerging technologies for improving the wireless link and mitigating interference in a multiple access system [God97], [Win98]. Many multiuser communication systems such as cellular radio networks had, until the end of the 20th century, operated below their designed capacity. As the market for these systems has grown and matured, the network traffic has grown as well. Smart antenna technology is seen as a cheap and effective solution for mitigating the problem of network congestion.

A directional antenna at a receiver or transmitter drastically changes the channel characteristics. Channel models that once applied to omnidirectional antennas must be modified and improved to account for the new space–time distortion of the channel by the directional antenna.

1.3.4 Faster, Smaller, Cheaper Hardware

Over the years, basic research in wireless communications has produced a plethora of modulation, multiple-access, and signal-processing innovations that combat the distortions introduced by a wireless channel. Only a small subset of these innovations are used in practice, since many algorithms and techniques do not have a feasible realization in hardware.

Radio frequency and digital signal-processing technology continues to develop, however. The computational power of baseband chipsets is increasing. The radio-frequency integrated circuits are operating at higher power levels and at higher frequencies. Above all, these transmitter and receiver components are becoming cheaper and cheaper to fabricate. As a result, many algorithms and techniques that are not feasible to implement today will become feasible tomorrow.

The added capabilities of future radio receivers, therefore, will be able to combat the detrimental effects of the multipath channel in new and innovative ways. With added functionality, receivers of the future need more than just an ad hoc approximation about the radio channel. Future receiver designs will require models that mimic the detailed dispersion, time-varying, and space-varying characteristics of a realistic wireless channel.

1.3.5 Frequency Congestion

Bandwidth is a finite resource. As wireless systems with wider and wider bandwidths continue to deploy, frequency congestion becomes a problem. One solution is to move outside of common frequency bands and into higher, uncrowded frequency in the upper microwave and mm-wave bands. Propagation at these higher frequencies presents an entirely different set of problems. Channel models developed around the 1 GHz microwave bands are inadequate to characterize wireless systems where both the carrier frequency and signal bandwidth are one or two orders of magnitude greater.

1.3.6 Multiple-Input, Multiple-Output Systems

Perhaps one of the most interesting trends in wireless communications is the proposed use of multiple-input, multiple-output (MIMO) systems. A MIMO system uses multiple transmitter antennas *and* multiple receiver antennas to break a multipath channel into several individual spatial channels. Such a system employs *space–time coding* to increase the link capacity [Fos96].

New MIMO systems represent a huge change in how wireless communications systems are designed. This change reflects how we view multipath in a wireless system:

> **The Old Perspective:** The ultimate goal of wireless communications is to combat the distortion caused by multipath in order to approach the theoretical limit of capacity for a band-limited channel.
>
> **The New Perspective:** Since multipath propagation actually represents *multiple channels* between a transmitter and receiver, the ultimate goal of wireless communications is to *use* multipath to provide higher total capacity than the theoretical limit for a conventional band-limited channel.

This philosophical reversal implies that many of the engineering design rules of thumb that were based on pessimistic, worst-case scenario channel models have now become unrealistically optimistic. Design of such systems will require new space–time channel models.

1.4 About This Book

This section presents an overview of the content in *Space–Time Wireless Channels.*

1.4.1 The Basic Disciplines

The primary purpose of this book is to provide a unique instruction or research reference for wireless researchers who require knowledge of the space–time wireless channel. The target reader is a researcher with limited exposure to the three basic disciplines of space–time channel modeling: random process theory (probability theory), electromagnetic propagation, and communications. Any reader with an undergraduate degree (or equivalent) in electrical engineering should have little difficulty understanding the mathematics and physics presented throughout the book.

All space–time channel concepts grow out of the three theoretical disciplines, listed in Figure 1.10. Communication theory, besides providing the general theory of information transmission, is the primary field of application in this text. All of the modulation, multiple access, coding, and signal processing algorithms of wireless communications depend, in part, on the space–time channel description.

Some basic concepts in random process theory are also vital to the study of space–time wireless channels. Since the type of channels experienced by a real radio link are varied and unpredictable, only stochastic modeling with random processes provides the engineer with a bridge from analytical description to realistic performance. Basic electromagnetic theory, one of the most neglected areas in practice, is important for understanding the physical properties of a space–time wireless channel.

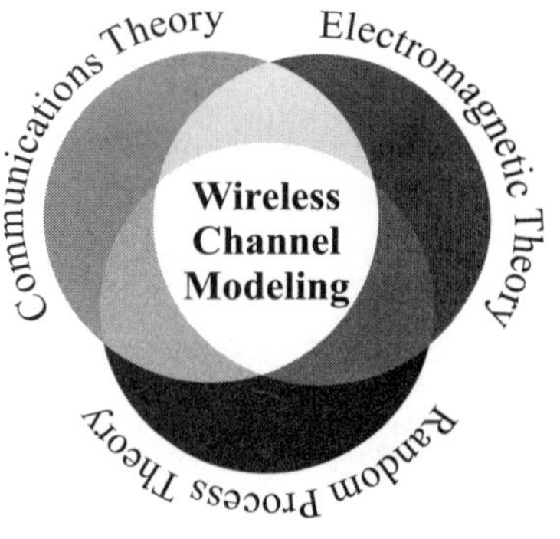

Figure 1.10 Wireless channel modeling theory is an amalgam of three different theoretical disciplines.

1.4.2 Contents

This book conveniently divides into three roughly equivalent parts. Chapters 2 to 4 are the basic principles of the three disciplines (communications, electromagnetics, and random process theory), presenting the basic framework of *Space–Time Wireless Channels*. Chapters 5 to 7 present the development of theory based on this basic framework. Chapters 8 to 10 present an overview of space–time applications, focusing on multiple antenna techniques for wireless communications. The following summarizes the contents of each chapter:

- Chapter 1: Introduction to the field of channel modeling.
- Chapter 2: Theory of transmission through space, time, and frequency.
- Chapter 3: Random process theory for space, time, and frequency.
- Chapter 4: Electromagnetic description of space–time channels.
- Chapter 5: First-order statistics of fading channels.
- Chapter 6: Angle spectrum concepts and applications.
- Chapter 7: Second-order statistics of fading channels.
- Chapter 8: Overview of diversity techniques.
- Chapter 9: Overview of space–time signal processing.
- Chapter 10: Design rules for antenna arrays in multipath channels.

In addition to the main text, the following appendices are included:

- Appendix A: List of special functions used in the text.
- Appendix B: Tables and examples of Fourier transforms.

- Appendix C: Review of definitions and concepts in random process theory.
- Appendix D: Glossary of mathematical conventions and acronyms.

By the conclusion of the book, the reader should have an understanding of general information transmission through many types of space–time wireless channels.

1.4.3 Features of This Book

In matters of scholarship and presentation, this text attempts to be as practical and as readable as possible. Since communications, stochastic process theory, and electromagnetic wave theory are vast subjects in themselves, advanced topics are introduced with thorough background information. Extensive mathematical derivations have been removed from the main text and placed in appendices at the end of the chapters. Important *theorems* are proven with pragmatic methods that emphasize understanding instead of mathematical rigor. *Examples* are used to illustrate concepts wherever possible. Each subsequent chapter is concluded with a set of problems, many of which relate theory to practical issues in wireless link design.

> *Note: Supplemental Information*
>
> Throughout this book there are *noteboxes* (like this one) that supplement technical concepts in the main text. These noteboxes are used to clarify possible points of confusion, to justify a certain type of notation, to alert the reader to misconceptions that exist in the research literature, or even to provide some history behind a useful concept and its inventor. Such "editorials-in-miniature" help break up long (and admittedly tedious) technical expositions.

This book is meant for more than just a treatment of various channel modeling issues in wireless communications. More broadly, the text is really a grassroots introduction to the principles of *stereostochastics*, the study of random processes as functions of both time and space. The principles of *Space–Time Wireless Channels* have numerous applications outside the field of commercial wireless engineering, including radar, optics, acoustics, quantum mechanics, and any other field of study involving dynamic wave interactions.

Chapter 2

SIGNAL TRANSMISSION

Before we can understand wireless communications, we must first understand what is meant by the wireless *channel* and how it distorts transmitted information. The wireless channel imposes the fundamental limitations on the data rate between transmitter and receiver. Hence, the wireless channel imposes the ultimate limitation on what types of information services may be offered in a given frequency band.

Compared to other types of communication channels (wires, waveguides, optical fibers, etc.), the wireless channel is quite data-hostile. There are two reasons for this: low power and unpredictable fading. This chapter concentrates on the unpredictable fading that occurs in a wireless link. As we will see, wireless channels can fade as functions of time, frequency, and space *simultaneously*.

The goal of this chapter is to develop the terminology, definitions, and basic concepts of modeling a wireless channel that can be a function of time, frequency, and receiver position in space. The chapter is broken into the following sections:

- Section 2.1: Basic definitions for complex baseband channel representation.
- Section 2.2: Introduction to channel coherence and selectivity.
- Section 2.3: Discussion of complex baseband channel usage.
- Section 2.4: Summary of important concepts.

By the end of this chapter, the reader will be able to understand the different varieties of wireless channels, to represent them mathematically, and to analytically predict how transmitted data signals are distorted by fading channels. As a whole, this chapter represents a steppingstone to every subsequent concept developed in this book.

2.1 Baseband Representation

Development of a *baseband* representation for modulated signals is the cornerstone of channel modeling and analysis. A baseband representation essentially removes the dependence of a passband radio channel from its carrier frequency, which both generalizes and simplifies channel modeling. This section discusses the mathematics behind switching between passband and baseband representations of radio signals and channels.

2.1.1 Signal Spectrum

Every realistic communications signal has a Fourier transform, or *spectrum* that defines its mathematical properties in the frequency domain. For every time-domain signal, $x(t)$, there is a frequency-domain signal, $X(f)$, given by the forward transform definition:

$$X(f) = \int_{-\infty}^{+\infty} x(t)\exp(-j2\pi ft)\,dt \qquad (2.1.1)$$

Every Fourier pair, $x(t)$ and $X(f)$, is unique so that the original time-domain signal may be recovered from the frequency spectrum using the following inverse transform:

$$x(t) = \int_{-\infty}^{+\infty} X(f)\exp(j2\pi ft)\,df \qquad (2.1.2)$$

So the spectrum, $X(f)$, contains all of the same information as $x(t)$. The information is just organized in a different way that helps us to perform certain types of signal analysis. Appendix B contains a discussion of Fourier transforms with tables of common Fourier transform pairs and properties.

The Fourier transform, as defined by Equation (2.1.1) and Equation (2.1.2), may be applied to *any* complex time-domain signal. What is mathematically permissible is not always physically meaningful, though. If the signal $x(t)$ represents a physical quantity (a time-domain voltage at the terminals of an antenna, for example), then it must be real-valued. Unless the time-domain signal is a contrived mathematical function, the spectrum of most real-valued time-domain functions is usually complex-valued.

Because of its complex values, a frequency-domain function, $X(f)$, is difficult to sketch on the page of a book. Since this graph would include a dependent frequency axis, a real function part, and an imaginary function part, the best way to graph a spectrum would be on "three-dimensional paper." With a little perspective and imagination, Figure 2.1 illustrates what a signal spectrum looks like in three dimensions. The spectrum in (a) is the Fourier transform of a *sinc pulse*, $\text{sn}(t)$ - which is a simple box spectrum - while the spectrum in (b) is the Fourier transform of a more realistic time-domain signal.

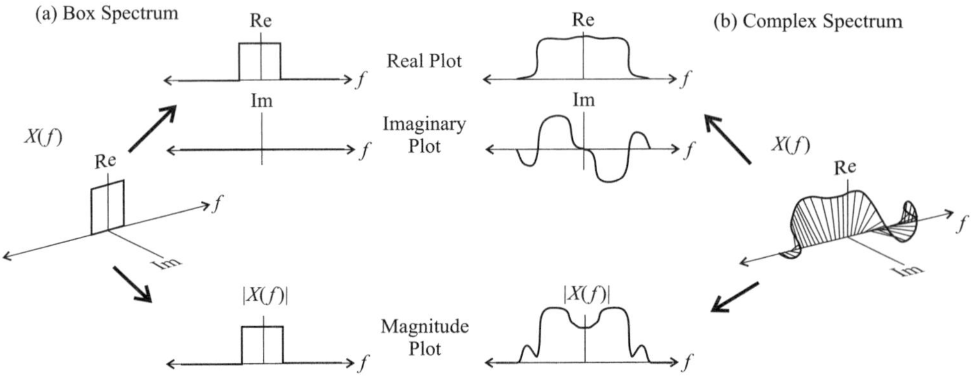

Figure 2.1 A simple box spectrum (a) and a complex-valued spectrum (b) are sketched in two-dimensions as either two real and imaginary plots or as a magnitude plot of $|X(f)|$.

As shown by Figure 2.1, we usually do one of two things to make the spectrum appear "flat" on a page. We either break the spectrum into a real and imaginary part or we simply sketch the *magnitude* of the spectrum, $|X(f)|$, and omit the phase information.

2.1.2 Signal Modulation

One of the most fundamental operations in radio communications is the act of modulating a carrier wave by a band-limited data signal. Modulation converts a *baseband* signal to a *passband* signal. To represent the act of modulation, we will use the modulation operator $\mathcal{M}\{\cdot\}$ to denote transformation of a baseband signal, $\tilde{x}(t)$, to a passband signal, $x(t)$, that modulates a carrier wave. Thus, using this notation we write

$$x(t) = \mathcal{M}\{\tilde{x}(t)\} \tag{2.1.3}$$

The tilde ($\tilde{\ }$) above a function will be the convention used throughout this book to denote baseband representation of a signal or channel.

It is easiest to view modulation in the frequency domain using the spectrum of baseband and passband signals: $\tilde{X}(f)$ and $X(f)$ respectively. The Fourier transform of the passband signal may be calculated from the baseband signal, $\tilde{X}(f)$, as follows:

$$X(f) = \frac{1}{2}\tilde{X}(f - f_c) + \frac{1}{2}\tilde{X}^*(-f - f_c) \tag{2.1.4}$$

where $*$ denotes complex conjugate. In the frequency domain, $X(f)$ is simply a copy of the spectrum $\tilde{X}(f)$ shifted to a center frequency of $f = +f_c$ and a conjugate mirror copy of $\tilde{X}(f)$ shifted to a center frequency of $f = -f_c$.

The operation of modulation can be defined directly in the time domain. Given a carrier frequency, f_c,

$$\mathcal{M}\left\{\tilde{x}(t)\right\} \doteq \underbrace{\text{Real}\{}_{\substack{\text{mirror}\\ \text{image}}}\ \tilde{x}(t)\ \underbrace{\exp(j2\pi f_c t)}_{\substack{\text{frequency}\\ \text{shift}}}\ \} \tag{2.1.5}$$

The complex exponential in Equation (2.1.5) shifts the baseband signal, $\tilde{x}(t)$, up to a carrier frequency of f_c, and the Real$\{\cdot\}$ operator produces the conjugate mirror-image spectrum at $-f_c$.

At this point, it is necessary to define the *bandwidth*, B, of a baseband signal. There are many different ways to define bandwidth, as illustrated in Figure 2.2. In this book, unless noted otherwise, we will use the largest-valued definition of bandwidth, *nonzero bandwidth.*

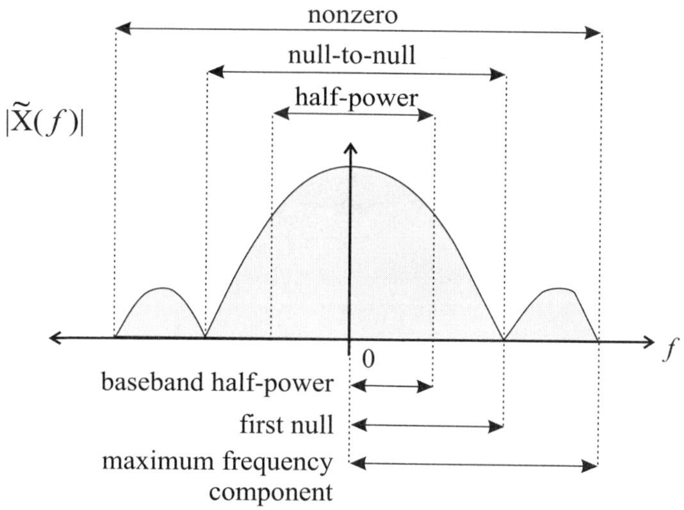

Figure 2.2 The many different bandwidths defined for a baseband signal spectrum.

2.1.3 Inverse Modulation

The inverse operation of modulation - converting a passband signal, $x(t)$, back to a baseband signal, $\tilde{x}(t)$ - also has a time domain definition:

$$\begin{aligned}\tilde{x}(t) &= \mathcal{M}^{-1}\left\{x(t)\right\}\\ &= \underbrace{[x(t)\ \exp(-j2\pi f_c t)]}_{\substack{\text{frequency}\\ \text{shift}}} \otimes \underbrace{[2B\text{sn}(Bt)]}_{\substack{\text{low-pass}\\ \text{filter}}}\end{aligned}$$

$$= 2B \int_{-\infty}^{+\infty} x(\zeta) \exp(-j2\pi f_c \zeta) \, \mathrm{sn}(B[t-\zeta]) \, d\zeta \tag{2.1.6}$$

where $\otimes$ denotes convolution and $\mathrm{sn}(\cdot)$ is the sinc function, $\mathrm{sn}(x) = \frac{\sin(\pi x)}{\pi x}$. The complex exponential term in Equation (2.1.6) shifts the passband signal spectrum, $X(f)$, by an amount f_c so that the copy of $\tilde{X}(f)$ lies centered at $f = 0$ and its conjugate mirror image lies at $f = -2f_c$. The convolution with the sinc function - a low-pass filter - then removes the high-frequency mirror image so that only $\tilde{X}(f)$ remains. The complete operation of carrier modulation and demodulation is shown in Figure 2.3.

> *Note: Definition of Convolution*
>
> The convolution operation, $\otimes$, is defined to be
>
> $$x(t) \otimes y(t) = \int_{-\infty}^{+\infty} x(\zeta)\, y(t-\zeta)\, d\zeta \qquad \textbf{or} \qquad x(t) \otimes y(t) = \int_{-\infty}^{+\infty} x(t-\zeta)\, y(\zeta)\, d\zeta$$
>
> Each integral evaluates to an identical result, regardless of $x(t)$ and $y(t)$.

If the modulated signal, $x(t)$, is to represent a physically realizeable transmission, then it must be a real-valued function. No such restriction is placed on the baseband signal, however, as *any* complex-valued function that modulates a carrier according to Equation (2.1.5) will produce a real-valued function. This difference between baseband and passband representations stems from the conjugate mirror image in the passband spectrum, $X(f)$. Thus, $X(f)$ has twice as much nonzero bandwidth as the baseband signal. A *complex* time-domain function is the easiest way to accommodate this extra information so that nothing is lost in the baseband representation of a modulated signal. Consider the following example:

Example 2.1: Amplitude and Phase

Problem: Passband signals are commonly written in the form

$$x(t) = V(t)\cos(2\pi f_c t + \varphi(t))$$

where $V(t)$ is amplitude and $\varphi(t)$ is phase, both of which are real-valued; $x(t)$ is a band-limited function with nonzero bandwidth B, centered at the carrier frequency f_c, and implicitly satisfying the relationship $B \leq 2f_c$. Find an expression for the baseband representation, $\tilde{x}(t)$, of this type of signal.

Solution: The first step is to plug the expression for $x(t)$ into the demodulation operation of Equation (2.1.6):

$$\tilde{x}(t) = 2B \int_{-\infty}^{+\infty} V(\zeta)\cos(2\pi f_c \zeta + \varphi(\zeta)) \exp(-j2\pi f_c \zeta)\, \mathrm{sn}(B[t-\zeta])\, d\zeta$$

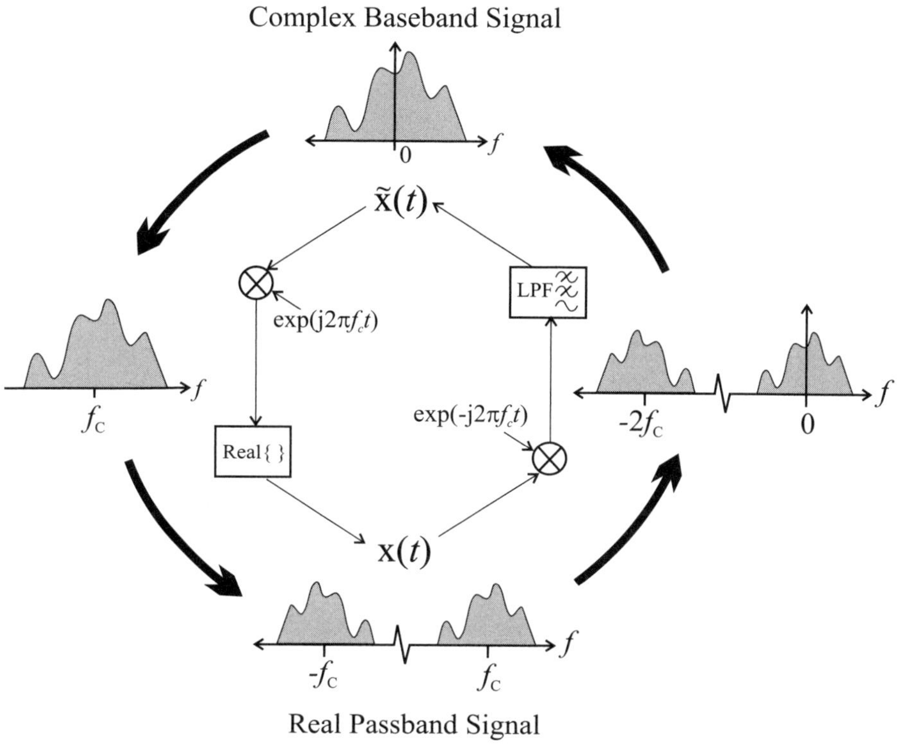

Figure 2.3 Baseband–passband transformations in the time domain (inner cycle) and the frequency domain (outer cycle).

Using the Euler relation for cos(·), it is possible to rewrite this integral as

$$\tilde{x}(t) = \int_{-\infty}^{+\infty} \frac{\{\overbrace{V(\zeta)\exp(j\varphi(\zeta))}^{\text{low frequency}} + \overbrace{V(\zeta)\exp(-j4\pi f_c\zeta + j\varphi(\zeta))}^{\text{high frequency}}\}\sin(\pi B[t-\zeta])\,d\zeta}{\pi(t-\zeta)}$$

which may be evaluated by inspection. The left term inside the braces is a low-frequency component, which passes through the integration unchanged. The right term inside the braces is a high-frequency component, which evaluates to zero upon integration. Thus,

$$\tilde{x}(t) = V(t)\exp(j\varphi(t))$$

which is the complex baseband representation of the real-valued transmitted signal.

Note: Removing the Carrier

From Example 2.1, it is clear that inverse modulation simply removes the oscillations of the carrier wave while retaining amplitude and phase information. The act of removing carrier oscillations is a staple of applied sciences. For example, when amplitude and phase are constant with respect to time ($V(t) = V_0$ and $\varphi(t) = \varphi_0$), the passband signal is said to be *time-harmonic*. The definitions for modulation and inverse modulation then degenerate into the commonly used *phasor transforms*.

2.1.4 The Baseband Channel

Three passband functions are required to represent the operation of the simplest wireless communication system: a transmitted signal, $x(t)$, a received signal, $y(t)$, and a channel, $H(t)$. If the channel is linear and time-invariant, then it is possible to relate these three quantities using convolution:

$$y(t) = x(t) \otimes H(t) \tag{2.1.7}$$

However, it is more convenient to analyze these functions using their baseband representation, since they become independent of carrier frequency. Using the following relationships for baseband and passband signals,

$$x(t) = \mathcal{M}\left\{\tilde{x}(t)\right\}, \qquad y(t) = \mathcal{M}\left\{\tilde{y}(t)\right\}, \qquad H(t) = \mathcal{M}\left\{\tilde{H}(t)\right\}$$

it is possible to write Equation (2.1.7) as a convolution of baseband signals:

$$\tilde{y}(t) = \frac{1}{2}\left[\tilde{x}(t) \otimes \tilde{H}(t)\right] \tag{2.1.8}$$

Keep in mind that while Equation (2.1.8) captures all of the behavior of the passband channel, it is a convenient representation and is *not* a physical process. An engineer must always return to passband in order to get the actual functions for the transmitted and received signals and the radio channel. However, all theory and development presented in this book deals with the baseband representation of signals and channels.

Note: About That $\frac{1}{2}$ Fudge Factor

Using the conventional definitions for baseband equivalent analysis in Equation (2.1.6), $\tilde{x}(t)$ and $\tilde{y}(t)$ have twice the total signal power as their passband counterparts, $x(t)$ and $y(t)$. If $H(t)$ and $\tilde{H}(t)$ are defined with the same baseband–passband transforms, then the baseband convolution with $\tilde{H}(t)$ will artificially amplify the output by a factor of 2. So the $\frac{1}{2}$ in Equation (2.1.8) adjusts the result of the baseband equivalent convolution so that it is consistent with the passband convolution.

The baseband and passband channel models for *single input, single output* (SISO) transmission are pictured in Figure 2.4. The *single input* is the transmitted

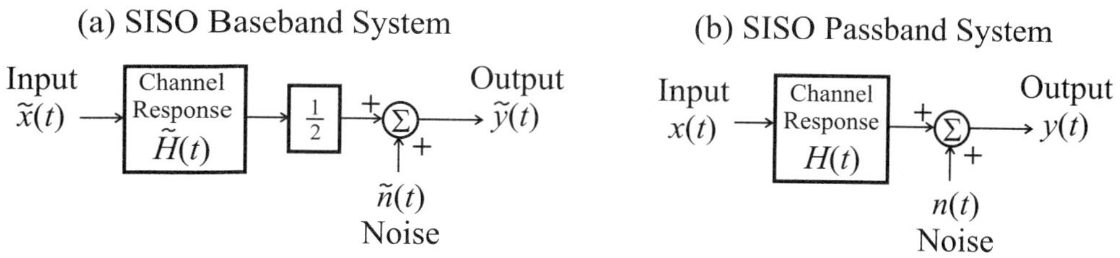

Figure 2.4 Block diagram of baseband and passband channel models for SISO transmission.

signal, and the *single output* is the received signal. Of course, a realistic communication system must contend with additive noise, $n(t)$. This noise can be thermal noise, impulsive noise, multiple-access interference, jamming - virtually any type of undesired signal that coexists in the passband of the desired signal. Additive noise modeling is a well-understood practice. For the purposes of this book, the additive noise function, $n(t)$ or $\tilde{n}(t)$, will be ignored until we develop some applications where the effects of noise or interference is important.

2.1.5 Time-Invariant Versus Time-Varying Channels

It is a hallmark of linear, time-invariant systems that the convolution of a transmitted signal and a channel may be written as a product in the frequency domain. Consider the following Fourier transform pairs:

$$\tilde{x}(t) \longleftrightarrow \tilde{X}(f), \qquad \tilde{y}(t) \longleftrightarrow \tilde{Y}(f), \qquad \tilde{H}(t) \longleftrightarrow \tilde{h}(f)$$

Using these definitions, Equation (2.1.7) and Equation (2.1.8) may be written respectively as

$$Y(f) = X(f)\,h(f) \tag{2.1.9}$$

$$\tilde{Y}(f) = \frac{1}{2}\left[\tilde{X}(f)\,\tilde{h}(f)\right] \tag{2.1.10}$$

These equations provide insight into the equivalence of dealing with passband signals and their baseband representations. Figure 2.5 illustrates the parallel nature of these operations in baseband and passband.

Note: Communications Versus Channel Modeling Convention

In communications theory, the time domain is always the base domain in which all transforms are referenced. In channel modeling theory, for reasons that will become clear later, it is more powerful and convenient to view the *frequency* domain as a base domain. To avoid confusion between the conventions of communications and space-time channel modeling, the same Fourier transform definitions are used to move back and forth between time and frequency for both signals and channels. However, *frequency* is chosen as the base domain for baseband channels, which receive the lowercase $\tilde{h}$ in this domain.

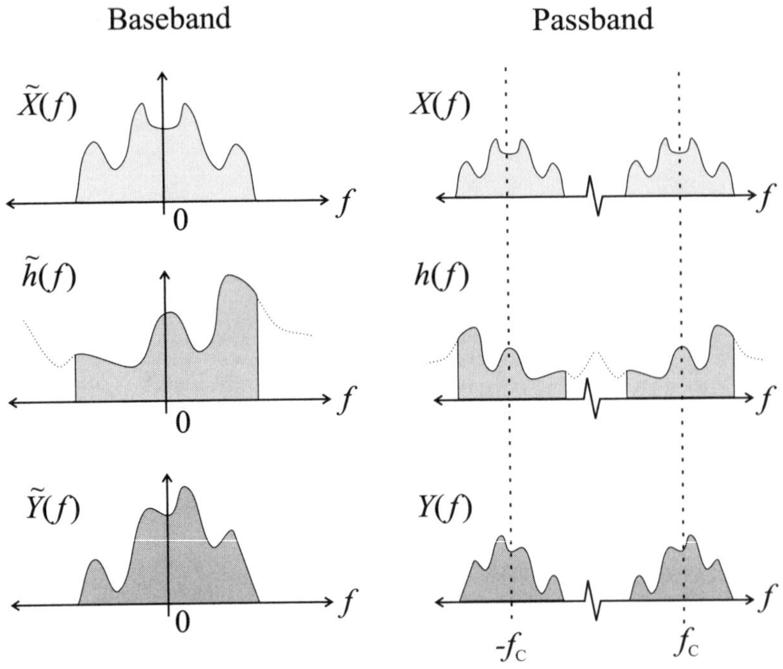

Figure 2.5 Spectral diagram of baseband and passband signals and channel.

If a channel is time-varying, neither convolution nor frequency-domain multiplication can be used to calculate signal transmission through the channel. For baseband transmission in a linear time-varying channel, the following input-output relationship must be used:

$$\tilde{y}(t) = \frac{1}{2} \int_{-\infty}^{+\infty} \tilde{X}(f)\, \tilde{h}(f,t)\, \exp(j2\pi f t)\, df \tag{2.1.11}$$

Equation (2.1.11) is basically the inverse Fourier transform of Equation (2.1.10) with a frequency-domain transfer function, $\tilde{h}(f,t)$, that is now also a function of time.

The input–output relationship of Equation (2.1.11) causes a new problem in the characterization of $\tilde{h}(f,t)$: The values of frequency, f, and time, t are no longer valid transform domain pairs as they were in the time-invariant system. For example, using the inverse Fourier transform on $\tilde{h}(f,t)$ produces a pure time-domain function that is useless for calculating the transmitted signal:

$$\times \quad \textbf{Incorrect:} \quad \tilde{h}(t) = \frac{1}{2} \int_{-\infty}^{+\infty} \tilde{h}(f,t)\, \exp(j2\pi f t)\, df$$

To circumvent this problem, the frequency and time dependencies of the channel must always be kept separate, even when performing spectral analysis on one or the other. This separation is discussed in greater detail in Section 2.3.

2.1.6 Detection Terminology

Detection is the operation performed by a receiver as it demodulates a passband signal. There are many ways to implement detection. For example, noncoherent receivers essentially take the magnitude of the received complex voltage signal $\tilde{y}(t)$ and throw away the phase information. Coherent receivers simultaneously detect the real and imaginary components of $\tilde{y}(t)$ so that both are available for baseband signal processing.

Regardless of technique, detection is described in using one or more real-valued functions, based on the complex baseband representation of a signal. Summarized in Table 2.1, there are four such functions:

In-Phase Component: This is the real part of a complex baseband voltage signal. Physically, the in-phase component arises by mixing the passband signal with $\cos(2\pi f_c t)$ and then low-pass filtering the result.

Quadrature Component: This is the imaginary part of a complex baseband voltage signal. The quadrature component arises by mixing the passband signal with $\sin(2\pi f_c t)$ and then low-pass filtering the result. Many modern receivers demodulate in-phase and quadrature components simultaneously. This type of receiver is called an IQ detector [Cou93].

Voltage Envelope: The voltage envelope is defined to be the magnitude of the complex baseband signal. A receiver operating on this principle is called a *noncoherent* receiver and uses an *envelope detector* to strip away the carrier and phase information from the signal.

Received Power: Similar to voltage envelope, the received power is defined to be the magnitude-squared of the complex baseband signal. Defined this way, received power has units of *Volts*^{2}. Received power is used in the calculation of a signal-to-noise or signal-to-interference ratio, which ultimately determines the theoretical limit of wireless channel capacity [Cov91].

These terms are used to analyze both signals and channels at baseband.

> *Note: Units of Power*
>
> As defined in Table 2.1, $P(t)$ is best described as *proportional* power, since it is proportional to the power of a signal, but does not have physical units of *Watts*. For example, if the signal is expressed in units of volts, then $P(t)$ would need to be multiplied by a constant impedance in order to produce units of *Watts*. This generic description of power rescues us from having to specify impedances, which would be highly specific to an application.

Table 2.1 Received Signal Functions Used in Complex Baseband Analysis

In-Phase Component:	$I(t)$	$=$	$\text{Real}\{\tilde{y}(t)\}$
Quadrature Component:	$Q(t)$	$=$	$\text{Imag}\{\tilde{y}(t)\}$
Voltage Envelope:	$R(t)$	$=$	$\lvert\tilde{y}(t)\rvert$
Received Power:	$P(t)$	$=$	$\lvert\tilde{y}(t)\rvert^2$

2.2 Channel Coherence

The most important concept in describing the wireless channel is *channel coherence.* This section presents an overview of the many types of coherence that a wireless channel exhibits.

2.2.1 Coherence Versus Selectivity

Fading is a general term used to describe a wireless channel affected by some type of *selectivity.* A channel has selectivity if it varies as a function of either time, frequency, or space. The opposite of selectivity is *coherence.* A channel has coherence if it does not change as a function of time, frequency, or space over a specified "window" of interest.

Indeed, wireless channels may be functions of time, frequency, *and* space. The most fundamental concept in channel modeling is classifying the three possible channel dependencies of time, frequency, and space as either coherent or selective. In order to keep track of these dependencies in the wireless channel, we will discuss each individually before developing the mathematics of joint characterization.

2.2.2 Temporal Coherence

A wireless channel has temporal coherence if the envelope of the unmodulated carrier wave does not change over a time window of interest. Mathematically, we express this condition in terms of a narrowband (no frequency dependence), fixed (no spatial dependence) channel, $\tilde{h}(t)$:

$$\left|\tilde{h}(t)\right| \approx V_0, \qquad \text{for } |t - t_0| \leq \frac{T_c}{2} \tag{2.2.1}$$

where V_0 is some constant voltage, T_c is the size of the time window of interest, and t_0 is some arbitrary moment in time. The largest value of T_c, on average, for which Equation (2.2.1) holds is called the *coherence time* and is the approximate time window over which the channel appears static. Figure 2.6 illustrates these definitions.

Note that the channel in Equation (2.2.1) is in complex phasor form and is independent of carrier frequency. Naturally, a transmitted wave will produce sinu-

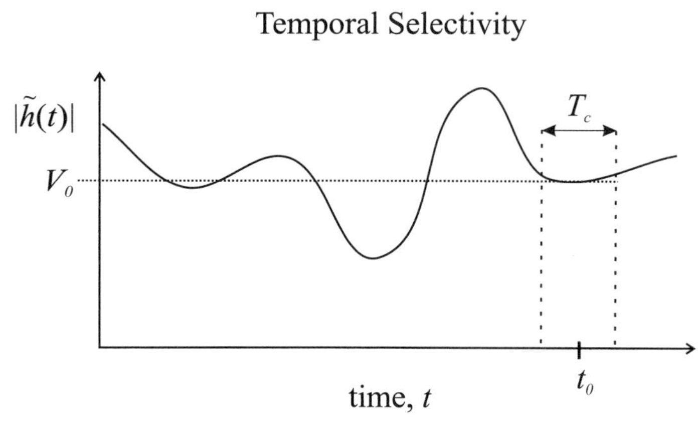

Figure 2.6 A time-varying channel.

soidal oscillations as a function of time, but the definition of temporal coherence is concerned with the *envelope* of those oscillations.

> *Note: Formal Definitions of Coherence*
> As is the case for each type of coherence defined in this chapter, the definition in Equation (2.2.1) is subjective, since the condition $\left|\tilde{h}(t)\right| \approx V_0$ is open to interpretation; it neither defines a formal metric for channel selectivity nor presents a threshold for characterizing a truly coherent channel. A formal treatment of coherence, however, requires much more background in stochastic channel modeling than has been presented thus far.

In the microwave and millimeter frequency regime the most common cause of temporal incoherence is motion by either the transmitter or the principle scatterers in the propagation environment. Temporal channel fading can degrade the performance of a wireless communication system. If the transmitted data rate is comparable to the temporal coherence, it becomes extremely difficult for the receiver to demodulate the transmitted signal reliably. Fluctuations due to data modulation and fluctuations due to the time-varying channel occur at the same time scale, causing catastrophic distortion.

One method for reliable digital communications in a time-varying channel is to transmit data using symbols that are much larger than the channel coherence time and rely on long-period averaging to filter out the fluctuations of the carrier from each symbol. When the envelope of the carrier wave fluctuates at a rate faster than the transmitted symbol rate, the channel is said to be *fast fading*.

Another method for reliable communications in a time-varying channel is to transmit data using symbols that are much smaller than the channel coherence time. For this case, the time-varying channel appears static over the short symbol period. When the envelope of the carrier wave fluctuates at a rate slower than the transmitted symbol rate, the channel is said to be *slow fading*. An illustration of the differences between fast and slow fading is shown in Figure 2.7.

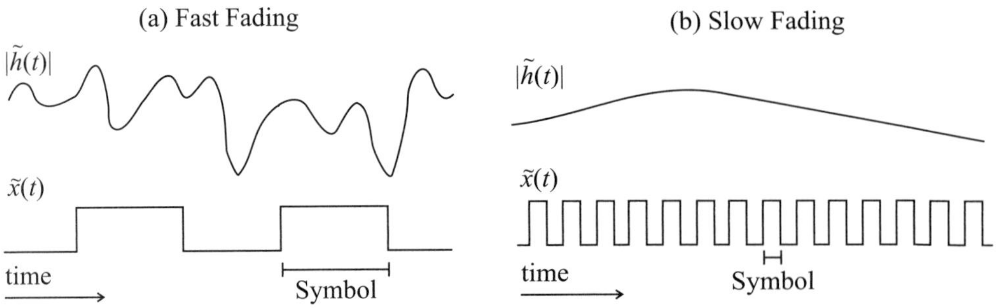

Figure 2.7 Fast and slow fading for simple square pulse symbols.

2.2.3 Frequency Coherence

A wireless channel has frequency coherence if the magnitude of the carrier wave does not change over a frequency window of interest. This window of interest is usually the bandwidth of the transmitted signal. Mathematically, we express the condition of frequency coherence in terms of the static (no time dependence), fixed channel, $\tilde{h}(f)$:

$$\left|\tilde{h}(f)\right| \approx V_0, \qquad \text{for } |f_c - f| \leq \frac{B_c}{2} \tag{2.2.2}$$

where V_0 is some constant amplitude, B_c is the size of the frequency window of interest, and f_c is the center carrier frequency. The largest value of B_c for which Equation (2.2.2) holds is called the *coherence bandwidth* and is the approximate range of frequencies over which the channel appears static. Figure 2.8 illustrates these definitions.

The loss of frequency coherence in a wireless communications system is caused by the dispersion of multipath propagation. Since each received multipath wave has traveled a different path from the transmitter, the same transmitted signal will arrive at the receiver as a cluster of symbols, each with a unique time delay. Thus, in the time domain, a dispersive channel introduces *intersymbol interference.* In the frequency domain, a dispersive channel has peaks and valleys across the bandwidth of interest. This behavior in the frequency domain gives rise to two distinct classifications of fading in wireless communications. A wireless channel with a coherence bandwidth that is less than the bandwidth of a transmitted signal is said to have *frequency-selective fading.* A channel with a coherence bandwidth that is greater than the transmitted signal bandwidth is said to have *frequency-flat fading.* A sketch of frequency-selective and frequency-flat channels is found in Figure 2.9.

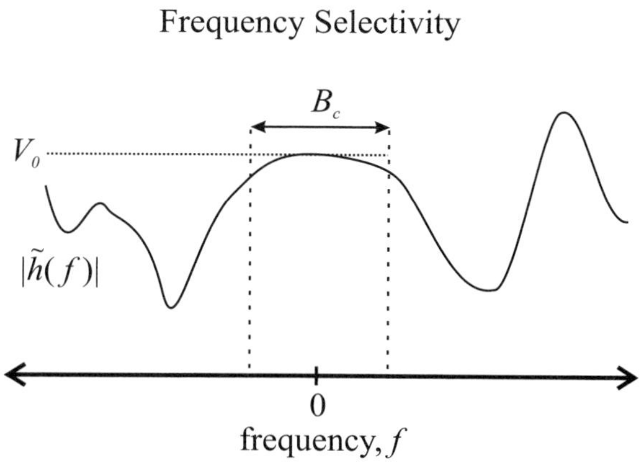

Figure 2.8 A frequency-varying channel.

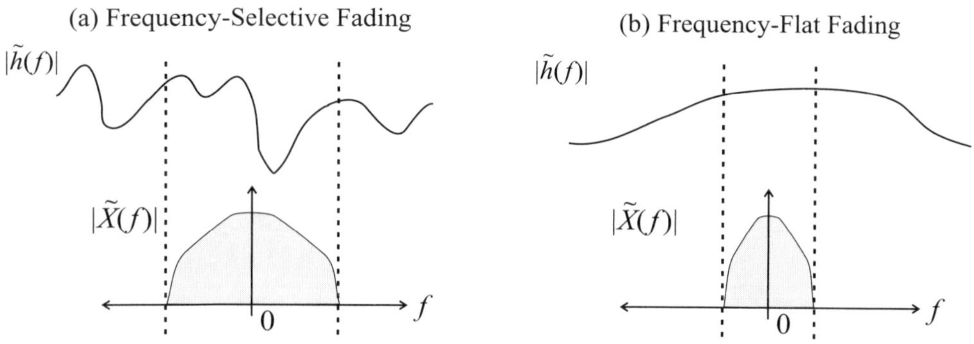

Figure 2.9 Frequency-selective and frequency-flat fading for a sample transmitted signal spectrum, $\tilde{X}(f)$.

2.2.4 Spatial Coherence

A wireless channel has spatial coherence if the magnitude of the carrier wave does not change over a spatial displacement of the receiver. Once again, we express the condition of spatial coherence in terms of the static narrowband channel, $\tilde{h}(r)$, which is a function of one-dimensional (1D) space, r:

$$\left|\tilde{h}(r)\right| \approx V_0, \qquad \text{for } |r - r_0| \leq \frac{D_c}{2} \tag{2.2.3}$$

where V_0 is some constant amplitude, D_c is the size of the position displacement, and r_0 is an arbitrary position in space. The largest value of D_c for which Equation (2.2.3) holds is called the *coherence distance* and is the approximate distance that a wireless receiver can move with the channel appearing to be static. Figure 2.10 illustrates these definitions.

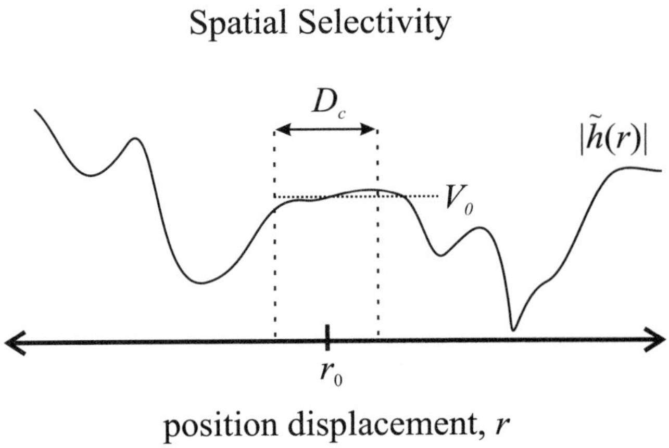

Figure 2.10 A space-varying channel (using one-dimensional position).

Note that for a wireless receiver moving in three-dimensional space, the coherence distance is a function of the direction that the receiver travels, since r in Equation (2.2.3) is scalar position. The three dimensions of free space make the study of spatial coherence more difficult than the scalar quantities of temporal or frequency coherence. We introduce three-dimensional spatial channels in Section 3.2.3.

While frequency incoherence is a result of multipath waves arriving with many different time delays, spatial incoherence is caused by multipath waves arriving from many different directions in space. These multipath waves create pockets of constructive and destructive interference so that the received signal power does not appear to be constant over small changes in receiver position. Thus, this type of channel exhibits *spatial selectivity.* If the distance traversed by a receiver is greater than the coherence distance of the channel, we say that the channel experiences *small-scale fading.*

Large-Scale Versus Small-Scale Fading

The converse of small-scale fading, *large-scale fading,* refers to fluctuations in spatially-averaged received power due to shadowing and scattering of objects in the propagation environment. Typically, small-scale fluctuations occur when a receiver moves a distance comparable to the size of the electromagnetic wavelength of the carrier. Large-scale fluctuations occur when the receiver moves over many wavelengths. An example of small-scale and large-scale fading for a received signal level as a function of position is illustrated in Figure 2.11.

It is quite possible in a small-scale fading channel to experience near-zero received power levels at points in space even if the large-scale power level is high. This type of dip in the power level is called a *null.* Conversely, a point in space that leads to maxima in the received power levels is called a *peak.*

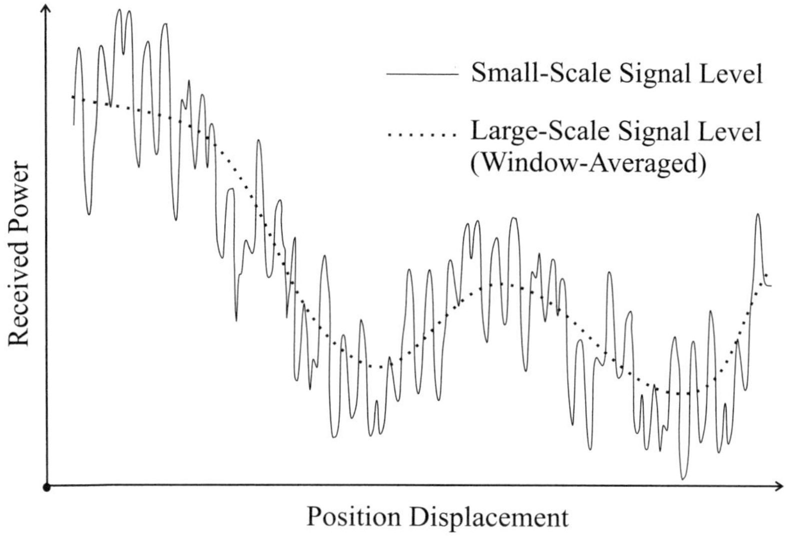

Figure 2.11 Small-scale and large-scale spatial channel fading.

Mobile Fading

For *mobile* wireless communications, spatial incoherence in a channel leads directly to temporal incoherence. As a receiver moves through space with constant velocity, the fading may be treated as a function of time instead of space by relating the position r and the time t by a simple equation of motion:

$$r = r_0 + vt \tag{2.2.4}$$

where r_0 is the position at time $t = 0$ (often set to zero for simplicity) and v is the speed of the moving receiver. Thus, small-scale fading causes temporal incoherence for wireless systems with mobile receivers.

> *Note: Not-So-Fast Fading*
>
> Some authors have confused the terms used for temporal and spatial incoherence and will refer to small-scale fading as *fast fading*. This is particularly unfortunate, since the motion-induced temporal fading that arises from small-scale fading is almost never really fast fading: The coherence time of the temporal fades is *always* larger than the symbol period of modern communications systems with useful data rates.

2.3 Using the Complete Baseband Channel

The complete wireless baseband channel, $\tilde{h}(f, r, t)$, is a function of frequency, position, and time. This section discusses the representations of the baseband channel with joint dependencies and how they related to the wireless transmission of signals.

2.3.1 Spectral Domain Representations

The definitions of Fourier transforms are useful tools in channel analysis. Since the full channel representation, $\tilde{h}(f, r, t)$, is a function of frequency, space, and time, transforms may be defined for each of these domains. There are three possible spectral domains:

delay domain: The spectral domain of frequency, f, is the *delay* domain, with dependence denoted by τ. Since f has units of *frequency*, the reciprocal units of delay, τ, must be *time.*

wavenumber domain: The spectral domain of position, r, is the *wavenumber* domain, with dependence denoted by k. Since position r has units of *distance*, wavenumber k must have reciprocal units of *radians per unit distance.*

Doppler domain: The spectral domain of time, t, is the *Doppler frequency* domain, or just the *Doppler* domain. This dependence is denoted by Doppler frequency variable ω.

The three spectral domains are summarized in Table 2.2 along with the mathematical definitions for their Fourier transformations. To perform a transformation into a spectral domain, insert the channel function into the brackets of the expression in the column of Table 2.2 marked *Transform.* To transform out of the spectral domain, insert the channel function into the brackets of the expression in the column marked *Inverse Transform.* Appendix B presents examples of forward and inverse transforms of these dependencies.

Table 2.2 Fourier Transform Definitions for Each Channel Dependency

DOMAIN PAIRS	TRANSFORM	INVERSE TRANSFORM
frequency, $f \longleftrightarrow$ delay, τ	$\int_{-\infty}^{+\infty} \{\} \exp(j2\pi\tau f) df$	$\int_{-\infty}^{+\infty} \{\} \exp(-j2\pi\tau f) d\tau$
position, $r \longleftrightarrow$ wavenumber, k	$\int_{-\infty}^{+\infty} \{\} \exp(-jkr) dr$	$\frac{1}{2\pi} \int_{-\infty}^{+\infty} \{\} \exp(jkr) dk$
time, $t \longleftrightarrow$ Doppler, ω	$\int_{-\infty}^{+\infty} \{\} \exp(-j\omega t) dt$	$\frac{1}{2\pi} \int_{-\infty}^{+\infty} \{\} \exp(j\omega t) d\omega$

> *Note: 2π or not 2π?*
>
> A keen observer will note that the frequency-delay Fourier pairs (f and τ), are defined differently than either time-Doppler (t and ω) or space-wavenumber (r and k). While the latter are *units* and *radian units*$^{-1}$ Fourier pairs, frequency and delay are defined as *units* and *units*$^{-1}$ Fourier pairs. This definition was chosen not only to follow conventions in the wireless literature but also to emphasize the difference between delay-frequency and time-Doppler relationships, which are easily confused.

The transformations of Table 2.2 may operate on $\tilde{h}(f, r, t)$ in any order and combination. When one or more of the channel dependencies have been transformed into the spectral domain, a *transfer function* has been created. A transfer function will be written in the following form throughout this book:

$$\tilde{H}(\text{transformed dependencies; untransformed dependencies})$$

where the transformed dependencies are either τ, k, or ω and the untransformed dependencies are either f, r, or t. For example, the transfer function $\tilde{H}(\tau, \omega; r)$ represents a Fourier transform of the channel, $\tilde{h}(f, r, t)$, with respect to both frequency, f, and time, t.

Various types of transfer functions are commonly used in the wireless literature and have developed specific terminology. Several of these specific transfer functions are listed below:

- $\tilde{H}(\tau; r, t)$: This transfer function is called the *channel impulse response* (CIR). It is the received baseband signal after sending a very short modulated pulse at the transmitter [Rap02a].
- $\tilde{H}(k; f, t)$: This transfer function of this general form is called the *radio channel* by the International Telecommunications Union (ITU). It is the formal definition used by the organization to characterize wireless propagation [Fle00].
- $\tilde{H}(\tau, k, \omega)$: This transfer function is a Fourier transform of $\tilde{h}(f, r, t)$ for all three dependencies. We define this to be the *complete transfer function.* Complete transfer functions are useful in stochastic channel modeling when defining a join power spectral density.

2.3.2 General Signal Transmission

Armed with the complete definition of the baseband space–time channel, it is now possible to study the wireless transmission of a signal, $\tilde{x}(t)$, through a channel that varies with space, time, and frequency. For a linear, time-varying channel, the received signal as a function of time and space is the result of the following integration:

$$\tilde{y}(t, r) = \frac{1}{2} \int_{-\infty}^{+\infty} \tilde{H}(\tau; t, r)\tilde{x}(t - \tau)\, d\tau \tag{2.3.1}$$

The result in Equation (2.3.1) may also be calculated using the untransformed channel, $\tilde{h}(f, r, t)$, rather than its delay transform, $\tilde{H}(\tau; r, t)$:

$$\tilde{y}(t, r) = \frac{1}{2} \int_{-\infty}^{+\infty} \tilde{h}(f, t, r) \tilde{X}(f) \exp(j2\pi f t)\, df \tag{2.3.2}$$

These equations are the most general input–output relationships for signal transmission through a wireless channel with a single transmitter antenna and a single receiver antenna. Theorem 2.1 sketches a proof of Equation (2.3.1) and Equation (2.3.2).

Theorem 2.1: Space–Time Transmission

Statement: Equation (2.3.1) and Equation (2.3.2) are both valid transmission equations for the baseband space–time channel.

Proof: Let us use Equation (2.1.11) as a starting point to show that Equation (2.3.2) is a straightforward extension of the time-varying, wideband channel; we simply add the spatial dependency, r:

$$\tilde{y}(t) = \frac{1}{2} \int_{-\infty}^{+\infty} \tilde{X}(f) \tilde{h}(f, t) \exp(j2\pi f t)\, df \longrightarrow \tilde{y}(t, r) = \frac{1}{2} \int_{-\infty}^{+\infty} \tilde{X}(f) \tilde{h}(f, t, r) \exp(j2\pi f t)\, df$$

Once Equation (2.3.2) is established, we can perform the following manipulations:

$$\begin{aligned}
\tilde{y}(t, r) &= \frac{1}{2} \int_{-\infty}^{+\infty} \underbrace{\left[\int_{-\infty}^{+\infty} \tilde{x}(\zeta) \exp(-j2\pi f \zeta)\, d\zeta \right]}_{\tilde{X}(f)} \tilde{h}(f, t, r) \exp(j2\pi f t)\, df \\
&= \frac{1}{2} \int_{-\infty}^{+\infty} \tilde{x}(\zeta) \left[\int_{-\infty}^{+\infty} \tilde{h}(f, t, r) \exp(j2\pi f [t - \zeta])\, df \right] d\zeta \\
&= \frac{1}{2} \int_{-\infty}^{+\infty} \tilde{x}(\zeta) \tilde{H}(t - \zeta; t, r)\, d\zeta
\end{aligned}$$

After making the change-of-variable $\zeta = t - \tau$, this expression reduces to Equation (2.3.1).

2.3.3 Time-Invariant Channel Transmission

A wireless channel may be broken into two parts: a *static* portion that does not vary with time and a *transient* portion that does. This may be written as

$$\tilde{h}(f, r, t) = \underbrace{\tilde{h}(f, r)}_{\text{static}} + \underbrace{\delta\tilde{h}(f, r, t)}_{\text{transient}} \tag{2.3.3}$$

In many types of channel, the transient portion is nonexistent or negligible, and the engineer may use only the static portion. This is the *time-invariant* channel.

The linear, time-invariant wireless channel depends only on frequency, f, and position, r. For such a channel, the general transmission equation of Equation (2.3.1) reduces to

$$\tilde{y}(t, r) = \frac{1}{2} \int_{-\infty}^{+\infty} \tilde{H}(\tau; r)\tilde{x}(t - \tau)\, d\tau \tag{2.3.4}$$

Note that Equation (2.3.4) is in the form of a convolution integral. Therefore, the relationship for the received signal, $\tilde{y}(t, r)$, may be written as the convolution of transmitted signal and channel:

$$\tilde{y}(t, r) = \frac{1}{2} \left[\tilde{H}(\tau; r) \Big|_{\tau = t} \otimes \tilde{x}(t) \right] \tag{2.3.5}$$

Of course, this convolution may also be written as the product of the Fourier pairs of Equation (2.3.5):

$$\tilde{Y}(f; r) = \frac{1}{2} \left[\tilde{h}(f, r)\tilde{X}(f) \right] \tag{2.3.6}$$

The static channel arises repeatedly in channel analysis either when the significant scatterers in the propagation environment are genuinely static or when the time variations are present but occur on a scale that does not affect the operation of the receiver.

2.3.4 Mobile Receiver Transmission

Another special case of the basic channel input–output relationships is that of a mobile receiver. The motion of a receiver at any given instant may be characterized by a velocity, v, which represents the speed and direction of the receiver. Now position, r, becomes a function of time, t, according to the simple equation of constant-velocity motion: $r = vt$. Therefore, the *mobile* received signal is expressed solely as a function of time, t:

$$\tilde{y}(t) = \tilde{y}(t, r) \Big|_{r = vt} \tag{2.3.7}$$

which may be calculated by

$$\tilde{y}(t) = \frac{1}{2} \int_{-\infty}^{+\infty} \left[\tilde{H}(\tau; r, t) \Big|_{r = vt} \right] \tilde{x}(t - \tau)\, d\tau \tag{2.3.8}$$

The mobile case emphasizes the flexibility of the full-dependency baseband channel representation, $\tilde{h}(f, r, t)$. As a function of position, the channel representation may be used to study receivers that operate with antenna arrays with elements at various positions in space. After the time-velocity substitution, however, the channel representation may be used to study single-antenna receivers moving through space. The original space and time dependencies of the channel cannot be recovered once the substitution is made.

2.4 Chapter Summary

This chapter introduced the terminology and mathematics used in the baseband analysis of space–time wireless channels. The following topics were discussed:

- Complex baseband representation is a useful analytical tool for studying communications systems that modulate signals onto carriers.
  - ▷ Baseband equivalent signals have a useful spectral interpretation.
  - ▷ Baseband analysis is independent of carrier frequency.
  - ▷ Complex baseband signals contain all the useful detection information.
- A channel may vary as a function of time, space, or frequency. Fluctuations as a function of each dependency may be characterized as follows:
  - ▷ A channel that does not change significantly has *coherence.*
  - ▷ A channel that changes significantly has *selectivity.*
  - ▷ A *fading* channel has selectivity that affects communications.
- There are three Fourier pairs that assist in channel analysis:
  - ▷ Frequency, $f \longleftrightarrow$ Delay, τ
  - ▷ Position, $r \longleftrightarrow$ Wavenumber, k
  - ▷ Time, $t \longleftrightarrow$ Doppler Frequency, ω

The theory presented thus far has been mathematical and generic. Except for the space-varying analysis, the concepts in this chapter are valid for analyzing *any* communications system with a stochastic time-varying channel. However, the wireless channel has additional characteristics and structure related to the physics of free-space radio wave propagation. Chapter 4 creates a link between generic channel modeling theory and the physics of the small-scale channel.

PROBLEMS

1. Sketch the real, imaginary, and magnitude portions of $\tilde{X}(f)$ for the following time-domain signals (like Figure 2.1):

   a. $\tilde{x}(t) = tu\left(t - \frac{T}{2}\right)$

b. $\tilde{x}(t) = j\text{sn}\left(\frac{t}{T}\right)$

c. $\tilde{x}(t) = \exp(-\alpha t - j\beta t)\text{u}(t)$

d. $\tilde{x}(t) = t\exp(-\alpha t)\text{u}(t)$

2. For each given complex baseband equivalent signal, calculate the real-valued passband signal, $\mathcal{M}\{\}$:

a. $\tilde{x}(t) = f(t) + jg(t)$

b. $\tilde{y}(t) = \cos(20\pi t)$

c. $\tilde{h}(t, r) = V(t)\exp(j\phi(t) - jk_0 r)$

d. $\tilde{x}(r) = f(r) + jg(r)$

e. $\tilde{h}(t) = 7 + j7$

3. For each given passband signal modulated to a carrier f_c, calculate the complex baseband equivalent signal, $\mathcal{M}^{-1}\{\}$:

a. $y(t) = f(r)\cos(2\pi f_c t + \frac{\pi}{4})$

b. $x(t) = \text{sn}(t)\cos(2\pi f_c t)$

c. $h(t, r) = 10\cos(2\pi f_c t - k_0 r - \frac{\pi}{3})$

d. $h(t, r) = \sin(\pi r)\sin(2\pi f_c t)$

4. Sketch both the baseband and passband frequency-domain representations of the following baseband signals:

a. $\tilde{x}(t) = \text{sn}(t)$

b. $\tilde{z}(t) = \text{J}_0(t)$

c. $\tilde{y}(t) = \text{sn}^2(10t)$

d. $\tilde{y}(t) = \text{sn}(t - 1/f_c)$

5. For a band-limited signal, $x(t)$, one useful metric for the amount of information carried by the signal is its total amount of nonzero spectrum. Use this simple rule of thumb to explain why the baseband representation of a real-valued modulated signal must be complex to avoid loss of information. Hint: think of the baseband representation as two real-valued, band-limited functions, $\tilde{x}(t) = x_a(t) + jx_b(t)$.

6. For many simple types of modulation, only in-phase signal components are transmitted. In other words, the baseband input signal, $\tilde{x}(t)$, is real-valued. Does this imply that the channel output, $\tilde{y}(t)$, will also be real-valued? Explain why or why not.

7. A wireless local area network (WLAN) is operating inside an office building. For the *downlink* of a typical WLAN, a fixed wireless access point in the building transmits a signal to a user terminal (a laptop computer or handheld data device). In this scenario, which type of downlink channel fading is caused by the following physical phenomenon?

a. People moving through an office building, opening and shutting doors.

b. Furniture in a room that scatters radio waves in many different *directions*.

c. Interior walls create multipath waves that arrive at the receiver with different *time delays*.

d. The receiver is moving through the room described in (b).

8. Suppose we have a channel, $\tilde{h}(f,t,r)$, that varies with time, frequency, and space. Explain the possible **errors** or **limitations** of the following channel transformations:

a. The frequency domain is transformed to the time domain:

$$\tilde{h}(t,r) = \int_{-\infty}^{+\infty} \tilde{h}(f,t,r)\exp(j2\pi ft)\,df$$

b. The time domain is converted to the frequency domain:

$$\tilde{H}(f,r) = \int_{-\infty}^{+\infty} \tilde{h}(f,t,r)\exp(-j2\pi ft)\,dt$$

c. A velocity substitution is made to represent a mobile receiver channel:

$$\tilde{h}(f,t) = \tilde{h}(f,t,r)\Big|_{r=vt}$$

9. Below is an expression for a time-varying, frequency-selective channel that has exponentially decaying dispersion and an oscillating phase:

$$\tilde{h}(\tau;t) = \exp(-\alpha\tau + j\beta t)\,\mathrm{u}(\tau)$$

A box pulse of width T, $x(t) = \mathrm{u}\left(t - \frac{T}{2}\right)$, is sent through this radio channel. Calculate the following:

a. The in-phase portion of the received signal.

b. The quadrature portion of the received signal.

c. The received signal *envelope*.

d. The power of the received signal.

10. There are some types of communication systems that operate at baseband without modulation to a carrier frequency. If we are modeling a physical channel with $f_c = 0$, what does that imply about $\tilde{h}(f,t,r)$ and $h(f,t,r)$?

Chapter 3

RANDOM FADING CHANNELS

Groves of trees rustling in the wind scramble received power from one second to the next in a point-to-point microwave link. Or a cellular handset drops a call after moving just a few centimeters from an operable location. Or the tap-delay line filter of a linear equalizer becomes unstable, incapable of canceling the intersymbol interference experienced by a wireless receiver. While the causes and effects of each channel-related problem are varied, the channel analysis is nearly identical for each case - *if rigorous stochastic channel modeling is employed.*

The goal of this chapter is to develop the terminology, definitions, and basic concepts of modeling a random wireless channel that can be a function of time, frequency, and receiver position in space. While the task of joint characterization may seem daunting, only a few basic concepts are required. In fact, this chapter applies the concept of *duality* to show that understanding random fading in one dependency leads immediately to understanding in others.

The chapter is broken into the following sections:

- Section 3.1: Concept of *correlation* in a random channel process.
- Section 3.2: Definition of a random process *power spectral density.*
- Section 3.3: Representation of random channels with multiple dependencies.
- Section 3.4: Definition of RMS spectral spreads.
- Section 3.5: Summary of important concepts.

By the conclusion of the chapter, the reader will be familiar with most of the terms and constructs of random process theory used to describe unpredictable space–time wireless channels. A short review of random process basics is also included in Appendix C for reference.

3.1 Channel Correlation

This section introduces the principle of stochastic channel *correlation*. Autocorrelation functions are then defined for the complex baseband channel in frequency, time, and space.

3.1.1 The Meaning of Correlation

In probability theory, *correlation* is a measure of conditional predictability, usually made between two observations of a random event. When we compare two random variables, X and Y, we say that X and Y are *dependent* if an observation of X provides some predictive information about an observation of Y, and vice versa. *Correlation* is one measure of dependency between random variables. Increased correlation between random events implies increased predictability. We would expect a strong correlation, for example, between random events such as the amount of sunshine and the average temperature of a given day. After all, sunny days are usually warmer than cloudy days.

If X and Y are *uncorrelated*, then knowing the value of X does not provide predictive information about Y, and vice versa. We would expect no correlation, for example, between random events such as the amount of sunshine and the monetary winnings of a game of poker played on the same day. (Those two events are also likely to be *independent* as well.) We can define the condition for uncorrelated random variables with more mathematical rigor:

$$\mathrm{E}\{XY\} = 0 \text{ for uncorrelated } X \text{ and } Y \tag{3.1.1}$$

Thus, if the above ensemble average evaluates to 0, we say that X and Y are uncorrelated. Equation (3.1.1) is only valid for zero-mean random variables.

The concept of correlation is also useful for describing the evolution of random processes - even *complex* random processes. Consider the time-varying random process, $\tilde{x}(t)$, in Figure 3.1, used here to describe correlation in qualitative terms. While we may not know the values of each realization of the random process at times t_1 and t_2, we do know that t_1 and t_2 are close in time. Thus, we expect that knowing either $\tilde{x}(t_1)$ or $\tilde{x}(t_2)$ provides a close estimate of the other - they are highly *correlated*. Sample values of $\tilde{x}(t)$ taken at t_1 and t_3 are farther apart in time and less correlated. Sample values of $\tilde{x}(t)$ taken at t_1 and t_4 are very far apart in time and probably uncorrelated, since knowledge of $\tilde{x}(t_1)$ gives virtually no information of $\tilde{x}(t_4)$; the random process changes a great deal over the interval $[t_1, t_4]$. As a rule of thumb, correlation between samples in a random process decreases as the time or distance separating them increases, though the decrease is not always monotonic.

The relationship between sample correlation and sample separation provides the starting point for characterizing the behavior of random process evolution. This type of analysis is a study in self-correlation or *autocorrelation*. The next section provides the rigorous definition of random process autocorrelation.

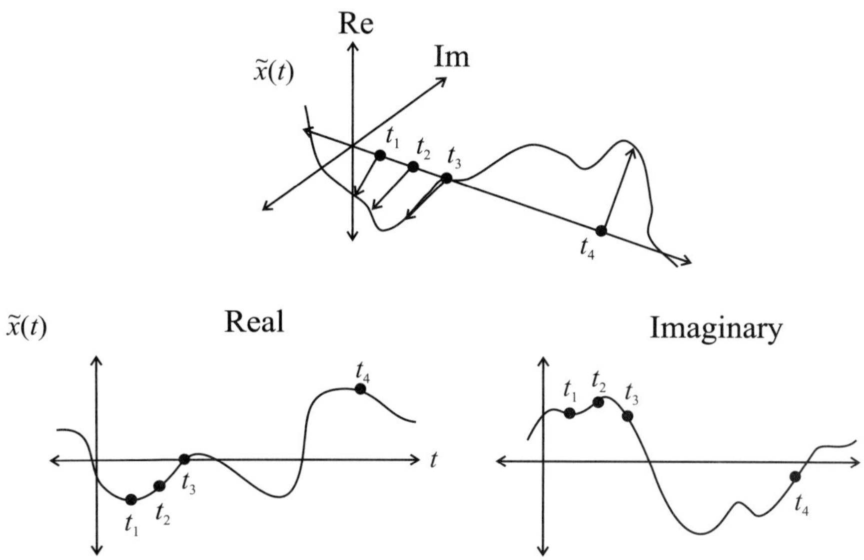

Figure 3.1 Correlation between samples with different separations in a complex random process (sketch of single realization).

3.1.2 Autocorrelation Relationships

The most common way to characterize the development of a stochastic process is by calculating the autocorrelation of a function. The definition for an autocorrelation, $C_{\tilde{h}}(t_1, t_2)$, of a time-varying stochastic channel, $\tilde{h}(t)$, is

$$C_{\tilde{h}}(t_1, t_2) \doteq \mathrm{E}\left\{\tilde{h}(t_1)\tilde{h}^*(t_2)\right\} \tag{3.1.2}$$

Equation (3.1.2) captures the time-evolution of $\tilde{h}(t)$ by averaging the products of all samples in the random process ensemble at two different points in time, t_1 and t_2. Thus, $C_{\tilde{h}}(t_1, t_2)$ is a snapshot of the typical correlation behavior for a random channel $\tilde{h}(t)$.

Most of the stochastic processes studied in this book and in channel modeling theory are *wide-sense stationary* (WSS). The autocorrelation of a WSS stochastic process, by definition, depends only on the difference in time between t_1 and t_2. In other words, the correlation behavior is invariant of absolute time:

$$C_{\tilde{h}}(t_1, t_2) = C_{\tilde{h}}(t_0 + t_1, t_0 + t_2) \quad \text{for all } t_0 \tag{3.1.3}$$

Therefore, a WSS autocorrelation is usually written as the function of one time variable, Δt, which is equal to the difference, $t_1 - t_2$. This WSS definition for autocorrelation is shown below:

$$C_{\tilde{h}}(\Delta t) = \mathrm{E}\left\{\tilde{h}(t_1)\tilde{h}^*(t_1 - \Delta t)\right\} \tag{3.1.4}$$

Similar WSS autocorrelation definitions exist for stochastic channels that are functions of frequency, f, and space, r.

There is a second condition that must hold for a stochastic process to be truly WSS. In addition to being *autocorrelation stationary*, a stochastic process must also be *mean stationary*. Using the time-varying baseband channel as an example, mean stationarity holds if the value $\mathrm{E}\left\{\tilde{h}(t)\right\}$ is *not* a function of time, t. Most processes encountered in real-life modeling fail the WSS test due to their autocorrelation statistics. Be aware, however, that certain processes exist having nonstationary means *and* stationary autocorrelations.

> *Note: Order of a Statistic*
>
> An autocorrelation function is considered a *second-order* statistic because it characterizes behavior between two samples within the random process. The term *order* refers to the number of samples involved in the computation of the statistic. Here are some example statistics:
>
> $$\overbrace{\mathrm{E}\left\{|\tilde{h}(t)|\right\}}^{\text{first-order}} \qquad \overbrace{\mathrm{E}\left\{\tilde{h}(t_1)\tilde{h}(t_2)\right\}}^{\text{second-order}} \qquad \overbrace{\mathrm{E}\left\{\tilde{h}(t_1)|\tilde{h}(t_2)||\tilde{h}(t_3)|^2\right\}}^{\text{third-order}}$$

3.1.3 Autocovariance

There are several useful variations on the definition of an autocorrelation function. First, if the process is a zero-mean process, then the autocorrelation is said to be an *autocovariance*. Using the time-varying channel as an example, if $\mathrm{E}\{\tilde{h}(t)\} = 0$, then the function $C_{\tilde{h}}(\Delta t)$ is an autocovariance function. Otherwise, the following definition for autocovariance, $\bar{C}_{\tilde{h}}(\Delta t)$, may be used, which removes the mean value, $\tilde{\mu}$, of the WSS process:

$$\begin{aligned}\bar{C}_{\tilde{h}}(\Delta t) &= \mathrm{E}\left\{\left[\tilde{h}(t_0) - \tilde{\mu}\right]\left[\tilde{h}^*(t_0 + \Delta t) - \tilde{\mu}^*\right]\right\}, \text{ where } \tilde{\mu} = \mathrm{E}\left\{\tilde{h}(t)\right\} \\ &= C_{\tilde{h}}(\Delta t) - |\tilde{\mu}|^2\end{aligned} \tag{3.1.5}$$

For many autocorrelation functions of the complex baseband channel, $\tilde{\mu} = 0$ and the autocorrelation *is* an autocovariance. The definition in Equation (3.1.5) is most useful when studying random processes of envelope and power, which have a positive mean value.

3.1.4 Unit Autocovariance

A second useful definition, the *unit autocovariance*, is an autocovariance function that has been normalized against the mean power of the process. In terms of the time-varying stochastic channel, mean power is equal to the autocorrelation evaluated at $\Delta t = 0$:

$$C_{\tilde{h}}(0) = \mathrm{E}\left\{\tilde{h}(t)\tilde{h}^*(t)\right\} = \mathrm{E}\left\{|\tilde{h}(t)|^2\right\} \tag{3.1.6}$$

Thus, the full definition for a unit autocovariance, hereafter denoted $\varrho_{\tilde{h}}(\Delta t)$, is given in terms of the autocorrelation by

$$\varrho_{\tilde{h}}(\Delta t) = \frac{C_{\tilde{h}}(\Delta t) - |\tilde{\mu}|^2}{C_{\tilde{h}}(0) - |\tilde{\mu}|^2}, \text{ where } \tilde{\mu} = \mathrm{E}\left\{\tilde{h}(t)\right\} \tag{3.1.7}$$

The normalization of the unit autocovariance makes it the most convenient, dimensionless measure of correlation within a random process. It can be shown that $|\varrho_{\tilde{h}}(\Delta t)| \leq 1$ for all values of Δt. A unit autocovariance of 1 indicates perfect correlation, while a value of 0 indicates the absence of correlation - regardless of the magnitude of $\tilde{h}(t)$.

The definition of unit autocovariance will arise in the study of random *envelope* processes; most formalized definitions of channel coherence are based on the unit autocovariance of received voltage envelope.

3.2 Power Spectral Density (PSD)

In the previous section, we defined the autocorrelation method for describing how a random process evolves in time (or frequency, or space). In this section, we present an alternative way of analyzing a random process by using Fourier analysis. It turns out that the autocorrelation method and the power spectral density method in this section are related by the Wiener-Khintchine theorem, providing different perspectives of the same information.

3.2.1 Correlation in Spectral Domains

Whenever a mathematical operation is performed on a random process, a new random process describes the result. This truth applies to the Fourier transform; the Fourier transform of a random process is itself a random process. Thus, transforming a random time-varying channel, $\tilde{h}(t)$, into a random frequency-varying channel produces a new random process, $\tilde{H}(\omega)$, that we can analyze with ensemble statistics.

One possible analysis is to define an autocorrelation, $C_{\tilde{H}}(\omega_1, \omega_2)$, for characterizing the development of $\tilde{H}(\omega)$ with respect to Doppler. The definition of this would be

$$C_{\tilde{H}}(\omega_1, \omega_2) \doteq \mathrm{E}\left\{\tilde{H}(\omega_1)\tilde{H}^*(\omega_2)\right\} \tag{3.2.1}$$

Equation (3.2.1) must have a particular form when describing a random process that, in this form, is WSS in the time domain. The spectral components of $\tilde{H}(\omega)$ must be *uncorrelated.* Mathematically, we can write this condition as

$$C_{\tilde{H}}(\omega_1, \omega_2) = 2\pi S_{\tilde{h}}(\omega_1)\delta(\omega_1 - \omega_2) \tag{3.2.2}$$

which clearly evaluates to 0 whenever $\omega_1 \neq \omega_2$. A proof of this may be found in Theorem 3.1.

Theorem 3.1: Uncorrelated Spectrum for WSS

Statement: A random process is WSS if and only if its spectral components are uncorrelated.

Proof: Start with the definition of temporal autocorrelation in the time domain - Equation (3.1.2) - and substitute the inverse Fourier transforms for $\tilde{h}(t_1)$ and $\tilde{h}(t_2)$:

$$\begin{aligned} C_{\tilde{h}}(t_1, t_2) &= \mathrm{E}\left\{\tilde{h}(t_1)\tilde{h}^*(t_2)\right\} \\ &= \frac{1}{4\pi^2}\int_{-\infty}^{+\infty}\int_{-\infty}^{+\infty} \underbrace{\mathrm{E}\left\{\tilde{H}(\omega_1)\tilde{H}^*(\omega_2)\right\}}_{C_{\tilde{H}}(\omega_1,\omega_2)} \exp(j[\omega_1 t_1 - \omega_2 t_2])\, d\omega_1\, d\omega_2 \end{aligned}$$

Now consider the result for uncorrelated spectral components: $C_{\tilde{H}}(\omega_1, \omega_2)$ evaluates to zero for all $\omega_1 \neq \omega_2$. For this case, we could then write the complex exponential function as $\exp(j\omega_1 [t_1 - t_2])$. The entire right-hand side then satisfies the condition for WSS, since it depends only on the time difference $t_2 - t_1$.

The converse - WSS leads to uncorrelated spectral components - is necessarily true as well. There is no mathematical way to force $\exp(j[\omega_1 t_1 - \omega_2 t_2])$ into a function expressible in terms of $t_2 - t_1$ unless its multiplier, $C_{\tilde{H}}(\omega_1, \omega_2)$, is zero for all $\omega_1 \neq \omega_2$. This is only true if the spectral components are all uncorrelated.

Note the special form used in Equation (3.2.2), where the delta function forces the spectral correlation to zero for all $\omega_1 \neq \omega_2$. The function $S_{\tilde{h}}(\omega_1)$ in Equation (3.2.2) is called the *power spectral density* (PSD). It describes how the spectral power of the random channel $\tilde{h}(t)$ is distributed in the Doppler domain. The PSD is the most important spectral-domain tool for analyzing WSS random processes.

In channel modeling circles, the condition of complete spectrum decorrelation in Equation (3.2.2) is often referred to as *uncorrelated scattering* (US) - especially when discussing the channel *delay* spectrum.

Note: Power Versus Energy Signals

Fourier transforms are only rigidly defined for *energy signals*. All truly WSS random processes, however, are *power signals*. Thus, the spectrum autocorrelation, $C_{\tilde{H}}(\omega_1, \omega_2)$, has infinite values. We avoid this problem in the future by using the finite-valued PSD $S_{\tilde{h}}(\omega)$ in all spectral analysis. Through a mathematical sleight-of-hand, the delta function in Equation (3.2.2) "absorbs" the infinite values, which is why $S_{\tilde{h}}(\omega)$ is called a density.

3.2.2 The Wiener-Khintchine Theorem

The usefulness of a PSD arises from the Wiener-Khintchine theorem [Cou93, p. 70]. This theorem states that the PSD and the autocorrelation of a WSS random process are Fourier transform pairs:

$$S_{\tilde{h}}(\omega) = \int_{-\infty}^{+\infty} C_{\tilde{h}}(\Delta t) \exp(-j\omega\Delta t)\, d\Delta t \tag{3.2.3}$$

$$C_{\tilde{h}}(\Delta t) = \frac{1}{2\pi} \int_{-\infty}^{+\infty} S_{\tilde{h}}(\omega) \exp(j\omega\Delta t)\, d\omega \tag{3.2.4}$$

The Wiener-Khintchine theorem implies that studying autocorrelation properties in the base domain is *equivalent* to studying average signal power in the spectral domain. Thus, there are two ways to characterize the same second-order statistics of a WSS random process. Theorem 3.2 proves the Wiener-Khintchine theorem.

Theorem 3.2: Wiener-Khintchine Theorem

Statement: The PSD and autocorrelation of a WSS random process are Fourier transform pairs.

Proof: Recall from Theorem 3.1 that the autocorrelation function is written as

$$C_{\tilde{h}}(t_1, t_2) = \frac{1}{4\pi^2} \int_{-\infty}^{+\infty}\int_{-\infty}^{+\infty} C_{\tilde{H}}(\omega_1, \omega_2) \exp(j[\omega_1 t_1 - \omega_2 t_2])\, d\omega_1\, d\omega_2$$

which, for a WSS random process, becomes

$$\begin{aligned} C_{\tilde{h}}(t_1, t_2) &= \frac{1}{2\pi} \int_{-\infty}^{+\infty}\int_{-\infty}^{+\infty} S_{\tilde{h}}(\omega_1)\delta(\omega_1 - \omega_2) \exp(j[\omega_1 t_1 - \omega_2 t_2])\, d\omega_1\, d\omega_2 \\ &= \frac{1}{2\pi} \int_{-\infty}^{+\infty} S_{\tilde{h}}(\omega_2) \exp(j\omega_2[t_1 - t_2])\, d\omega_2 \end{aligned}$$

Making the cosmetic substitutions $\omega = \omega_2$ and $\Delta t = t_1 - t_2$ produces the Fourier transform relationship of Equation (3.2.4).

3.2.3 Statistics with Three-Dimensional Space

In our discussion of spatial selectivity, we have dealt with only scalar space - movement along a linear spatial variable, r. Of course, reality dictates that wireless receivers are capable of operating in three-dimensional space, which requires adding dependencies to the spatial representation as well as its Fourier spectrum. The Fourier transform of a function of three-dimensional position is actually a threefold transform over scalar coordinates. Thus, we write the transform pairs as

$$\tilde{h}(x, y, z) \longleftrightarrow \tilde{H}(k_x, k_y, k_z)$$

where x, y, and z are the Cartesian position coordinates and k_x, k_y, and k_z are their corresponding wavenumbers in the spectral domain. The three-fold Fourier transform therefore requires three integrations, which are written as

$$\tilde{H}(k_x, k_y, k_z) = \int_{-\infty}^{+\infty}\int_{-\infty}^{+\infty}\int_{-\infty}^{+\infty} \tilde{h}(x, y, z) \exp\left(-j\left[k_x x + k_y y + k_z z\right]\right) dx\, dy\, dz \qquad (3.2.5)$$

The notation of Equation (3.2.5) is cumbersome. Throughout this work, a set of vector shorthand notation is used to simplify the equations and emphasize the concepts rather than the raw mathematics.

First, the dependencies of both scalar positions and wavenumbers are collapsed into a three-dimensional position vector wavevector, respectively:

$$\vec{\mathrm{r}} \doteq x\hat{\mathrm{x}} + y\hat{\mathrm{y}} + z\hat{\mathrm{z}}$$

$$\vec{\mathrm{k}} \doteq k_x\hat{\mathrm{x}} + k_y\hat{\mathrm{y}} + k_z\hat{\mathrm{z}}$$

where the $\hat{\cdot}$ denotes a unit vector. Next, the threefold integrations over position variables or wavenumber variables is abbreviated into one integration over a vector partial, $d\vec{\mathrm{r}}$ or $d\vec{\mathrm{k}}$. These single integrations are defined to be

$$\int_{-\infty}^{+\infty} d\vec{\mathrm{r}} \doteq \int_{-\infty}^{+\infty}\int_{-\infty}^{+\infty}\int_{-\infty}^{+\infty} dx\, dy\, dz$$

$$\int_{-\infty}^{+\infty} d\vec{\mathrm{k}} \doteq \int_{-\infty}^{+\infty}\int_{-\infty}^{+\infty}\int_{-\infty}^{+\infty} dk_x\, dk_y\, dk_z$$

With these substitutions, the set of forward and inverse Fourier transforms for functions of vector position is given by the following:

$$\tilde{H}(\vec{\mathrm{k}}) = \int_{-\infty}^{+\infty} \tilde{h}(\vec{\mathrm{r}}) \exp(-j\vec{\mathrm{k}}\cdot\vec{\mathrm{r}})\, d\vec{\mathrm{r}} \qquad (3.2.6)$$

$$\tilde{h}(\vec{\mathrm{r}}) = \frac{1}{(2\pi)^3}\int_{-\infty}^{+\infty} \tilde{H}(\vec{\mathrm{k}}) \exp(j\vec{\mathrm{k}}\cdot\vec{\mathrm{r}})\, d\vec{\mathrm{k}} \qquad (3.2.7)$$

where $(\cdot)$ is a scalar product ($\vec{\mathrm{k}}\cdot\vec{\mathrm{r}} = xk_x + yk_y + zk_z$).

The simple vector notation makes it almost trivial to define a 3D spatial autocorrelation function and a wavevector PSD, since the definitions follow logically from their scalar spatial counterparts. The 3D spatial autocorrelation function (assuming a WSS spatial channel) is defined as

$$C_{\tilde{h}}(\Delta\vec{r}) = \mathrm{E}\left\{\tilde{h}(\vec{r})\tilde{h}^*(\vec{r}+\Delta\vec{r})\right\} \tag{3.2.8}$$

The wavevector PSD for this random channel process is given by the Fourier transform:

$$S_{\tilde{h}}(\vec{k}) = \int_{-\infty}^{+\infty} C_{\tilde{h}}(\Delta\vec{r}) \exp\left(-j\vec{k}\cdot\Delta\vec{r}\right)\, d\Delta\vec{r} \tag{3.2.9}$$

Although it is easier to introduce spatial channel modeling concepts with scalar space dependencies r and k, a full 3D spatial channel model will be important in subsequent chapters.

3.2.4 Summary of Single-Dependency PSDs

The spectral definitions for time-Doppler channels are equally valid for use with other channel dependencies. Thus, the Doppler PSD is used for time-varying channels, the delay PSD is used for frequency-varying channels, and the wavenumber PSD is used for space-varying channels. Engineers often refer to these PSDs as simply the Doppler spectrum, the delay spectrum, and the wavenumber or wavevector spectrum (although this is, technically, a little ambiguous). Table 3.1 summarizes all of the relevant autocorrelation-spectrum relationships for the WSS complex baseband channel.

Table 3.1 Transform Definitions for Autocorrelation and Power Spectral Densities

TIME AUTOCORRELATION $C_{\tilde{h}}(\Delta t) = \frac{1}{2\pi}\int_{-\infty}^{+\infty} S_{\tilde{h}}(\omega)\exp(j\omega\Delta t)\,d\omega$	DOPPLER SPECTRUM $S_{\tilde{h}}(\omega) = \int_{-\infty}^{+\infty} C_{\tilde{h}}(\Delta t)\exp(-j\omega\Delta t)\,d\Delta t$
FREQUENCY AUTOCORRELATION $C_{\tilde{h}}(\Delta f) = \int_{-\infty}^{+\infty} S_{\tilde{h}}(\tau)\exp(-j2\pi\tau\Delta f)\,d\tau$	DELAY SPECTRUM $S_{\tilde{h}}(\tau) = \int_{-\infty}^{+\infty} C_{\tilde{h}}(\Delta f)\exp(j2\pi\tau\Delta f)\,d\Delta f$
SCALAR SPACE AUTOCORRELATION $C_{\tilde{h}}(\Delta r) = \frac{1}{2\pi}\int_{-\infty}^{+\infty} S_{\tilde{h}}(k)\exp(jk\Delta r)\,dk$	WAVENUMBER SPECTRUM $S_{\tilde{h}}(k) = \int_{-\infty}^{+\infty} C_{\tilde{h}}(\Delta r)\exp(-jk\Delta r)\,d\Delta r$
VECTOR SPACE AUTOCORRELATION $C_{\tilde{h}}(\Delta\vec{r}) = \frac{1}{(2\pi)^3}\int_{-\infty}^{+\infty} S_{\tilde{h}}(\vec{k})\exp(j\vec{k}\cdot\Delta\vec{r})\,d\vec{k}$	WAVEVECTOR SPECTRUM $S_{\tilde{h}}(k) = \int_{-\infty}^{+\infty} C_{\tilde{h}}(\Delta\vec{r})\exp(-j\vec{k}\cdot\Delta\vec{r})\,d\Delta\vec{r}$

3.3 Joint Statistics

Autocorrelation functions and PSDs are not just for characterizing random channels with a single dependency; these techniques are every bit as useful for multiple-dependency channels. The only difficult part in analyzing a random channel with multiple dependencies is visualizing the distinct behavior. To help with this, transform maps are presented in this section along with the definitions of joint autocorrelation and spectrum.

3.3.1 Joint Autocorrelation and Spectrum

To accommodate all the random dependencies of a channel, it is possible to define a joint PSD as a function of Doppler, delay, and wavenumber. Specifically, a random process describing the full channel with respect to time, frequency, and space has the following Fourier transform pair:

$$\tilde{h}(t, f, r) \longleftrightarrow \tilde{H}(\omega, \tau, k)$$

Just as in the single-dependency case, we can study the correlation of spectral components by writing

$$C_{\tilde{H}}(\omega_1, \omega_2, \tau_1, \tau_2, k_1, k_2) \doteq \mathrm{E}\left\{\tilde{H}(\omega_1, \tau_1, k_1)\tilde{H}^*(\omega_2, \tau_2, k_2)\right\} \tag{3.3.1}$$

Once again, if the spectral components are uncorrelated with one another, we may write Equation (3.3.1) as

$$C_{\tilde{H}}(\omega_1, \omega_2, \tau_1, \tau_2, k_1, k_2) = 4\pi^2 S_{\tilde{h}}(\omega_2, \tau_2, k_2)\delta(\omega_1 - \omega_2)\delta(\tau_1 - \tau_2)\delta(k_1 - k_2) \tag{3.3.2}$$

The multidependency random process described by Equation (3.3.2) is called a *wide-sense stationary, uncorrelated scattering* (WSS-US) process.

> *Note: Shaded Meaning of Uncorrelated Scattering*
>
> In seminal work on WSS-US time-varying channels, the term WSS refers to *wide-sense stationary with respect to time*, and the term US refers to *wide-sense stationary with respect to frequency* [Bel63]. This terminology gets confusing as we add dependencies to the random space–time channel. In this text, US emphasizes the ability to write the PSD in the form of Equation (3.3.2) - spectral components are jointly uncorrelated across *all* spectral domains.

Logically, we may also define an autocorrelation function for the multidependent WSS-US channel and apply the Wiener-Khintchine theorem. The definition follows from the single-dependency autocorrelation of Equation (3.1.4):

$$C_{\tilde{h}}(\Delta t, \Delta f, \Delta r) = \mathrm{E}\left\{\tilde{h}(t, f, r)\tilde{h}^*(t + \Delta t, f + \Delta f, r + \Delta r)\right\} \tag{3.3.3}$$

The Wiener-Khintchine theorem for WSS-US processes then leads to the following Fourier transform relationship between autocorrelation function and PSD:

$$C_{\tilde{h}}(\Delta t, \Delta f, \Delta r) \longleftrightarrow S_{\tilde{h}}(\omega, \tau, k)$$

See Example 3.1 for a demonstration of how to apply these definitions to a WSS-US channel with multiple dependencies.

Example 3.1: WSS-US Autocorrelation

Problem: Given the following autocorrelation function for a WSS-US channel,

$$C_{\tilde{h}}(\Delta t, \Delta f, \Delta r) = \frac{S_0 \sigma_\tau \cos(k_0 \Delta r)}{1 + j2\pi \Delta f \sigma_\tau} \delta(\Delta t)$$

derive an expression for the PSD, $S_{\tilde{h}}(\omega, \tau, k)$.

Solution: The solution follows mechanically by taking the Fourier transform of the autocorrelation function:

$$\begin{aligned}
S_{\tilde{h}}(\omega, \tau, k) &= \int_{-\infty}^{+\infty}\int_{-\infty}^{+\infty}\int_{-\infty}^{+\infty} C_{\tilde{h}}(\Delta t, \Delta f, \Delta r) \;\times \\
&\qquad \exp(j[2\pi\tau\Delta f - \omega\Delta t - k\Delta r])\, d\Delta t\, d\Delta f\, d\Delta r \\
&= S_0 \underbrace{\int_{-\infty}^{+\infty} \delta(\Delta t) \exp(-j\omega\Delta t)\, d\Delta t}_{=1} \underbrace{\int_{-\infty}^{+\infty} \frac{\sigma_\tau \exp(j2\pi\tau\Delta f)}{1 + j2\pi\Delta f \sigma_\tau}\, d\Delta f}_{=\exp\left(-\frac{\tau}{\sigma_\tau}\right)\mathrm{u}(\tau)} \;\times \\
&\qquad \underbrace{\int_{-\infty}^{+\infty} \cos(k_0\Delta r) \exp(-jk\Delta r)\, d\Delta r}_{=\frac{1}{2}[\delta(k-k_0)+\delta(k+k_0)]} \\
&= \frac{S_0}{2}[\delta(k - k_0) + \delta(k + k_0)] \exp\left(-\frac{\tau}{\sigma_\tau}\right) \mathrm{u}(\tau)
\end{aligned}$$

> *Note: Why Uncorrelated Scattering?*
>
> It is possible for a random channel to be WSS with respect to each dependency *individually*, but not *jointly* (i.e. WSS-US as used by this text). To understand how this can happen, consider the following example of joint spectral correlation with respect to Doppler and delay dependencies:
>
> $$C_{\tilde{H}}(\omega_1, \omega_2, \tau_1, \tau_2) = 2\pi S_{\tilde{h}}(\omega_2, \tau_2)\delta(\omega_1 - \omega_2 + \tau_1 - \tau_2)\delta(\tau_1 - \tau_2)$$
>
> This spectral correlation results in an autocorrelation function that depends only on $t_1 - t_2$ and $\omega_1 - \omega_2$. Yet the cross-dependent term causes correlated behavior *between* Doppler and delay, which adds complexity to a second-order analysis.

3.3.2 Time–Frequency Transform Map

To truly understand the concept of joint spectra and autocorrelation functions, it is best to eliminate some of the dependencies from the full space–time–frequency channel model. Consider the case of a fixed, single-antenna receiver operating in a channel that is only a function of time, t, and frequency, f; this channel has no spatial dependency. For this type of channel, Figure 3.2 is a roadmap through the various stochastic relationships [Bel63], [Gre62]. The stochastic, second-order behavior of this channel may be characterized by its joint autocorrelation, $C_{\tilde{h}}(\Delta t, \Delta f)$, or its PSD, $S_{\tilde{h}}(\omega, \tau)$. In a WSS channel, knowledge of only one is necessary, since they are two-dimensional Fourier transform pairs.

If the channel is *narrowband* (signal bandwidth is less than the coherence bandwidth), then the fading is flat about the carrier frequency, and it suffices to characterize the stochastic channel with an autocorrelation that is only a function of time or a PSD that is only a function of Doppler. These one-dimensional functions may be calculated from their two-dimensional counterparts as

$$C_{\tilde{h}}(\Delta t) = C_{\tilde{h}}(\Delta t, \Delta f)\big|_{\Delta f = 0} \tag{3.3.4}$$

$$S_{\tilde{h}}(\omega) = \int_{-\infty}^{+\infty} S_{\tilde{h}}(\omega, \tau)\, d\tau \tag{3.3.5}$$

This calculation is shown on the left side of Figure 3.2. The remaining one-dimensional functions - the temporal autocorrelation and Doppler spectrum - are Fourier transform pairs.

The right-hand side of Figure 3.2 represents characterization of a *static* channel that does not vary with time. A one-dimensional frequency autocorrelation is calculated from the joint time–frequency autocorrelation by setting Δt equal to 0. A one-dimensional delay spectrum is calculated from the joint Doppler–delay spectrum by integrating out the Doppler frequency, ω. And, of course, the resulting frequency autocorrelation and delay spectrum are Fourier transform pairs.

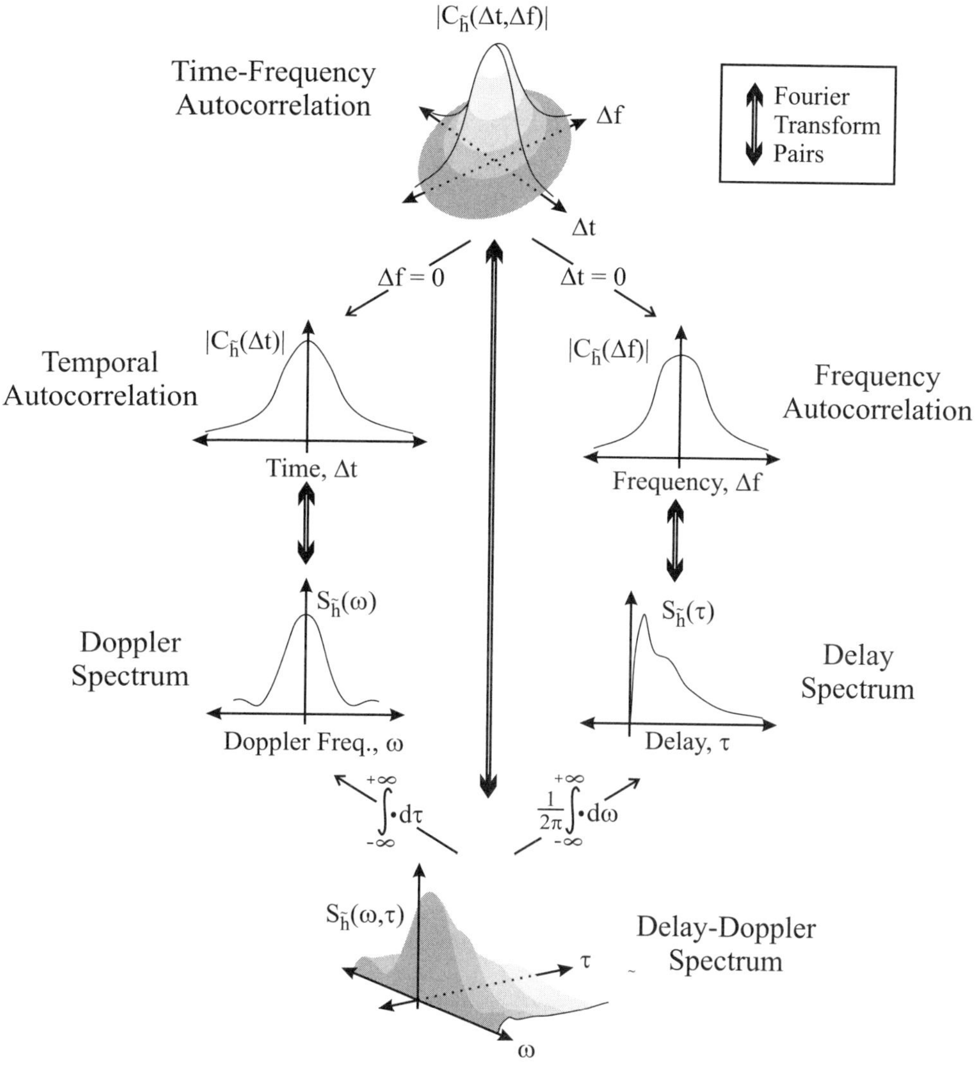

Figure 3.2 Autocorrelation function and PSD relationships for time and frequency.

Many researchers have used Figure 3.2 to understand and characterize the stochastic behavior of linear, time-varying channels. It was first presented by [Gre62] in the context of radio astronomy measurements. For this application, an antenna dish is located in a fixed position and receives signals that vary with time and frequency from the heavens. The concept may be extended to include spatial relationships as well.

> *Note: Well, Sort of Radio Astronomy*
>
> The real impetus for the original work in [Gre62] had more to do with spying on radio transmissions by the former Soviet Union using moon bounce measurements. Soviet radio signals originating from the other side of the earth would bounce off the revolving and rotating moon, producing a radio signal with both dispersion (frequency-selective fading) and Doppler spreading (temporal fading). Spy satellites eventually led to a much more effective way to monitor the radio signals of other countries.

3.3.3 Space–Frequency Transform Map

For a static channel, we may extend the time-frequency transform map in Figure 3.2 for the space–frequency relationships shown in Figure 3.3. In this diagram, joint space–frequency autocorrelation and joint wavenumber–delay spectrum are shown to be two-dimensional Fourier transform pairs. One-dimensional spatial or frequency autocorrelations may be calculated by zeroing out Δf or Δr respectively in the joint autocorrelation. One-dimensional wavenumber or delay spectra may be calculated by integrating out τ or k respectively in the joint PSD.

3.3.4 Complete Transform Map

Of course, all three dependencies - space, time, and frequency - may be studied jointly. The combined relationships of Figure 3.2 and Figure 3.3 are shown in one complete transform map, Figure 3.4. This figure is best viewed as a single "tile" in a transform map that continues with identical tiles in all directions. For example, the conversion from a joint space–time autocorrelation to a one-dimensional temporal autocorrelation is made by wrapping around from the right edge of Figure 3.4 to the left edge. Likewise, the complete autocorrelation of the stochastic channel is shown to be the three-domain Fourier transform pair of the complete PSD by wrapping around from the top edge of Figure 3.4 to the bottom edge.

Although Figure 3.4 contains numerous channel dependencies and definitions, navigation through the relationships is crucial for understanding the stochastic characterization of the multivariable wireless channel. When one considers vector space instead of scalar space, the characterization is even more complicated. Remember, however, that each dependency can be removed to isolate and understand the fading behaviors in a space–time channel. Use Table 3.2, which summarizes how to remove the dependencies, along with the transform map of Figure 3.4 to analyze the random space–time channel.

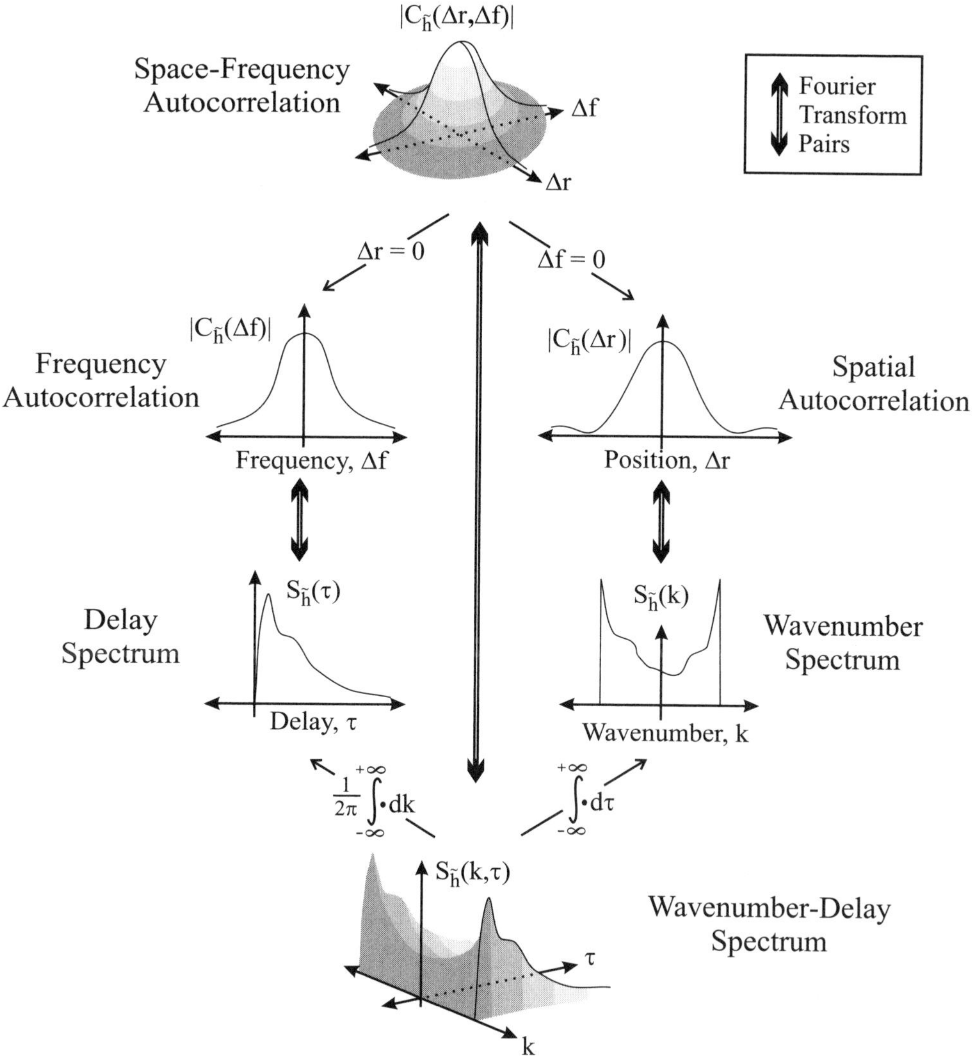

Figure 3.3 Autocorrelation function and PSD relationships for space and frequency.

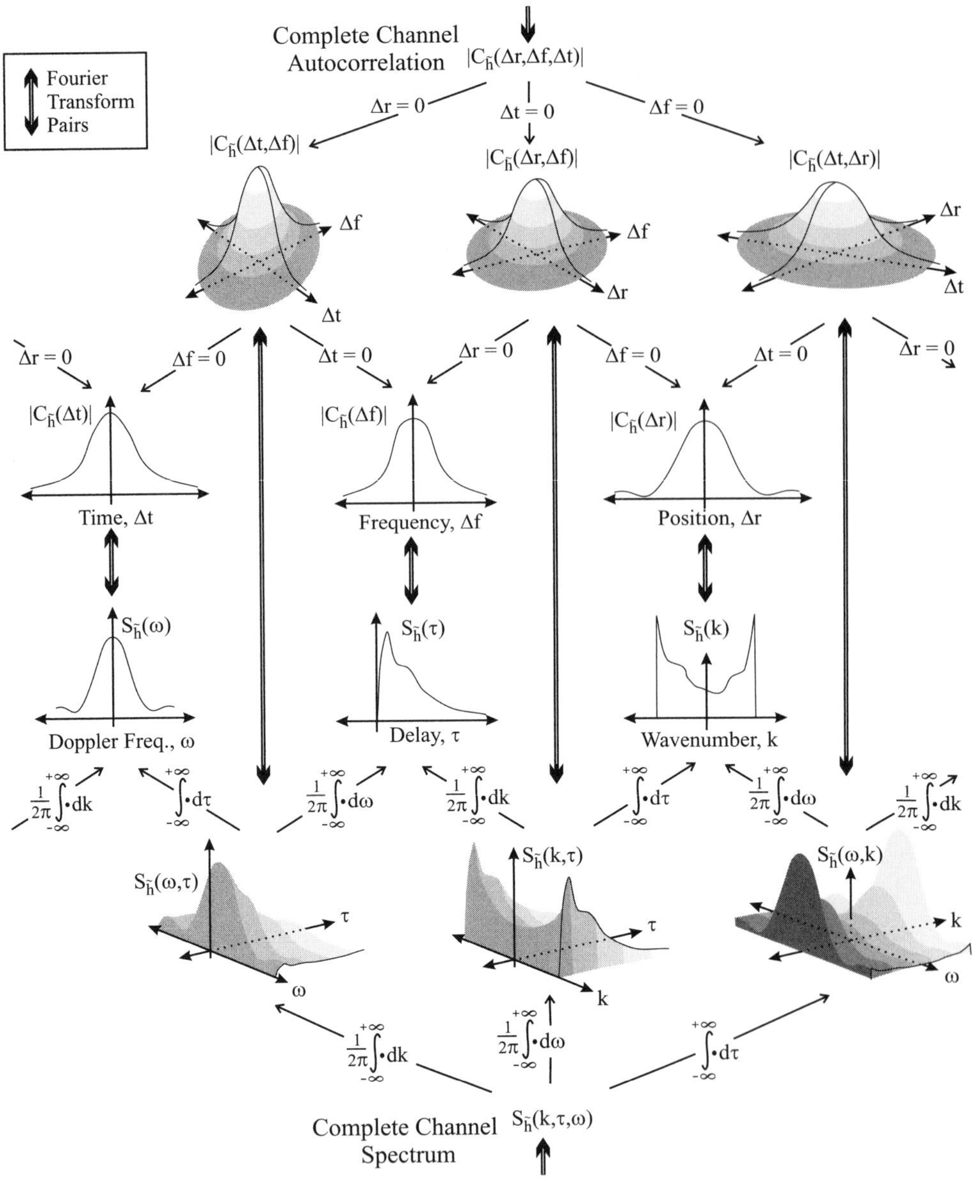

Figure 3.4 Autocorrelation function and power spectrum relationships for space, time, and frequency.

Table 3.2 How to Remove a Dependency in a Random Space–Time Channel Model.

DEPENDENCY TO REMOVE	FROM AUTOCORRELATION	FROM PSD
Time	$\Delta t = 0$	$\frac{1}{2\pi}\int_{-\infty}^{+\infty} \cdot\, d\omega$
Frequency	$\Delta f = 0$	$\int_{-\infty}^{+\infty} \cdot\, d\tau$
Scalar Space	$\Delta r = 0$	$\frac{1}{2\pi}\int_{-\infty}^{+\infty} \cdot\, dk$
Vector Space	$\Delta \vec{\mathrm{r}} = \vec{0}$	$\frac{1}{(2\pi)^3}\int_{-\infty}^{+\infty} \cdot\, d\vec{\mathrm{k}}$

3.4 Width of the PSD

It is a basic principle in Fourier analysis that as a function becomes narrower in one domain, its Fourier transform pair becomes wider in the other domain. Therefore, the *width* of a PSD can provide insight into the coherence of a channel: As the PSD widens in the transform domain, its autocorrelation becomes narrower and coherence decreases. The most common (and useful) definition of a channel's PSD width is the root-mean-square (RMS) width. RMS widths may be defined for PSDs as a function of delay, Doppler, or wavenumber.

3.4.1 RMS Delay Spread

The RMS delay spread is defined from the delay spectrum of a stochastic channel. Mathematically, this delay spread is the second centered moment of this spectrum, defined as

$$\sigma_\tau^2 = \overline{\tau^2} - (\overline{\tau})^2, \quad \text{where } \overline{\tau^n} = \frac{\int_{-\infty}^{+\infty} \tau^n S_{\tilde{h}}(\tau)d\tau}{\int_{-\infty}^{+\infty} S_{\tilde{h}}(\tau)d\tau} \tag{3.4.1}$$

A larger delay spread implies increased frequency selectivity and a smaller coherence bandwidth. Many wireless engineers calculate the coherence bandwidth using the rule of thumb $B_C \approx \frac{1}{5\sigma_\tau}$.

Example 3.2: Exponential Delay Spectrum

Problem: The received power of multipath components typically decreases exponentially as a function of time delay. As a result, many engineers approximate the delay spectrum as an exponential-shaped function. Develop expressions for delay spectrum and autocorrelation as a function of RMS delay spread, σ_τ.

Solution: If the RMS delay spread of an exponential-shaped spectrum is σ_τ, then the delay spectrum must take the form

$$S_{\tilde{h}}(\tau) = S_0 \exp\left(-\frac{\tau}{\sigma_\tau}\right) u(\tau)$$

where S_0 is an arbitrary constant and $u(\tau)$ is the unit step function. Computing the inverse Fourier transform of this function produces the frequency autocorrelation:

$$C_{\tilde{h}}(\Delta f) = \frac{S_0 \sigma_\tau}{1 + j2\pi \Delta f \sigma_\tau}$$

The shape and widths of the delay spectrum and frequency autocorrelation are shown below:

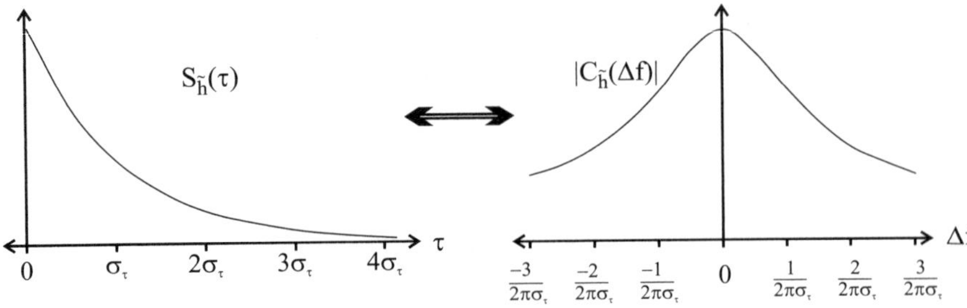

Note the inverse proportionality between the widths of the spectrum and the autocorrelation as the RMS delay spread, σ_τ, increases.

3.4.2 RMS Doppler Spread

In definition, the RMS Doppler spread is nearly identical to RMS delay spread. This delay spread is calculated as the second centered moment of the Doppler spectrum:

$$\sigma_\omega^2 = \overline{\omega^2} - (\overline{\omega})^2, \quad \text{where } \overline{\omega^n} = \frac{\int_{-\infty}^{+\infty} \omega^n S_{\tilde{h}}(\omega) d\omega}{\int_{-\infty}^{+\infty} S_{\tilde{h}}(\omega) d\omega} \tag{3.4.2}$$

An increased Doppler spread implies a channel with faster temporal fluctuations and a smaller coherence time.

Example 3.3: Gaussian Doppler Spectrum

Problem: The power in the Doppler spectrum of many types of time-varying channels is concentrated heavily at $\omega = 0$ and quickly diminishes for larger values of $|\omega|$. One model for this type of behavior is a Gaussian-shaped Doppler spectrum. Develop expressions for Doppler spectrum and autocorrelation as a function of Doppler spread, σ_ω.

Solution: If the RMS Doppler spread of a Gaussian-shaped spectrum is σ_ω, then the Doppler spectrum must take the following form:

$$S_{\tilde{h}}(\omega) = S_0 \exp\left(-\frac{\omega^2}{2\sigma_\omega^2}\right)$$

where S_0 is some arbitrary constant. The temporal autocorrelation, $C_{\tilde{h}}(\Delta t)$, is the inverse Fourier transform:

$$C_{\tilde{h}}(\Delta t) = \frac{S_0 \sigma_\omega}{\sqrt{2\pi}} \exp\left(-\frac{\Delta t^2 \sigma_\omega^2}{2}\right)$$

The shape and relative widths of the Doppler spectrum and temporal autocorrelation are shown below:

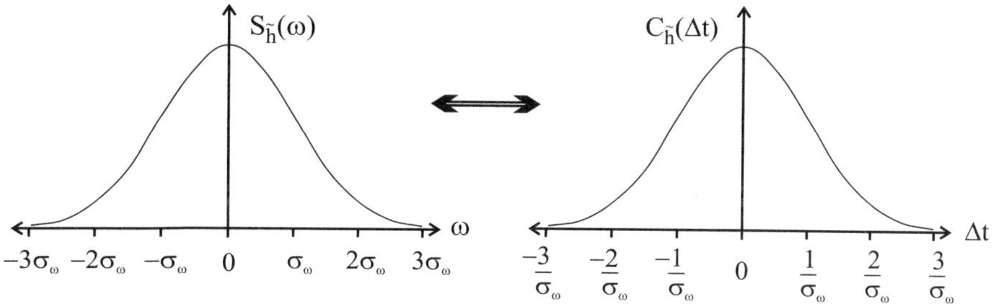

Note the inverse proportionality between the widths of the spectrum and the autocorrelation as the RMS Doppler spread, σ_ω, increases.

3.4.3 RMS Wavenumber Spread

The definition for RMS wavenumber spread follows immediately from the definitions of the previous RMS spreads. The definition for wavenumber spread is then given by

$$\sigma_k^2 = \overline{k^2} - (\overline{k})^2, \quad \text{where } \overline{k^n} = \frac{\int_{-\infty}^{+\infty} k^n S_{\tilde{h}}(k)dk}{\int_{-\infty}^{+\infty} S_{\tilde{h}}(k)dk} \tag{3.4.3}$$

Increased wavenumber spread implies a channel with more rapid spatial fluctuations and a smaller coherence distance.

Example 3.4: Omnidirectional Wavenumber Spectrum

Problem: A common model for wavenumber spectrum is the Clarke omnidirectional spectrum, which arises in cluttered outdoor environments when multipath power arrives from the horizon in every direction [Cla68]. The wavenumber spectrum for this case is

$$S_{\tilde{h}}(k) = \frac{S_0}{\sqrt{k_0^2 - k^2}} \quad \text{for } |k| \leq k_{\max}$$

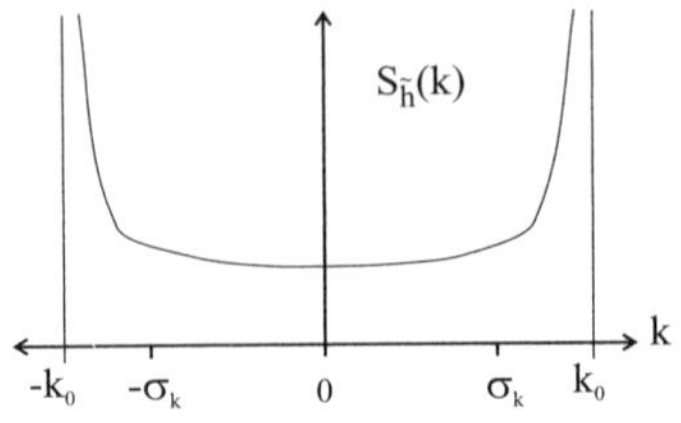

where k is wavenumber, S_0 is an arbitrary constant, and k_0 is the maximum free-space wavenumber. Find expressions for the RMS wavenumber spread and spatial autocorrelation.

Solution: The wavenumber spread may be calculated in terms of k_0 using Equation (3.4.3):

$$\sigma_k = \frac{k_0}{\sqrt{2}}$$

The spatial autocorrelation of this wavenumber spectrum is

$$C_{\tilde{h}}(\Delta r) = \frac{S_0}{2}\, \mathrm{J}_0(k_0 \Delta r)$$

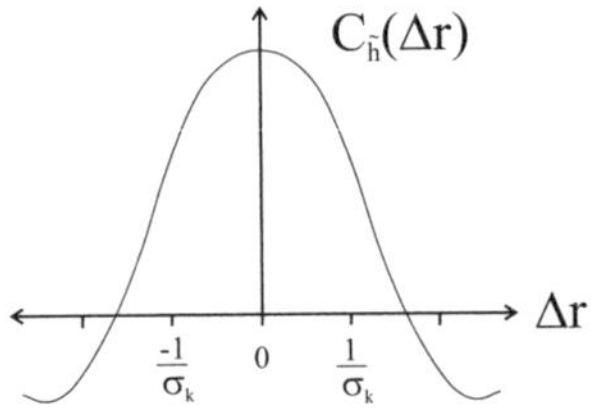

where $\mathrm{J}_0(\cdot)$ is a zero-order Bessel function.

Note: A Familiar Spectrum

The U-shaped spectrum in Example 3.4 is a famous classical result in mobile radio propagation, although it is usually presented in textbooks as a *Doppler* spectrum instead of a wavenumber spectrum. These textbooks also characterize spatial fading, although there is an implied distance = velocity × time ($r = vt$) substitution. This substitution converts a space-varying channel to a time-varying channel, as discussed in Chapter 2, Section 2.3.4.

3.4.4 Channel Duality Principle

By now it is obvious that a tremendous amount of similarity exists in the stochastic characterization of time, frequency, and space. An analytical technique developed for one dependency is equally applicable to the others, whether it is an autocorrelation function, a PSD, an RMS width, or other quantity. This is the principle of *duality* in channel modeling [Bel64].

The time, frequency, and spatial *duals* of the basic stochastic concepts discussed in this section are summarized by Table 3.3. Despite that each dependency characterizes a completely different aspect of the wireless channel, the concepts and terminology used to study each are the same. A generic summary of the random channel characterization presented thus far might go something like this:

> *The stochastic wireless channel as a function of* **(time, frequency, space)** *displacement may be characterized by an autocorrelation function. The Fourier transform of the autocorrelation produces a* **(Doppler, delay, wavenumber)** *spectrum. The width of this spectrum is characterized by its RMS* **(Doppler, delay, wavenumber)** *spread. As this spread increases, the* **(temporal, frequency, spatial)** *selectivity of the channel increases and the coherence* **(time, bandwidth, distance)** *decreases.*

The above statement may be read several times, substituting the first, second, and third words in each set with each respective reading.

Table 3.3 Channel Duality Relationships Between Time, Frequency, and Space.

	TIME	FREQUENCY	SPACE
Dependency	time, t	frequency, f	position, r
Coherence	time, T_c	bandwidth, B_c	distance, D_c
Spectral Domain	Doppler, ω	delay, τ	wavenumber, k
Spectral Width	Doppler spread, σ_ω	delay spread, σ_τ	w.n. spread, σ_k

Example 3.5: Coherence Time Duality

Problem: The RMS Doppler spread, σ_ω, of a time-varying, narrowband channel is known. Estimate the coherence time, T_c.

Solution: We have yet to formally define *coherence time*. However, it was stated in Section 3.4 that the coherence bandwidth, B_c, for a dispersive channel is inversely proportional to $5\sigma_\tau$. It follows by duality that a similar rule of thumb should work for time-varying channels. The following substitutions are made: $B_c \Longrightarrow T_c$ and $\sigma_\tau \Longrightarrow \frac{1}{2\pi}\sigma_\omega$ (the 2π is the difference between

Fourier transform definitions of the Doppler and delay domains). The "new" coherence time rule of thumb is now

$$T_c \approx \frac{2\pi}{5\sigma_\omega}$$

which agrees with rule-of-thumb estimates presented in [Rap02a].

3.4.5 Definition of a Rate Variance

Before continuing the discussion on angle spectra, it is important to introduce the concept of a *fading rate variance*. The fading rate variance is a key second-order statistic that determines how quickly a channel fluctuates as a function of time, space, or frequency.

Consider the snapshots of two stochastic processes in Figure 3.5. Both processes represent time-varying voltages and both have magnitudes that follow an identical Rayleigh PDF (which are introduced in Chapter 5). Clearly, they are not the same process, as Signal 1 fluctuates much faster than the process on the bottom. Their *second-order statistics*, which relate to development over time, are quite different.

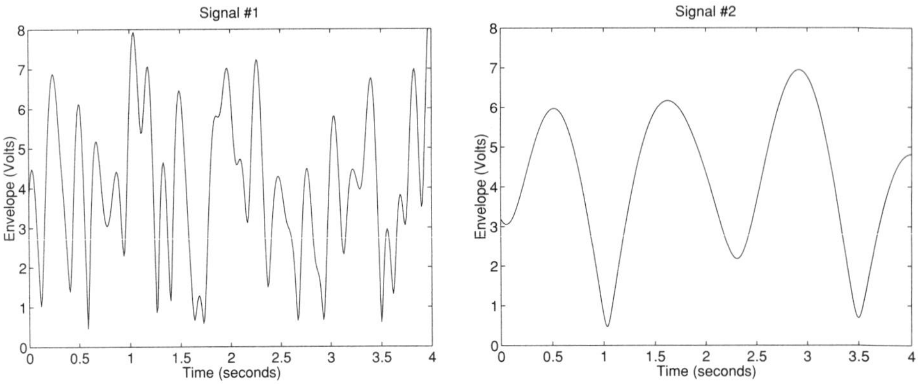

Figure 3.5 Time-varying, Rayleigh-distributed stochastic processes with different second-order statistics.

Although WSS second-order statistics are best characterized with either an autocorrelation function or a PSD, there are simpler measures of how rapidly a process evolves over time, space, or frequency. One such measure may be based on the *derivative* of the channel. Consider the narrowband, fixed, time-varying channel, $\tilde{h}(t)$. A useful measure of fluctuations could be based on $\frac{d\tilde{h}(t)}{dt}$. Simply taking the ensemble average of this quantity is not useful, since the mean derivative of all WSS processes is 0. Rather, taking the ensemble average of $\left|\frac{d\tilde{h}(t)}{dt}\right|^2$ produces a result which, true to our intentions, measures the amount of fluctuation of the stochastic time-varying process, $\tilde{h}(t)$.

There is one final and subtle adjustment that must be made to our fluctuation measure. For a complex-valued process such as a baseband channel, we would like to measure the fluctuations of a process that has *stationary phase*. For example, a WSS stochastic channel has a phase that is itself a stochastic process. This process may be written as

$$\Phi(t) = \arg\{\tilde{h}(t)\}$$

If the centroid of the Doppler spectrum, $S_{\tilde{h}}(\omega)$, is not zero, then the stochastic *phase* process will not be mean-stationary. Instead, the mean phase will be a function of time:

$$\mathrm{E}\left\{\Phi(t)\right\} = \phi_0 + \bar{\omega}t \qquad \text{where } \bar{\omega} = \frac{\int_{-\infty}^{+\infty} \omega\, S_{\tilde{h}}(\omega)\, d\omega}{\int_{-\infty}^{+\infty} S_{\tilde{h}}(\omega)\, d\omega}$$

The constant $\bar{\omega}$ is the centroid of the Doppler spectrum. The solution to removing this nonstationary phase is to multiply the channel, $\tilde{h}(t)$, by a complex exponential of the form, $\exp(-j\bar{\omega}t)$.

> *Note: The Utility of Phase Centering*
>
> Keep in mind that performance in a communication link depends on envelope - or, equivalently, power - which is independent of phase. Multiplication of any complex process by a linear phase taper does nothing to the envelope of the process. We use statistics based on a phase-centered complex channel because (as we will see in Chapter 7) they can be linked directly to average changes in the envelope process.

After making the final adjustment for nonstationary phase, the final measure of temporal channel fluctuation becomes

$$\sigma_t^2 = \mathrm{E}\left\{\left|\frac{d\left[\tilde{h}(t)\exp(-j\bar{\omega}t)\right]}{dt}\right|^2\right\} \tag{3.4.4}$$

The measure, σ_t^2, is called a *fading rate variance* because the time-derivative of the channel describes the rate of fading, and the mean-squared value represents the variance of that process [Dur98a]. Since the complex exponential term in Equation (3.4.4) corrects the nonstationary phase, the fluctuations measured by the temporal fading rate variance is related to the envelope fluctuations of $\tilde{h}(t)$, and not just the steady-state progression of phase.

Of course fading rate variances exist for other dependencies as well. For a measure of channel fluctuations as a function of frequency, it is possible to define a *frequency fading rate variance*, σ_f^2, for a static, fixed channel $\tilde{h}(f)$:

$$\sigma_f^2 = \mathrm{E}\left\{\left|\frac{d\left[\tilde{h}(f)\exp(-j2\pi\bar{\tau}f)\right]}{df}\right|^2\right\} \tag{3.4.5}$$

where $\bar{\tau}$ is the centroid of the delay spectrum. By duality, a *spatial fading rate variance*, σ_r^2, can be defined for a static, narrowband channel $\tilde{h}(r)$, where r represents displacement in a fixed direction in space:

$$\sigma_r^2 = \mathrm{E}\left\{\left|\frac{d\left[\tilde{h}(r)\exp(-j\bar{k}r)\right]}{dr}\right|^2\right\} \tag{3.4.6}$$

The value of $\bar{k}$ is the centroid of the wavenumber spectrum calculated for the direction of displacement in space. These fading rate variances are more than just an intuitive academic measure of fading fluctuations - there is a powerful relationship between fading rate variance and the RMS spread of a PSD.

3.4.6 Fundamental Spectral Spread Theorem

It is also possible to calculate the mean-squared derivative of a time-varying channel given its Doppler spectrum, $S_{\tilde{h}}(\omega)$. There is a basic theorem in stochastic process theory that relates the mean-squared derivatives of complex processes to PSDs [Pap91]:

$$\mathrm{E}\left\{\left|\frac{d^n\tilde{h}(t)}{dt^n}\right|^2\right\} = \frac{1}{2\pi}\int_{-\infty}^{+\infty} \omega^{2n}\, S_{\tilde{h}}(\omega)\, d\omega \tag{3.4.7}$$

For the case of fading rate variance, we are interested in the case of $n = 1$:

$$\mathrm{E}\left\{\left|\frac{d\tilde{h}(t)}{dt}\right|^2\right\} = \frac{1}{2\pi}\int_{-\infty}^{+\infty} \omega^{2}\, S_{\tilde{h}}(\omega)\, d\omega \tag{3.4.8}$$

To account for the modulation of the process by the factor $\exp(-j\bar{\omega}t)$, it is possible to shift the spectrum in Doppler frequency by an amount $-\bar{\omega}$ and adjust the limits of integration:

$$\sigma_t^2 = \frac{1}{2\pi}\int_{-\infty}^{+\infty} (\omega-\bar{\omega})^2\, S_{\tilde{h}}(\omega)\, d\omega \tag{3.4.9}$$

This equation may be rearranged into the following form:

$$\sigma_t^2 = \underbrace{\frac{1}{2\pi}\int_{-\infty}^{+\infty} S_{\tilde{h}}(\omega)\,d\omega}_{\mathrm{E}\{P(t)\}} \underbrace{\frac{\int_{-\infty}^{+\infty}(\omega-\bar{\omega})^2\,S_{\tilde{h}}(\omega)\,d\omega}{\int_{-\infty}^{+\infty} S_{\tilde{h}}(\omega)\,d\omega}}_{\sigma_\omega^2} \tag{3.4.10}$$

where $P(t) = |\tilde{h}(t)|^2$. Now the fading rate variance has become a simple function of two familiar measures: average power, $\mathrm{E}\{P(t)\}$, and RMS Doppler spread, σ_ω.

The relationship in Equation (3.4.10) is a fundamental result and holds for any fading rate variance and spectral spread. For stochastic wireless channels, the following results hold:

$$\text{Time:}\quad \sigma_t^2 = \mathrm{E}\{P(t)\}\,\sigma_\omega^2 \tag{3.4.11}$$

$$\text{Frequency:}\quad \sigma_f^2 = (2\pi)^2\,\mathrm{E}\{P(f)\}\,\sigma_\tau^2 \tag{3.4.12}$$

$$\text{Space:}\quad \sigma_r^2 = \mathrm{E}\{P(r)\}\,\sigma_k^2 \tag{3.4.13}$$

It becomes clear from equations (3.4.11), (3.4.12), and (3.4.13) why RMS spectral spreads provide such accurate measures of channel coherence. The RMS spectral spread, due to its direct relationship to fading rate variance, is an *excellent* measure of channel fluctuation in the base domain. Equations (3.4.11), (3.4.12), and (3.4.13) are valuable starting points when applying stochastic channel theory to real-world problems.

3.5 Chapter Summary

This chapter discussed the principles of modeling channels as random processes. Following is an outline of the major points presented:

- An autocorrelation function is used to characterize how a random channel evolves in time.
    - ▷ Temporal, frequency, and spatial autocorrelation definitions are possible.
    - ▷ A wide-sense stationary (WSS) process autocorrelation function depends only on the distance between correlated samples.
- A power spectral density (sometimes called a PSD) measures average spectral power in a random channel.
    - ▷ Doppler, delay, and wavenumber spectra are defined.

- ▷ The Wiener-Khintchine theorem states that a PSD and an autocorrelation function are Fourier transform pairs.
- ▷ The scalar spatial channel may be extended to three dimensions by using a vector autocorrelation function and a wavevector spectrum.

- Joint autocorrelation functions and PSDs characterize channels with multiple dependencies. Relationships between numerous dependencies are best viewed on a *transform map.*

- The RMS spread is a formal measure of PSD width.
  - ▷ Doppler, delay, and wavenumber spreads are defined.
  - ▷ RMS fading rates are measures of average channel fluctuation with respect to time, frequency, and space.
  - ▷ The fading rate variance and the squared RMS spectrum spread of a channel are proportional.
  - ▷ All stochastic modeling of the space–time wireless channel involves a symmetry among dependencies called *duality.*

With the discussions on baseband channel transmission and random channel characterization in our wake, the next step in channel modeling involves the physics of radio wave propagation and the implication for three-dimensional spatial channels. Chapter 4 discusses the unique form and properties that a spatial channel must have to satisfy *Maxwell's equations.*

PROBLEMS

1. Construct a random process from any number of random variables that have a stationary autocorrelation and a nonstationary mean.

2. Prove that the magnitude of a unit autocovariance of any random process is always less than or equal to 1.

3. Fill in the blanks of the following table:

AUTOCORRELATION	MEAN, μ_h	AUTOCOVARIANCE	UNIT AUTOCOVARIANCE
$S_0 \exp(-\alpha\lvert\Delta t\rvert)$	0		
	A	$S_0 \exp\left(-\frac{\Delta r^2}{2\sigma^2}\right)$	
$A^2 \cos^2\left(\pi\left[\frac{\Delta f}{B_0}\right]\right)$			$\cos\left(2\pi\left[\frac{\Delta f}{B_0}\right]\right)$
	C	$\frac{A}{1+j2\pi\Delta f\sigma_\tau}$	

4. Explain why the PSD for any random channel must be a function containing only real, nonnegative values.

5. Below are eight real-valued functions. Each may be a valid autocorrelation function (A), a valid PSD (P), both (B), or neither (N). Mark each function with either A, P, B, or N, based on the properties of PSDs and autocorrelations. (Assume that the vertical line in the center is the zero point.)

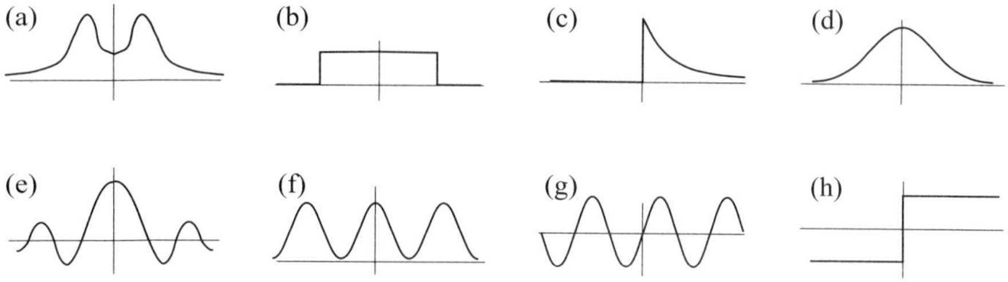

6. A WSS random process is said to be white if its PSD is a constant. Answer the following questions based on a time-varying random channel, $\tilde{h}(t)$, that is *white*:

   a. Prove that the random Doppler process $\tilde{H}(\omega)$ is also WSS with respect to ω.

   b. For the process described in (a), write an expression for the time-domain PSD of $\tilde{H}(\omega)$, $S_{\tilde{H}}(t)$.

7. Use the Fourier transform to complete the following table of autocorrelation and PSD functions of one dependency:

AUTOCORRELATION	PSD
$\mathrm{J}_0(2\pi\Delta t)\cos(2\pi\Delta t)$	
	$6\tau\exp(-\pi\tau)$
	$\cos(\pi\omega)\mathrm{u}(1-\lvert\omega\rvert)$
$\frac{\mathrm{sn}(\Delta r)\cos(\pi\Delta r)}{1-4\Delta r^2}$	
	$\sum_{i=-\infty}^{+\infty}\delta(k-10i)$

8. If we desire to construct an autocorrelation function for a time-varying *passband* channel, $h(t)$, we can use the following expression in terms of complex baseband channels, $\tilde{h}(t)$ [Bel63]:

$$C_h(t_1,t_2) = \frac{1}{2}\mathrm{Real}\Big\{C_{\tilde{h}}(t_1,t_2)\exp\left(j2\pi f_c[t_1-t_2]\right) + \mathrm{E}\left\{\tilde{h}(t_1)\tilde{h}(t_2)\right\}\exp\left(j2\pi f_c[t_1+t_2]\right)\Big\}$$

   a. Derive the above expression from first principles. For the general case, what information besides $C_{\tilde{h}}(t_1,t_2)$ is needed to calculate $C_h(t_1,t_2)$?

b. If we know that the passband channel is WSS, what simplifications can we make to this expression?

9. Consider a spectral autocorrelation for a time- and frequency-varying channel that takes the following form:

$$C_{\tilde{H}}(\omega_1, \omega_2, \tau_1, \tau_2) = 2\pi S_{\tilde{h}}(\omega_2, \tau_2)\delta(\omega_1 - \omega_2 + W(\tau_1, \tau_2))\delta(\tau_1 - \tau_2 + T(\omega_1, \omega_2))$$

Use this form of the spectral autocorrelation to answer the following questions:

a. If we desire only a narrowband channel analysis (no frequency selectivity), what is the criterion for a WSS time-varying channel?

b. If we desire only a static channel analysis (no temporal selectivity), what is the criterion for a WSS frequency-varying channel?

c. To simultaneously model frequency and temporal selectivity, what are the criteria for a WSS-US channel?

d. **Challenge:** Derive criteria that produce a WSS time–frequency autocorrelation function (depends only on $t_1 - t_2$ and $f_1 - f_2$) with *correlated* scattering. This involves specifying a relationship between the functions T and W and their dependencies.

10. Use the Fourier transform to complete the following table of autocorrelation and PSD functions of multiple dependencies:

AUTOCORRELATION	PSD
	$\cos(\pi\omega)\,\mathrm{u}(1-\lvert\omega\rvert)\,\mathrm{u}(1-\lvert\tau\rvert)\,\mathrm{u}(1-\lvert k\rvert)$
	$\exp(-a\omega - b\tau - ck)\,\mathrm{u}(\omega)\,\mathrm{u}(\tau)\,\mathrm{u}(k)$
$\exp(-j\,[a\Delta t + b\Delta f + c\Delta r])$	
$\exp\left(-\Delta f^2 - 2\lvert\Delta r\rvert\right)(1-\lvert\Delta t\rvert)\mathrm{u}(1-\lvert\Delta t\rvert)$	

11. Calculate the RMS spreads and fading rate variances for the random channels described by the following:

a. $S_{\tilde{h}}(\omega) = \mathrm{sn}^2(t_0\omega)$

b. $S_{\tilde{h}}(k) = 5\cos\left(\frac{\pi k}{2k_{\max}}\right)\,\mathrm{u}(k_{\max} - \lvert k\rvert)$

c. $C_{\tilde{h}}(\Delta f) = \mathrm{J}_0(\alpha f)$

d. $C_{\tilde{h}}(\Delta r) = 8\exp(-jk_0 r)$

e. $S_{\tilde{h}}(k, \omega) = \exp(-\pi\lvert k\rvert)[\mathrm{u}(\tau) - \mathrm{u}(\tau - 1)]$ (needs a pair of answers)

12. Not all random processes have well-defined RMS PSD widths. Consider a random channel with a triangular autocorrelation function, $C_{\tilde{h}}(\Delta t) = \left(1 - \frac{\lvert\Delta t\rvert}{T}\right)\mathrm{u}(T - \lvert\Delta t\rvert)$. Answer the following questions based on this random process:

a. Calculate the PSD for this process.

b. Set up the integral for calculating an RMS Doppler spread. How does this integral evaluate?

c. What does the result in (b) imply about the fading rate variance? What characteristics of a random process might imply such behavior?

13. In the text, we stated that fading rate variance was a second-order statistic. Yet, upon inspection, this statistic is calculated from only a single sample of the differentiated random process. Explain why this is still considered a second-order statistic.

14. Channel duality can be a powerful tool for understanding many signal processing techniques in wireless communications by constructing a *duality analogy.* For example, we can construct a sentence that uses duality to show similarity between the purposes of frequency-hopping and interleaving:

> The technique of **(frequency-hopping, interleaving)** transmits sequential digital symbols in pseudorandom **(frequency, time)** slots so that bursts of errors are avoided in a **(frequency, time)** selective channel.

Construct a duality analogy for each of the following concepts. (This may require some research.)

a. How multiple users share the common air interface in **(TDMA, FDMA)**. T/FDMA is Time/Frequency Division Multiple Access.

b. How fading is overcome using **(frequency-hopping, selection antenna diversity)**.

c. How increased spectral spreads lead to instability in **(a linear equalizer, adaptive gain control)**.

d. How to remove fading using **(switched-beam antenna diversity, a rake equalizer)**. Hint: Switching beams in a linear antenna array is equivalent to selecting only a portion of a wavenumber spectrum.

e. How digital symbols are modulated using **(QAM, OFDM)**. QAM: quadrature amplitude modulation, OFDM: orthogonal frequency division multiplexing.

f. Find two original concepts in digital modulation to create your own duality analogy.

Chapter 4

PHYSICS OF SMALL-SCALE FADING

This chapter addresses a subject that is too often overlooked in the analysis of small-scale radio channels: *A radio channel must obey Maxwell's equations for free-space propagation.* People often use stochastic process theory - such as that presented in Chapter 3 - and apply it to arbitrarily constructed channel models. Such channel models do not necessarily represent realistic propagation and will yield deceptive and incorrect results when applied to problems in wireless communications.

The analysis in this chapter focuses on the frequency-selective and spatially selective behavior of the static channel, $\tilde{h}(f, \vec{r})$. Scatterers in the propagation environment produce multipath waves with different time delays and wavevectors, causing both spatial and frequency selectivity. Temporal selectivity, on the other hand, is caused by the *motion* of scatterers and lacks the close relationship shared by spatial and frequency selectivity.

The principal goal of this chapter is to develop a fundamental understanding of what constitutes realistic small-scale channels. The following summarizes the topics of this chapter:

- Section 4.1: Overview of plane wave propagation in free-space.
- Section 4.2: Definition of a *local area.*
- Section 4.3: Classification of wave components into groups.
- Section 4.4: Presentation of the *stochastic local area channel model.*
- Section 4.5: Summary of important concepts.

These concepts make meaningful links between stochastic modeling and the physical realm.

4.1 Plane Wave Representation

This section describes the plane wave representations used in the analysis of small-scale wireless channels.

4.1.1 Electromagnetic Fields and Received Signals

The propagation of electromagnetic waves in space must be described using the vector quantities of electric or magnetic fields. However, a radio receiver does not receive a true vector signal. Rather, the input of a radio receiver is a scalar voltage or current emanating from the terminals of the receiver antenna. In essence, an antenna is a device that maps a space-varying, vector field quantity (with units of *Volts* m^{-1} or *Amps* m^{-1}) to a scalar voltage quantity (with units of *Volts* or *Amps*). This mapping is illustrated in Figure 4.1 for an electric field.

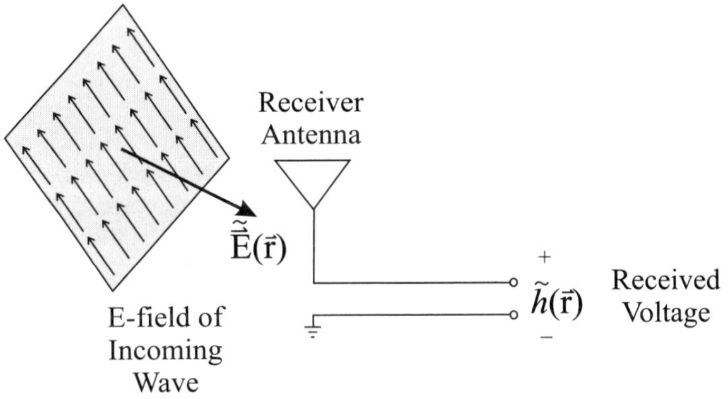

Figure 4.1 An antenna maps the complex electric field vector, $\tilde{\vec{E}}(\vec{r})$, to a scalar baseband channel voltage, $\tilde{h}(\vec{r})$.

To avoid the complications of vector notation, we often use scalar voltage or current of the antenna instead of the vector fields of free-space. This implies that the effects of the antenna - gain, phase change, and polarization mismatch - have already been accounted for in the voltage representation. All of these antenna effects for a propagating wave may be accounted for by a single *polarization vector*, $\tilde{\vec{a}}$. If the propagation of a single time-harmonic electromagnetic wave is described by an electric field vector, $\tilde{\vec{E}}(\vec{r})$, then the baseband voltage, $\tilde{h}(\vec{r})$, results from the following operation:

$$\tilde{h}(\vec{r}) = \tilde{\vec{E}}(\vec{r}) \cdot \tilde{\vec{a}} \tag{4.1.1}$$

where $\vec{r}$ is spatial translation of the receiver, or at least of its antenna. The dot product of the electric field and the polarization vector of the antenna produce the baseband voltage at the terminals of the receiver antenna.

The polarization vector, $\tilde{\vec{a}}$, of Equation (4.1.1) must have units of distance (meters) to be dimensionally correct. The magnitude of the polarization vector is proportional to the gain of the antenna. Since the polarization vector is complex, it is capable of modeling phase change of the incoming signal. The orientation of the vector is capable of modeling polarization mismatch. The only limitation regarding the polarization vector notation is that $\tilde{\vec{a}}$ changes as the incident angle of the impinging wave changes. For example, if multiple waves from different directions impinge upon the antenna, then a polarization vector must be calculated for each wave:

$$\tilde{h}(\vec{r}) = \underbrace{\tilde{\vec{E}}_1(\vec{r}) \cdot \tilde{\vec{a}}_1}_{\tilde{V}_1(\vec{r})} + \underbrace{\tilde{\vec{E}}_2(\vec{r}) \cdot \tilde{\vec{a}}_2}_{\tilde{V}_2(\vec{r})} + \underbrace{\tilde{\vec{E}}_3(\vec{r}) \cdot \tilde{\vec{a}}_3}_{\tilde{V}_3(\vec{r})} + \ldots \tag{4.1.2}$$

Since much of our analysis requires breaking complicated propagating wave solutions into individual wave components, this aspect of polarization vectors is not restrictive.

The Physical Channel Convention

The channel in Equation (4.1.2) has units of *Volts* and, as defined, represents the physical voltage excited by waves at the terminals of the antenna. With this definition, the terms *channel* and *received voltage* may be used interchangeably. This convention, which we call the *physical channel convention* (PCC), is convenient. If a unitless information signal, $\tilde{x}(t)$, is sent through the channel according to Figure 2.4, then the received baseband signal, $\tilde{y}(t)$, has the units of *Volts* and represents an actual signal processed by receiver hardware.

The Normalized Channel Convention

Another useful way of defining the radio channel is the *normalized channel convention* (NCC). Using this convention, we define the radio channel model as a normalized version of Equation (4.1.2), dividing out the RMS channel power:

$$\tilde{h}(\vec{r}) = \frac{\sum_i \tilde{V}_i(\vec{r})}{\sqrt{\left\langle \left| \sum_i \tilde{V}_i(\vec{r}) \right|^2 \right\rangle}} \tag{4.1.3}$$

where $< \cdot >$ is an averaging operator (ensemble or spatial). Equation (4.1.3) is useful for application analysis involving fading power levels. In the NCC case, the average power is treated as a multiplying constant that subsumes hardware-dependent constants such as impedance, digital signal filtering, or the $\frac{1}{2}$ multiplier discussed in Section 2.1.4. Most of our discussion is general to both the PCC and the NCC; if not, the text will specify which channel convention is appropriate.

4.1.2 The Maxwellian Basis

In order to describe complicated propagation in a bounded region of linear free-space, it is useful to break down the received voltage levels as a function of space into a *solution basis*. A solution basis is a set of elementary functions whose linear combinations span every possible solution for a set of differential equations - in our study, Maxwell's equations of free-space radio wave propagation. For example, we may postulate that a solution basis for any baseband received voltage is the set of all complex sinusoidal waves:

$$\tilde{h}(\vec{r}) = \sum_i V_i \exp(j[\phi_i - \vec{k}_i \cdot \vec{r}]) \tag{4.1.4}$$

where $\{V_i\}$ are real amplitudes, $\{\phi_i\}$ are real phases, and $\{\vec{k}_i\}$ are real vectors. According to basic Fourier analysis, *any* 3D function such as complex voltage, $\tilde{h}(\vec{r})$, may be written in the form of Equation (4.1.4).

The basis of Equation (4.1.4) is purely mathematical. In radio wave propagation, the solution basis for a channel must satisfy Maxwell's equations in addition to spanning all possible functions of received voltage. In fact, only a select few terms of the summation in Equation (4.1.4) will obey the fundamental laws of propagation. A complete solution basis in which every term solves Maxwell's equations is called a *Maxwellian basis* [Bro98].

One Maxwellian basis that may be used to construct any realizeable set of received voltages in bounded free-space is the set of all plane waves. This basis is very similar to Equation (4.1.4):

$$\tilde{h}(\vec{r}) = \sum_i V_i \exp(j[\phi_i - \vec{k}_i \cdot \vec{r}]) \qquad \text{where } \vec{k}_i \cdot \vec{k}_i = k_0^2 \tag{4.1.5}$$

In Equation (4.1.5), $\{V_i\}$ are real amplitudes, $\{\phi_i\}$ are real phases, and $\{\vec{k}_i\}$ are constant wavevectors [Bor80, p. 562]. The terms of Equation (4.1.5) are called *plane waves* due to the geometry of the equiphase surfaces. An equiphase surface of a propagating wave is defined as the set of points, $\vec{r}$, in three-dimensional space that satisfy the following equation:

$$\arg\{\tilde{h}(\vec{r})\} = \phi_0 \tag{4.1.6}$$

where ϕ_0 is some arbitrary phase constant. The equiphase surfaces of each term in Equation (4.1.5) form planes in three-dimensional space.

At first glance, Equation (4.1.5) appears to be a small subset of the Fourier basis in Equation (4.1.4). The condition $\vec{k}_i \cdot \vec{k}_i = k_0^2$ restricts the solution set to plane waves with wavevectors of a certain magnitude (see Theorem 4.1 for a derivation of this condition). But unlike the Fourier basis, *the wavevectors of Equation (4.1.5) may be complex-valued.* A complex $\vec{k}_i$ still solves Maxwell's time-harmonic equations and is needed to describe every possible instance of received voltage in free-space.

Theorem 4.1: Wavevector Criterion for free-space

Statement: The free-space wavevectors in a plane wave basis obey the condition in Equation (4.1.5).

Proof: The electric field of a plane wave propagating in free-space takes the following form:

$$\tilde{\vec{\mathrm{E}}}(\vec{\mathrm{r}}) = \tilde{\vec{\mathrm{E}}}_0 \exp(-j\vec{\mathrm{k}}_0 \cdot \vec{\mathrm{r}})$$

where $\tilde{\vec{\mathrm{E}}}_0$ is a complex, constant electric field vector and $\vec{\mathrm{k}}_0$ is a constant-valued wavevector. Any solution of this equation must solve the scalar wave equation for free-space:

$$(\nabla^2 + k_0^2)\tilde{\vec{\mathrm{E}}}(\vec{\mathrm{r}}) = 0$$

where k_0 is the free-space wavenumber of time-harmonic propagation and $\nabla^2 = \frac{\partial^2}{\partial x^2} + \frac{\partial^2}{\partial y^2} + \frac{\partial^2}{\partial z^2}$. We may then write

$$\begin{aligned} 0 &= (\nabla^2 + k_0^2)\, \tilde{\vec{\mathrm{E}}}_0 \exp(-j\vec{\mathrm{k}}_0 \cdot \vec{\mathrm{r}}) \\ &= \tilde{\vec{\mathrm{E}}}_0 \left[k_0^2 - \vec{\mathrm{k}}_0 \cdot \vec{\mathrm{k}}_0\right] \exp(-j\vec{\mathrm{k}}_0 \cdot \vec{\mathrm{r}}) \end{aligned}$$

which took advantage of the following vector relationship:

$$\nabla^2 \exp(-j\vec{\mathrm{k}}_0 \cdot \vec{\mathrm{r}}) = -\vec{\mathrm{k}}_0 \cdot \vec{\mathrm{k}}_0 \exp(-j\vec{\mathrm{k}}_0 \cdot \vec{\mathrm{r}})$$

Thus, the wavevector, $\vec{\mathrm{k}}_0$, must satisfy the following relationship:

$$\vec{\mathrm{k}}_0 \cdot \vec{\mathrm{k}}_0 = k_0^2$$

All plane waves must have wavevectors of this form in order to be a valid solution of Maxwell's equations in free space.

Note: Frequency, Wavelength, and Wavenumber

Recall the basic result from physics that all waves traveling in a linear medium satisfy $f_c\lambda = c$, where f_c is carrier frequency, λ is wavelength, and c is the speed of propagation (for electromagnetic propagation, this is the speed of light, 3.0×10^8 m/s). Since k_0 is the free-space wavenumber, we may write this simple relationship in an alternative form:

$$\frac{2\pi f_c}{k_0} = c \qquad \left(k_0 = \frac{2\pi}{\lambda}\right)$$

The terms of the Maxwellian plane wave basis neatly divide into two different types of plane waves. The first group consists of *homogeneous* plane waves with strictly real-valued wavevectors. The second group consists of *inhomogeneous* plane waves with complex-valued wavevectors. Thus, Equation (4.1.5) may be rewritten as

$$\tilde{h}(\vec{\mathrm{r}}) = \underbrace{\sum_l V_l \exp(j[\phi_l - \vec{\mathrm{k}}_l \cdot \vec{\mathrm{r}}])}_{\substack{\text{Homogeneous,} \\ \text{Real } \vec{\mathrm{k}}_l}} + \underbrace{\sum_m V_m \exp(j[\phi_m - \vec{\mathrm{k}}_m \cdot \vec{\mathrm{r}}])}_{\substack{\text{Inhomogeneous,} \\ \text{Complex } \vec{\mathrm{k}}_m}} \qquad (4.1.7)$$

Each classification of plane wave is discussed below.

> *Note: Definition of Free-Space*
>
> The term *free-space* in propagation implies the following three characteristics:
>
> 1. There are no current sources or charges present in the medium.
> 2. The propagation medium is *linear*, *isotropic*, and *lossless*.
> 3. The material parameters of permittivity and permeability are constants and are equal to that of vacuous space ($\epsilon = \epsilon_0$ and $\mu = \mu_0$).
>
> It is generally accepted that the atmosphere - despite the presence of gases - behaves as a free-space medium for short-range propagation.

4.1.3 Homogeneous Plane Waves

Homogeneous plane waves are also called *uniform* plane waves because their envelope is a constant value that does not depend on position in space. Since the wavevector of a homogeneous plane wave, $\vec{\mathrm{k}}$, is real-valued and satisfies the restriction noted in Equation (4.1.5), the plane wave may be written as

$$V(\vec{\mathrm{r}}) = V_0 \exp(j[\phi_0 - k_0 \hat{\mathrm{k}} \cdot \vec{\mathrm{r}}]) \tag{4.1.8}$$

where $\hat{\mathrm{k}}$ is a unit vector that points in the direction of propagation and k_0 is the free-space wavenumber. In a time-harmonic free-space analysis, k_0 is related to the wavelength of radiation, λ:

$$k_0 = \frac{2\pi}{\lambda} \tag{4.1.9}$$

Regardless of the direction of travel, all homogeneous plane waves propagate at the same speed with the same wavenumber. This is not true of inhomogeneous plane waves.

4.1.4 Inhomogeneous Plane Waves

Inhomogeneous plane waves are only slightly more difficult to understand than homogeneous plane waves. First, it helps to break the complex wavevector into a real and imaginary part:

$$\vec{\mathrm{k}} = \beta\hat{\beta} - j\alpha\hat{\alpha} \tag{4.1.10}$$

where $\hat{\beta}$ is a unit vector pointing in the direction of the real part of $\vec{\mathrm{k}}$ and $\hat{\alpha}$ is a unit vector pointing in the direction of the imaginary part of $\vec{\mathrm{k}}$. This forces α and β to be real quantities.

The form of Equation (4.1.10) illuminates some basic properties of inhomogeneous plane waves. Imposing the wavevector restriction of the Maxwellian basis from Equation (4.1.5) leads to two descriptive conditions for a vector of this form:

$$\vec{\mathrm{k}} \cdot \vec{\mathrm{k}} = k_0^2 \longrightarrow \begin{array}{ll} \text{Condition 1:} & \hat{\alpha} \cdot \hat{\beta} = 0 \\ \text{Condition 2:} & k_0^2 = \beta^2 - \alpha^2 \end{array}$$

First, we see that the unit vectors $\hat{\alpha}$ and $\hat{\beta}$ must be orthogonal to one another (their dot products are zero). Second, while α and β may take on any positive value, they are dependent on one another. Specifically, the value $\sqrt{\beta^2 - \alpha^2}$ must be equal to the free-space wavenumber, k_0.

Armed with this information, it is now useful to introduce the split wavevector of Equation (4.1.10) into the expression for a propagating plane wave:

$$\tilde{h}(\vec{r}) = V_0 \overbrace{\exp(-\alpha[\hat{\alpha} \cdot \vec{r}])}^{\text{Attenuation}} \underbrace{\exp\left(j[\phi_0 - \beta(\hat{\beta} \cdot \vec{r})]\right)}_{\text{Phase Progression}} \tag{4.1.11}$$

Now the physical meanings of α and β are apparent. The value of β is the actual wavenumber of the inhomogeneous plane wave, determining the rate of phase progression through space in the direction $\hat{\beta}$. The value of α represents the rate of amplitude decay. As Equation (4.1.11) shows, the amplitude of the plane wave attenuates in the direction of $\hat{\alpha}$. The direction of amplitude attenuation must always be orthogonal to the direction of propagation for an inhomogeneous plane wave.

Figure 4.2 illustrates a comparison of homogeneous and inhomogeneous plane waves. Notice the two key differences between the types of plane waves. The wavenumber, β, of the inhomogeneous plane wave is always greater than the free-space wavenumber, k_0. It is useful to refer to an "effective wavelength," λ_{eff}, for inhomogeneous plane waves, defined as

$$\lambda_{\text{eff}} = \frac{2\pi}{\beta} \tag{4.1.12}$$

The effective wavelength of an inhomogeneous plane wave is always smaller than the free-space wavelength of radiation. As a result, the equiphase fronts of the inhomogeneous plane wave are more densely packed than the homogeneous plane wave.

The second key difference between homogeneous and inhomogeneous plane waves is the exponential decay of amplitudes perpendicular to the direction of propagation. If the decay rate, α, is small, then the wavenumber, β, approaches the homogeneous free-space wavenumber, k_0. If the decay rate, α, is large, then the wavenumber, β, is much larger than k_0. This principle is crucial for later sections in this chapter that develop physically-based small-scale channel models.

4.1.5 Homogeneous Versus Inhomogeneous Plane Waves

If free-space were unbounded, having infinite extent in every direction, then inhomogeneous plane waves would not exist. A quick study of Equation (4.1.11) tells us why. A wave that decays exponentially in one direction actually grows exponentially in the opposite direction. Thus, in unbounded free-space, the power density of an inhomogeneous plane wave becomes infinite towards one side of space - clearly a

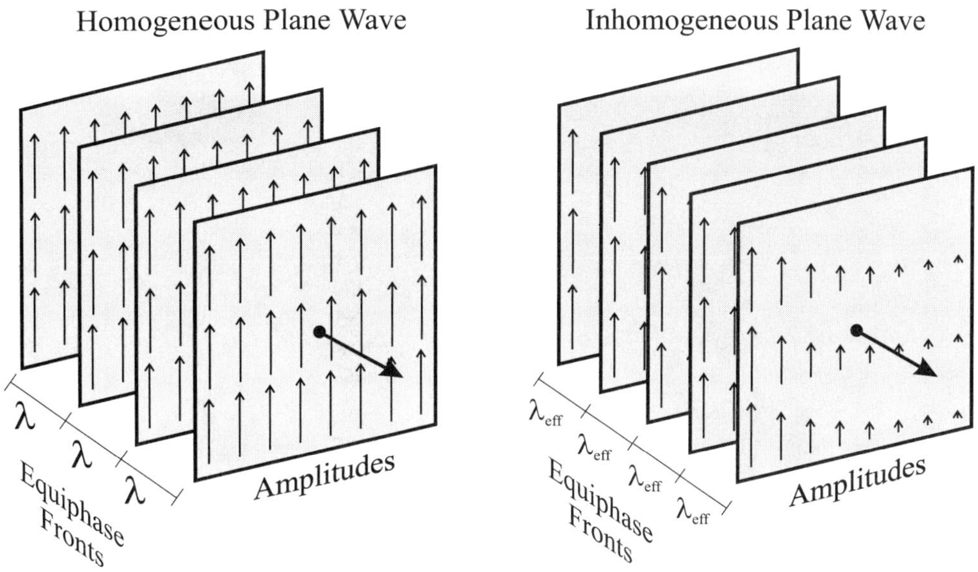

Figure 4.2 Homogeneous and inhomogeneous plane waves.

nonphysical result. Homogeneous waves have no such imbalance, having a constant, finite power density in all of space.

Although inhomogeneous plane waves solve Maxwell's equations in bounded free-space, these waves are caused by scatterers or sources in the propagation medium *outside* the bounded free-space region. The larger wavenumbers of inhomogeneous plane waves help meet the boundary conditions of electromagnetic fields close to material surfaces with fine spatial structure [Bor80, p. 563]. This physical characteristic leads to a rule of thumb for inhomogeneous plane wave propagation:

> *Boundary conditions near large scatterers typically introduce inhomogeneous plane waves into a region of bounded free-space. The direction of decay, $\hat{\alpha}$, for these waves tends to point away from the scatterers.*

The homogeneous waves follow a different rule of thumb:

> *The direction of propagation, $\hat{k}$, for homogeneous plane waves is typically away from the scatterers.*

These two rules of thumb are illustrated in Figure 4.3.

An Analogy From Circuit Theory

A useful analogy exists between free-space propagation and linear circuit theory that aids in understanding the nature of homogeneous and inhomogeneous plane

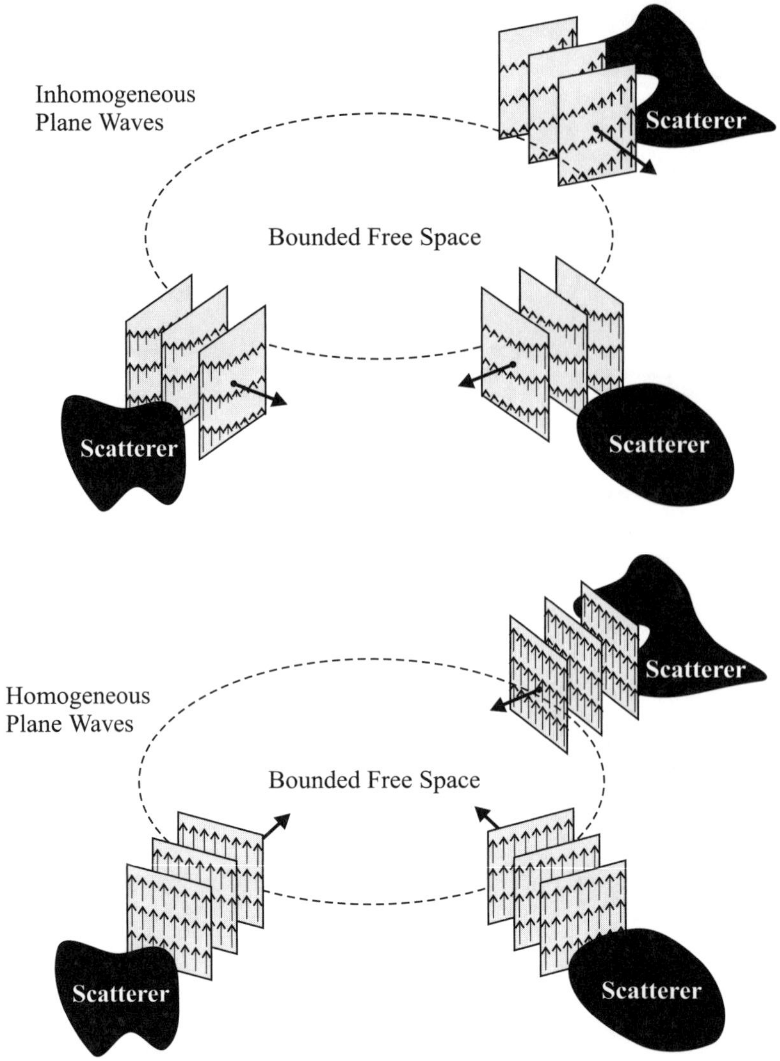

Figure 4.3 Rules of thumb for homogeneous and inhomogeneous plane wave propagation.

waves. Consider as an example the linear circuit of Figure 4.4. This circuit contains an AC voltage source, which switches on at $t = 0$. To solve for an output voltage, $V(t)$, as a function of time, we must solve the governing differential equation of the system. The order of a circuit differential equation is equal to the total number of inductors and capacitors. Thus, the example in Figure 4.4 has a seventh-order differential equation governing the input–output relationships.

While the source of Figure 4.4 is a time-harmonic AC source, the closing switch at time $t = 0$ introduces transient behavior into the circuit. The solution of this

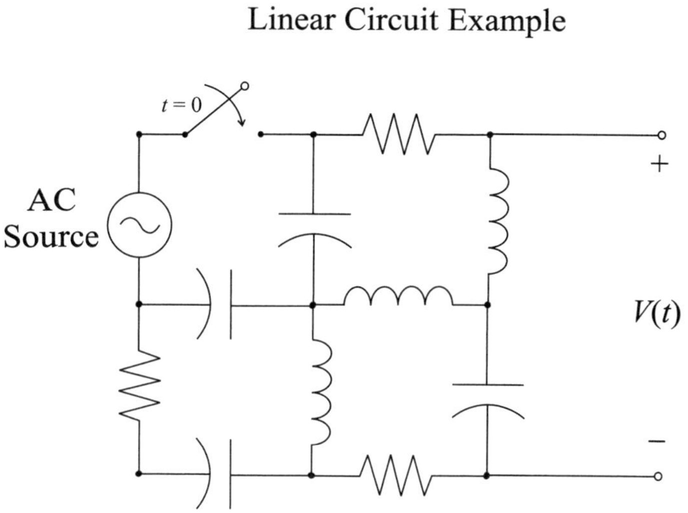

Figure 4.4 A linear circuit contains capacitors, inductors, resistors, and an AC source.

circuit for $t \geq 0$ may be written in the following form:

$$V(t) = [\text{Steady-State Solution}] + [\text{Transient Solution}] \quad \text{for } t \geq 0$$

A *steady-state solution* is a time-harmonic, sinusoidal voltage that eventually becomes the dominant solution for large values of t. The transient solution is an exponentially decaying set of voltages that transfers the circuit from its zero-excited state ($t < 0$) to its steady state ($t \gg 0$). The concept of steady-state and transient solutions are illustrated in Figure 4.5. If we make several modifications in terminology, then the linear circuit becomes a powerful analogy for understanding free-space propagation.

Analogy to Free-Space Plane Waves

The key difference between free-space propagation and the linear circuit example is that the plane wave propagation is a function of space rather than time. Homogeneous plane waves are like the steady-state solutions in the linear circuit. These plane waves are uniformly periodic (space-harmonic) with respect to space just as the steady-state solution is uniformly periodic (time-harmonic) with respect to time. Inhomogeneous plane waves are like the transient solutions in the linear circuit. The amplitudes of these plane waves decay exponentially with respect to space just as the transient solution decays in time.

In the propagation problem, the electromagnetic scatterer is like a "switch in space." The position of the scatterer in space is like the time that the circuit switch is thrown. Solutions for fields that are very close to the scatterer must pay attention

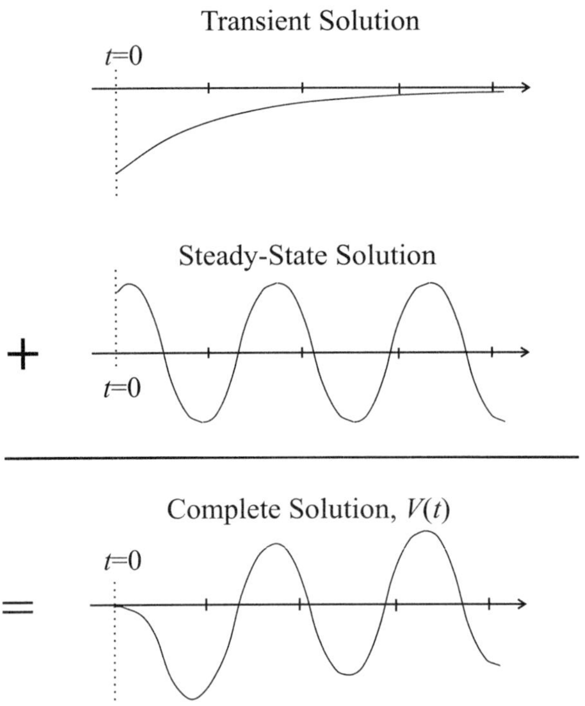

Figure 4.5 A linear circuit solution may be broken into a steady-state and transient solution.

to the behavior of inhomogeneous plane waves, just as the transient circuit solutions dominate for times close to the switching time of $t = 0$. Solutions for fields that are further from the scatterer may ignore the inhomogeneous plane waves, since they decay exponentially like the transient circuit solution.

There are numerous analogies to be made between the linear circuit of Figure 4.4 and the free-space propagation of plane wave solutions. Below is a summary of many of these parallel concepts:

LINEAR CIRCUIT CONCEPT	FREE-SPACE PROPAGATION CONCEPT
time	position
switch	scatterer
$t > 0$	bounded free-space
time elapsed since switch	distance from scatterer
steady-state solution	homogeneous plane waves
transient solution	inhomogeneous plane waves

The goal of this circuit analogy is, by using familiar engineering constructs, to make an esoteric concept (wave propagation in three dimensional space) easier to understand. Another motive is to help justify the key assumption in the definition of a local area, namely, the removal of inhomogeneous plane waves for receivers that operate away from significant scatterers.

> *Note: Analogy from Waveguide Theory*
>
> A reader familiar with dielectric waveguide theory may recognize inhomogeneous plane waves as the *evanescent* waves that propagate along a material interface of the waveguide and an open medium. Much like Figure 4.3, these waves are bounded to the material surface and decay into free-space exponentially away from the waveguide.

4.2 The Local Area

Used often in the wireless literature without definition, the term *local area* is one of the most important concepts in small-scale channel modeling. This section presents a definition of the local area and analyzes its usefulness and validity.

4.2.1 Definition of a Local Area

Small-scale fading due to spatial selectivity is the focus of a *local area* propagation analysis. Characterization of a channel autocorrelation function is just one example of a local area propagation analysis. This is in contrast to a *macro area* analysis, which focuses on the differences in propagation from one radio environment or location to the next. A comparison of average power received in the front of a shopping mall to average power received on the top story of a parking garage is an example of a macro area analysis.

Before proceeding, it is crucial to define exactly what is meant by a local area, since the definition can be quite arbitrary. There is no magical distance quantity that can be applied to any region of space and called a local area. Rather, the following definition will be applied throughout this work:

> A *local area* is the largest volume of free-space about a specified point ($\vec{r} = \vec{0}$) in which the wireless channel can be modeled accurately as the sum of homogeneous plane waves of the form
>
> $$\tilde{h}(f, \vec{r}) = \sum_i V_i \exp\left(j\left[\phi_i - \vec{k}_i \cdot \vec{r} - 2\pi f \tau_i\right]\right)$$

The justification for this definition of the local area channel is discussed next.

> *Note: A Multipath Wave*
>
> The term *multipath* is used by wireless engineers to describe the multiple wave components, each traveling different paths, that impinge upon a receiver antenna. Hence, *multiple* + *paths* = *multipath*. When the term *multipath wave* is used in this work, we are speaking of a single wave term in the local area representation. With this definition, it is possible that a single scatterer may produce more than one *multipath wave*. Therefore, be aware that the *multi* in multipath refers to the number of waves and not the number of scatterers.

4.2.2 Scatterer Proximity

First note that the local area definition does not include inhomogeneous plane waves in the description of the propagation. For the local area definition to be valid, all inhomogeneous plane waves must be either ignored or approximated by homogeneous plane waves. To rationalize this restriction, first group the inhomogeneous waves into two types:

Abruptly Decaying: The direction of decay for the inhomogeneous plane waves typically points into the local area (see Figure 4.3). If the decay rate is high, then the power carried by the plane wave extinguishes after the wave penetrates the local area by only one or two wavelengths. Thus, when one considers the total power carried by the fast-decaying wave with respect to all other multipath waves in the local area, it becomes insignificant and may be ignored.

Gently Decaying: For slowly decaying inhomogeneous plane waves, the value of decay rate α is small. As a result, the wavenumber β is approximately equal to the free-space wavenumber k:

$$\beta = \sqrt{k^2 + \alpha^2} \approx k$$

Thus, a slowly decaying inhomogeneous plane wave moves in space with a phase progression similar to a homogeneous plane wave and has only a gradual taper to its amplitude. A homogeneous plane wave is a valid approximation to this wave over a small region of space.

Of course, the size of the free-space region that satisfies these approximations varies from case to case and is largely a function of *scatterer proximity*. We know that inhomogeneous plane waves are caused by scatterers in free-space and that their contribution to propagation diminishes in regions away from the significant scatterer. Thus, the proximity of the local area to significant scatterers usually determines the acceptable size of the local area.

This concept is illustrated in Figure 4.6. When the local area is not close to a large scatterer, then the homogeneous plane waves dominate the description of propagation; the size of the region over which inhomogeneous plane waves may be ignored is large (like the steady-state solution in the circuit analogy). When the local area is close to a large scatterer, then inhomogeneous plane waves contribute significant power to the total power propagating through the space. The size of the valid local area is small, since the abruptly and gently decaying approximations fall apart after a few wavelengths of displacement.

4.2.3 A Wideband Plane Wave

The ideal wireless channel, from the standpoint of simplicity, results from a single uniform plane wave propagating through space. This plane wave impinges upon

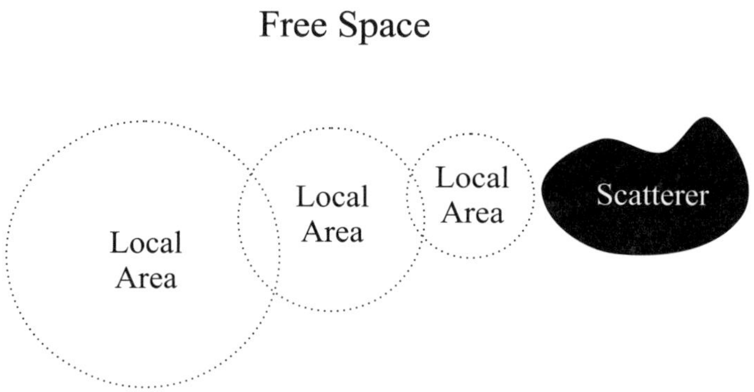

Figure 4.6 The size of a local area decreases closer to significant scatterers.

a receiver antenna and excites a constant-amplitude voltage at the receiver that does not change in magnitude as a function of position. The resulting complex, time-harmonic voltage at the receiver is given by

$$\tilde{h}(\vec{r}) = V \exp\Big(j[\phi - \vec{k} \cdot \vec{r} - k_0 L]\Big) \tag{4.2.1}$$

where V is a constant, real-valued amplitude and ϕ is a constant, real-valued phase. If the homogeneous plane wave exciting the voltage in Equation (4.2.1) satisfies Maxwell's equations for free-space propagation, then the real vector $\vec{k}$ points in the direction of propagation and has magnitude k_0. The value L is the total distance traveled from transmitter to receiver by the wave.

The phase terms in the complex exponential of Equation (4.2.1) each represent a different physical contribution to the total phase of the incoming wave, as illustrated in Figure 4.7. The three phase terms and their physical causes are listed below:

$$\arg\{\tilde{h}(\vec{r})\} = \phi - \vec{k} \cdot \vec{r} - k_0 L$$

ϕ	phase change accumulated by antenna and scatterer interaction
$k_0 L$	phase change due to total distance traveled
$\vec{k} \cdot \vec{r}$	phase change due to motion of the receiver

For a single homogeneous plane wave, the phase term contains most of the interesting physical behavior.

Hidden in Equation (4.2.1) is an inherent narrowband dependence on the carrier frequency, f_c. Since k_0 is related to wavelength, λ, and wavelength is related to carrier frequency ($\lambda = c/f_c$, where c is the free-space speed of light), it is possible to substitute $k_0 = \frac{2\pi f_c}{c}$ into Equation (4.2.1) to observe the carrier dependence:

$$\tilde{h}(\vec{r}) = V \exp\left(j\left[\phi - \frac{2\pi f_c}{c}\hat{k} \cdot \vec{r} - 2\pi f_c \tau\right]\right) \tag{4.2.2}$$

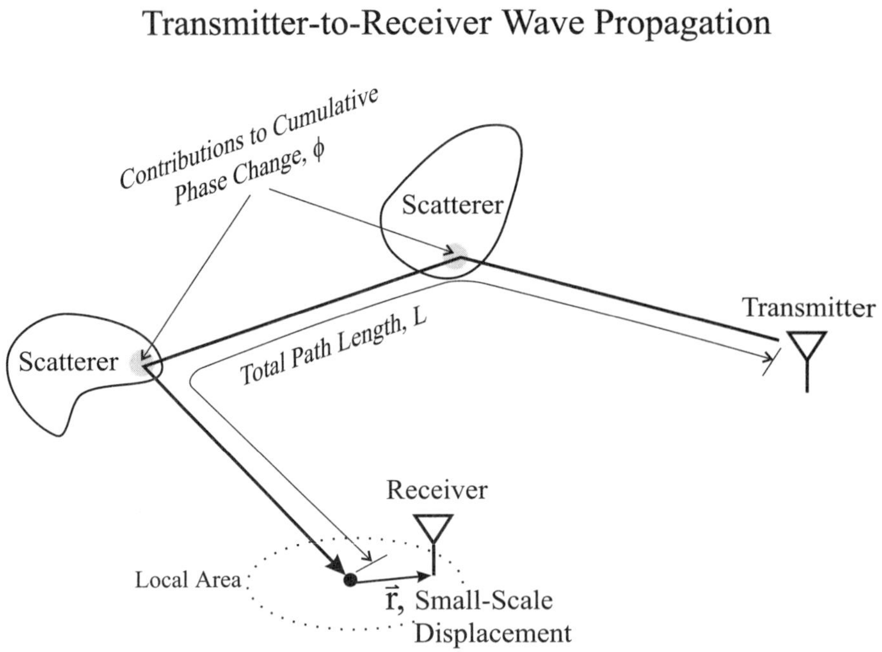

Figure 4.7 The basic quantities of time-harmonic wave propagation through a scattering environment.

where $\tau = L/c$, which is the propagation delay of the wave, the time it takes to travel from transmitter to receiver. In Equation (4.2.2), the value $\hat{\mathrm{k}}$ is a unit vector that points in the direction of propagation.

Adding Signal Bandwidth

If the plane wave has been modulated with a band-limited transmitted signal, then the narrowband characterization of Equation (4.2.2) becomes insufficient to characterize the channel. For a small frequency displacement, f, about the carrier frequency, Equation (4.2.2) may be rewritten as

$$\tilde{h}(\vec{\mathrm{r}}, f) = V \exp\left(j\left[\phi - \frac{2\pi(f_c + f)}{c}\hat{\mathrm{k}} \cdot \vec{\mathrm{r}} - 2\pi(f_c + f)\tau\right]\right) \tag{4.2.3}$$

The phase term of Equation (4.2.3) may be rearranged to produce

$$\tilde{h}(\vec{\mathrm{r}}, f) = V \exp\left(j\left[\phi_0 - \vec{\mathrm{k}} \cdot \vec{\mathrm{r}} - 2\pi f\tau - \frac{2\pi f}{c}\hat{\mathrm{k}} \cdot \vec{\mathrm{r}}\right]\right) \tag{4.2.4}$$

where $\phi_0 = \phi - 2\pi f_c \tau$. Equation (4.2.4) still retains the basic characteristics of a plane wave. The envelope of the channel is still a constant, independent of position, $\vec{\mathrm{r}}$, *and* frequency, f.

Adding More Bandwidth

The plane wave in Equation (4.2.4) is actually a small-frequency expansion about the carrier frequency. If a wideband signal modulates the plane wave, then the amplitude and phase of the single wave may also become functions of frequency, f. This frequency dependence results from the fine scattering mechanisms in the environment and sometimes even the electromagnetic interactions with receiver and transmitter antennas. Every interaction with a realistic material has some type of inherent frequency-dependence, which makes the excess phase change, ϕ_0, and the amplitude, V, a function of frequency.

To treat this case analytically, simply expand the amplitude and excess phase change in terms of a Fourier series of an arbitrary number of terms:

$$V(f)\exp\left(j\phi_0(f)\right) = \sum_i V_i \exp(j[\phi_i - 2\pi f \tau_i']) \tag{4.2.5}$$

Inserting this expansion into Equation (4.2.4), we arrive at

$$\tilde{h}(\vec{\mathrm{r}}, f) = \sum_i \underbrace{V_i \exp(j[\phi_i - 2\pi f \tau_i])}_{V(f)\exp[j\phi_0(f)]} \exp\left(-j\left[\vec{\mathrm{k}}\cdot\vec{\mathrm{r}} + \frac{2\pi f}{c}\hat{\mathrm{k}}\cdot\vec{\mathrm{r}}\right]\right) \tag{4.2.6}$$

where $\tau_i = \tau_i' + \tau$. Therefore, the full wideband expression of a modulated uniform plane wave will break the wave into the sum of smaller waves with various propagation delays, τ_i. Note, however, that each wave component in Equation (4.2.6) has the same form as a single homogeneous plane wave of the form:

$$V_i \exp\left(j\left[\phi_i - \vec{\mathrm{k}}\cdot\vec{\mathrm{r}} - 2\pi f \tau_i - \frac{2\pi f}{c}\hat{\mathrm{k}}\cdot\vec{\mathrm{r}}\right]\right) \tag{4.2.7}$$

The envelope of this wave is constant, independent of position, $\vec{\mathrm{r}}$, and frequency, f.

4.2.4 The Bandwidth-Distance Threshold

If time-harmonic propagation in a region of free-space may be broken into the sum of homogeneous plane waves, and each plane wave may be broken into components of the form Equation (4.2.7) as bandwidth is added, then we have almost validated the usefulness of the local area definition. There is one key difference between the type of waves specified in the local area definition and the type of wave specified by Equation (4.2.7). Consider the phase of the wave in Equation (4.2.7):

$$\exp\left(j\left[\phi_i - \overbrace{\vec{\mathrm{k}}\cdot\vec{\mathrm{r}}}^{\substack{\text{Space}\\ \text{Term}}} - \underbrace{2\pi f \tau_i}_{\substack{\text{Frequency}\\ \text{Term}}} - \overbrace{\frac{2\pi f}{c}\hat{\mathrm{k}}\cdot\vec{\mathrm{r}}}^{\substack{\text{Cross}\\ \text{Product}}}\right]\right) \tag{4.2.8}$$

The homogeneous wave terms in the local area definition contain the space and frequency terms, but do not contain the space–frequency cross-product term.

Thus, the local area representation makes the following approximation to the uniform plane wave equation:

$$\exp\left(-j\left[\vec{\mathrm{k}}\cdot\vec{\mathrm{r}}+2\pi f\tau_i+\frac{2\pi f}{c}\hat{\mathrm{k}}\cdot\vec{\mathrm{r}}\right]\right)\approx\exp\left(-j\left[\vec{\mathrm{k}}\cdot\vec{\mathrm{r}}+2\pi f\tau_i\right]\right) \tag{4.2.9}$$

The temptation is to justify this approximation by simply stating that the frequency and space terms are much larger than the cross-product term. After all, both frequency, f, and small-scale displacement, $\vec{\mathrm{r}}$, must change substantially for the cross product to affect the total phase.

Completely ignoring the cross-product term, however, is a poor approximation for a phase term. Phase terms *always* remain sensitive to small changes in a large net value due to the 2π periodicity of the complex exponential function. Therefore, an absolute limit on the contribution of the cross-product term must be imposed if the local area representation is accurate. We will restrict the cross-product term to always contribute less than a $\frac{\pi}{2}$-radian phase change in any single homogeneous plane wave component. This restriction may be written as

$$\frac{2\pi f}{c}\hat{\mathrm{k}}\cdot\vec{\mathrm{r}}<\frac{\pi}{2} \tag{4.2.10}$$

The maximum value for frequency displacement, f, in the baseband channel is one-half the RF signal bandwidth, B. The maximum value for the small-scale position displacement is one-half the size of the local area, L_A:

$$\hat{\mathrm{k}}\cdot\vec{\mathrm{r}}<|\hat{\mathrm{k}}||\vec{\mathrm{r}}|<\frac{L_A}{2} \tag{4.2.11}$$

Therefore, the largest possible value for the left-hand side in Equation (4.2.10) occurs when frequency and small-scale position displacement are at a maximum $\left(f\rightarrow\frac{B}{2},\ \hat{\mathrm{k}}\cdot\vec{\mathrm{r}}\rightarrow\frac{L_A}{2}\right)$:

$$\frac{BL_A}{c}<1 \tag{4.2.12}$$

which may be rewritten in terms of carrier frequency, f_c, and wavelength, λ:

$$\frac{L_A}{\lambda}<\frac{f_c}{B} \tag{4.2.13}$$

This relationship may be stated in words: The ratio of the local area size to the wavelength of radiation must be less than the ratio of the carrier frequency to the signal bandwidth. The relationship is illustrated in Figure 4.8

Note: Narrowband and Fixed-Receiver Analysis

The cross-product term of Equation (4.2.9) is unique to joint analysis of space–frequency channel selectivity. If the analysis were narrowband, then frequency displacement, f, about the carrier would always be 0 and the cross-product term would vanish. If the analysis were on a fixed receiver, the position vector, $\vec{r}$, would be 0 and the term would also vanish. Thus, the cross-product term and the resulting bandwidth-distance threshold is a unique artifact of full space–time channel modeling.

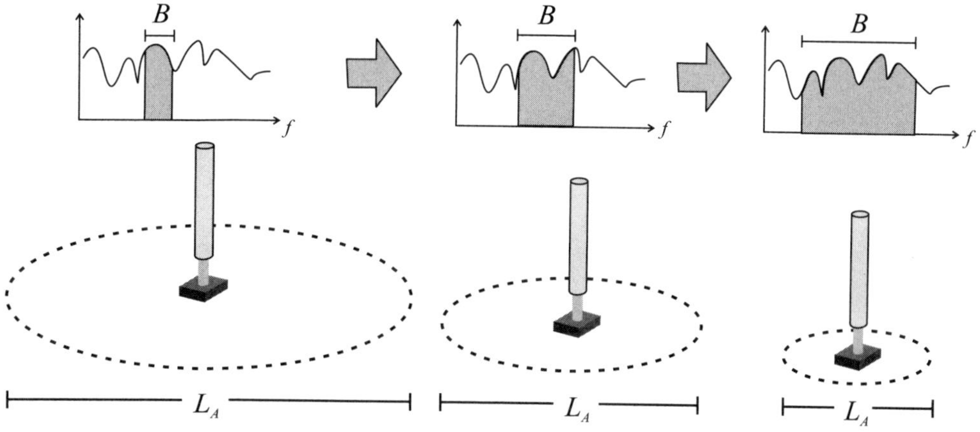

Figure 4.8 If we define a local area with smaller size, the bandwidth over which the local area approximation is valid increases.

Along with scatterer proximity, the bandwidth-distance threshold limits the size of a valid local area. This threshold is typically less restrictive than scatter proximity for the discussion of realistic wireless systems. Table 4.1 lists the carrier frequencies and bandwidths of several common wireless applications. The resulting local area size according to the bandwidth-distance threshold is large in terms of wavelengths - even for applications that the commercial wireless industry considers "wideband." Of course, the size of a wavelength decreases as the carrier frequency increases, so the absolute size of a local area does become small for mm-wave applications.

Note: Trying to Get US

The manipulations of this section are not just gratuitous mathematics. We desire a final form for a local area channel that not only consists of homogeneous plane waves, but of band-limited waves that have *uncorrelated scattering* between frequency and space. The bandwidth-distance threshold, along with the scatterer proximity assumption, helps us model baseband channels with WSS-US random processes.

Table 4.1 Maximum Size of a Local Area (only) According to the Bandwidth–Distance Threshold for Example Wireless Applications

Wireless Application	Carrier Frequency	Signal Bandwidth	Local Area Size
Analog Cellular Channel	840 MHz	30 kHz	28000 λ
PCS Spread-Spectrum Channel	1910 MHz	1.25 MHz	1528 λ
NII Campus Link	5.7 MHz	100 MHz	57 λ
LMDS Link (proposed)	28 GHz	500 MHz	56 λ

4.3 Wave Groupings for Multipath Components

In the study of multipath channels, it is useful to group waves into different types of multipath components. The terms specular, nonspecular, and diffuse are used to describe the electromagnetic waves propagating in a region of space.

4.3.1 Specular Wave Component

A *specular wave component* refers to a received multipath component that may be described by a single homogeneous plane wave. Mathematically, a specular component has the following form:

$$\tilde{h}(f, \vec{r}) = V_i \exp\left(j\left[\phi_i - \vec{k}_i \cdot \vec{r} - 2\pi f \tau_i\right]\right) \tag{4.3.1}$$

The envelope of a specular wave component does not change as a function of space and frequency. Clearly, if a local area wireless channel consisted of a single homogeneous plane wave component, then the channel would have perfect frequency coherence across the signal bandwidth and perfect spatial coherence across the local area.

There are numerous propagation mechanisms that may give rise to large specular components propagating through a local area. Examples of specular components include the direct line-of-sight wave from a transmitter antenna, the specular reflection from a flat surface in the propagation environment, or even a far-field wave diffracted from the edge of an object.

4.3.2 Nonspecular Wave Component

An *nonspecular wave component* refers to a voltage that may be written as the sum of many differently phased specular waves. For a local area, a nonspecular wave component is a collection of homogeneous plane waves that travel in different directions and arrive with different time delays. Mathematically, a nonspecular wave component has the following form:

$$\tilde{h}(f, \vec{r}) = \sum_i V_i \exp\left(j\left[\phi_i - \vec{k}_i \cdot \vec{r} - 2\pi f \tau_i\right]\right) \tag{4.3.2}$$

Given a position in frequency, f, and space, $\vec{r}$, the envelope of such a channel is random and unknown unless every phase variable in Equation (4.3.2) is known. The channel resulting from nonspecular waves clearly has frequency selectivity and spatial selectivity.

The distinction between nonspecular and specular wave components may seem arbitrary, since a nonspecular wave component may be expressed as the sum of specular wave components in a local area. In actuality, the grouping of multipath waves into nonspecular and specular components is quite useful. The scattering mechanisms in many wireless channel are so numerous and varied that characterizing each homogeneous plane wave may take thousands or even millions of terms. Often, it suffices to lump the majority of the waves into a nonspecular term and analyze it as a single component.

4.3.3 Diffuse Wave Component

One more important type of wave component grouping is the *diffuse* wave component. In a local area, a nonspecular wave component is said to be *diffuse* if its constituent homogeneous plane wave amplitudes, $\{V_i\}$, satisfy the following relationship:

$$[\max(V_1, V_2, V_3, \dots)]^2 \ll \sum_i V_i^2 \tag{4.3.3}$$

In other words, a nonspecular wave component is diffuse if each homogeneous plane wave carries power that is negligible compared to the total average power in the component.

Diffuse nonspecular wave propagation occurs in nearly every wireless environment. In a realistic propagation environment, there are numerous objects and surfaces that scatter power to the receiver and back and forth among each other. A wireless receiver is flooded with tiny waves that have negligible power individually, but by sheer quantity contribute a large portion to the overall received signal strength. There is always a collection of diffuse, nonspecular waves in every realistic wireless channel [Ogi91], [Bro97].

Note: For the Physics Buffs

In quantum mechanics, it is possible for an electromagnetic wave to become so weak that it loses its observable coherence and becomes an intermittent trickle of photons. Researchers Elitzur and Vaidman, working at the University of Tel-Aviv, showed in 1993 that even in this state, a wave can add constructively and destructively to other waves [Eli93]. Such a wave is, by itself, insignificant and invisible to observation, but together with millions of other similar waves can contribute significant power in a wireless channel.

4.3.4 Reduced Wave Grouping

Now that specular, nonspecular, and diffuse wave components have been defined, it is possible to discuss the concept of a *reduced wave grouping.* In a local area, it is possible to write the channel as the sum of a finite number of specular waves and a nonspecular component:

$$\tilde{h}(f,\vec{r}) = \underbrace{\sum_{i=1}^{N} V_i \exp\left(j\left[\phi_i - \vec{k}_i \cdot \vec{r} - 2\pi f \tau_i\right]\right)}_{\text{Specular Waves}} + \tilde{h}_{\text{dif}}(f,\vec{r}) \quad (4.3.4)$$

where N is the number of specular waves.

The grouping in Equation (4.3.4) is arbitrary. Any number of specular components may be lumped inside or outside the nonspecular channel component, $\tilde{h}_{\text{dif}}(f,\vec{r})$. The *reduced wave grouping*, however, refers to a specific combination of specular and nonspecular wave components. Specifically, the following definition will be used throughout this work:

> A *reduced wave grouping* of local area propagation is the representation of Equation (4.3.4) with the smallest value of N such that the nonspecular component is diffuse.

In other words, a reduced wave grouping reduces a local area channel of any arbitrary complexity to the sum of a few large specular components and a single nonspecular term of "leftover" components that are diffuse.

There is a simple recipe for constructing the reduced wave grouping of a channel if the voltage amplitudes of the constituent waves are known. The procedure for generating the reduced wave grouping is shown in Table 4.2. When this procedure is finished, the reduced wave grouping for the local area has been achieved. The grouping consists of N specular components and a nonspecular component. It is possible that, at the end of the procedure, there are no specular components ($N = 0$) or that nothing is left in the nonspecular component. We should also note that the criterion for whether a component is diffuse depends on the application.

Table 4.2 Reduced Wave Grouping Algorithm

Step 1:	Group *all* of the local area wave components into one single nonspecular component and set $N = 0$.
Step 2:	Is the nonspecular component diffuse? If so, then stop.
Step 3:	Remove the homogeneous plane wave from the nonspecular component with the largest amplitude and increment N.
Step 4:	Go to Step 2.

4.4 The SLAC Model

This section defines the *stochastic local area channel* (SLAC) model and explores several of its key properties. This model is a cornerstone of small-scale channel analysis.

4.4.1 Stochastic Model

Following from the definition of a local area, the stochastic model for small-scale propagation is written in the following form:

$$\tilde{h}(f, \vec{\mathrm{r}}) = \sum_{i=1}^{N} V_i \exp\left(j \left[\Phi_i - \vec{\mathrm{k}}_i \cdot \vec{\mathrm{r}} - 2\pi f \tau_i\right]\right) \tag{4.4.1}$$

In this model, $\{V_i\}$, $\{\vec{\mathrm{k}}_i\}$, and $\{\tau_i\}$ are all constants, determined by the propagation that characterizes the local area. The phases, $\{\Phi_i\}$, are the random variables that make Equation (4.4.1) a stochastic model [Oss64]. Equation (4.4.1) will be referred to as the SLAC model throughout the rest of this work.

As written in the form of Equation (4.4.1), the SLAC model appears to be a discrete sum of waves - but it would be incorrect to view this as a discrete model. No restriction is placed on the number of components, N, or the types of values assigned the amplitudes, $\{V_i\}$. It may be that some local area propagation requires an *infinite* number of terms, each with an infinitesimal amplitude value, in Equation (4.4.1) to model propagation accurately.

4.4.2 Random Phases

Random phases in Equation (4.4.1) create a vast ensemble of realizations in the SLAC model. This variety is useful for simulation or analytic study involving fading channels. Common scenarios that apply the SLAC model include

- **Measurement Supplement:** Quantities such as multipath amplitude, time-delay, and wavevector are much easier to measure than individual multipath phases. As a result, we can use the SLAC model to extract channel realizations from measurements without phase data.

- **Channel Template:** If, through measurement or careful large-scale modeling, we have obtained realistic sets of amplitudes, delays, and wavevectors, the SLAC model provides a useful *template* for generating many different channel realizations.

- **Position Uncertainty:** Even if every piece of Equation (4.4.1) is determined by measurement, it is unlikely that a receiver will ever operate exactly at the measurement location. Position uncertainty, in the context of the SLAC model, is equivalent to scrambling the phases of multipath waves.

- **Frequency Uncertainty:** Even if every piece of Equation (4.4.1) is determined by measurement, it is unlikely that a receiver will ever operate exactly at the same carrier frequency, especially in multiple access wireless systems that allocate different portions of radio spectrum to different users. Frequency uncertainty, in the context of the SLAC model, is also equivalent to scrambling the phases of multipath waves.

Most types of local area analysis fall within these categories. A stochastic model never has more meaning than that endowed by the engineer, so always be careful to rigorously define its application. Otherwise, the old computer maxim holds: garbage in, garbage out.

4.4.3 Other Random Quantities

In stochastic channel modeling, other quantities in Equation (4.4.1) that are fixed constants in the SLAC model are sometimes treated as random variables. The physical implications of random amplitudes, wavevectors, or delays in Equation (4.4.1) lead to a *stochastic macro-area channel* (SMAC) model. For example, if the $\{V_i\}$ of Equation (4.4.1) are random variables, then each channel realization in the ensemble will represent a set of plane waves with different amplitudes. Thus, each channel realization in the ensemble will represent a completely different local area.

This book concentrates mostly on SLAC modeling because this is the perspective most useful for describing fading effects in a wireless link. At any given moment, a wireless receiver performs all of its operations within a single local area, not an ensemble of local areas simultaneously.

> *Note: SMAC or SLAC?*
>
> The difference between SLAC and SMAC models is an endless source of confusion. The classical presentation of small-scale fading statistics by Jakes used a model similar to Equation (4.4.1) with random phases *and* random amplitudes (a SMAC model) [Jak74]. Such a model is valid for deriving small-scale fading statistics only when the receiver operates in a diffuse, high-scatter channel - the type presented by Jakes. Since then, however, many channel modeling researchers indiscriminately randomize variables in Equation (4.4.1) with little regard to the physics of what is actually modeled.

4.4.4 Random Phase Models

For a SLAC model, the essence of the randomness boils down to the distribution of phases, $\{\Phi_i\}$, in Equation (4.4.1). The distribution of phases is characterized by a *probability density function* (PDF). Without exception, the random phases in the SLAC model are considered to be uniformly distributed over the interval $[0, 2\pi)$. Mathematically, the PDF for the ith phase, Φ_i, is written as

$$f_{\Phi_i}(\phi_i) = \frac{1}{2\pi} \qquad 0 \leq \phi_i < 2\pi \tag{4.4.2}$$

The justification for this distribution comes from a simple physical argument. A typical radio wave travels a long distance to reach a local area - usually the distance of hundreds or thousands of wavelengths. Since all effects of phase are periodic, only the 2π-modulus of the phase is essential for characterizing a propagating wave. The 2π-modulus of a very large absolute phase value is unpredictable and therefore may be treated as a uniformly distributed random variable.

While the PDF of Equation (4.4.2) is useful for describing the individual distribution of phase values, the true nature of any SLAC model relates to how random phases relate to one another in an ensemble of realizations. This behavior is described by the *joint PDF* of the random phases. Two useful joint characterizations for phases are discussed below.

Uncorrelated Phases

If the phases of a SLAC model are *uncorrelated*, then we refer to this representation of local area propagation as a U-SLAC model. The mathematical definition for uncorrelated phases - Φ_l and Φ_m, for example - is slightly different than the conventional definition for uncorrelated random variables:

$$\begin{aligned} \text{Conventionally Uncorrelated:} \quad & \mathrm{E}\{\Phi_l \Phi_m\} - \mathrm{E}\{\Phi_l\}\,\mathrm{E}\{\Phi_m\} = 0 \\ \text{Uncorrelated Phases:} \quad & \mathrm{E}\{\exp(j[\Phi_l - \Phi_m])\} = 0 \end{aligned} \tag{4.4.3}$$

The former expression is the common definition for random variable decorrelation. Throughout this book, the term *uncorrelated phases* refers to the latter definition. This subtle distinction accounts for the periodic nature of phase that is absent in other types of random variables.

Formally, we state that a SLAC model is a U-SLAC model if the following condition holds for all values of phase in Equation (4.4.1):

$$\mathrm{E}\{\exp(j[\Phi_l - \Phi_m])\} = 0 \quad \text{for } l \neq m \tag{4.4.4}$$

The physical argument for uncorrelated phase is straightforward. Different phases describe the behavior of differently scattered waves. Therefore, a weak, uncorrelated relationship should govern the selection of random phases with respect to one another in the SLAC model.

As we will show, uncorrelated phases lead to very useful behavior in a SLAC model. The condition *uncorrelated* still does not produce an expression for the joint phase PDF, however.

Independent Phases

If the random phases of a SLAC model are *independent*, then we refer to the channel model as an I-SLAC model. Imposing *independence* on the random variables of a SLAC model is much stronger than simply stating that the phases are uncorrelated. Thus, all I-SLAC models are U-SLAC models, but the converse is not necessarily true.

By definition, independence implies that a joint PDF may be written as the product of individual PDFs. For the case of random phases, the joint PDF is

$$f_{\vec{\Phi}}(\vec{\phi}) = f_{\Phi_1}(\phi_1)\, f_{\Phi_2}(\phi_2) \ldots f_{\Phi_N}(\phi_N), \qquad \vec{\phi} = \begin{bmatrix} \phi_1 \\ \phi_2 \\ \vdots \\ \phi_N \end{bmatrix} \tag{4.4.5}$$

Since the individual PDFs of each phase is a uniform distribution, the final form of the joint PDF for an I-SLAC model is

$$f_{\vec{\Phi}}(\vec{\phi}) = \frac{1}{(2\pi)^N} \quad 0 \leq \vec{\phi} < 2\pi \tag{4.4.6}$$

Unlike the U-SLAC model, phase-independence in the I-SLAC model leads to a specific joint PDF. This powerful feature of the I-SLAC model will make it the target of much of our analysis in Chapter 5. Figure 4.9 illustrates the relationship between the different types of SLAC channel models with a Venn diagram.

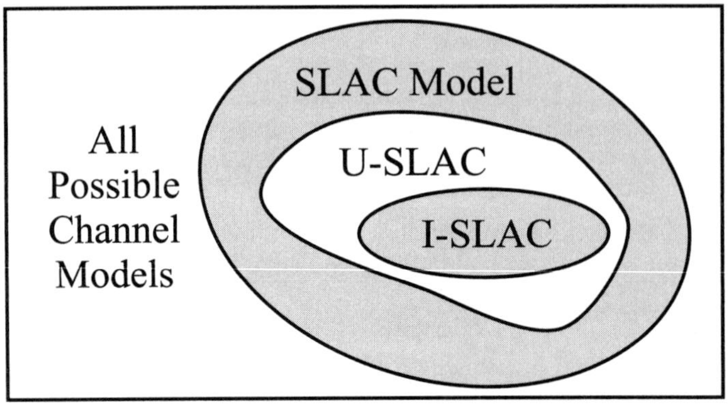

Figure 4.9 Venn diagram of various local area channel models.

The independence of phases is justified in a manner similar to the argument for uncorrelated phases in a channel model. The argument for independent phase simply states that each individual wave arrives from a different set of scattering mechanisms. As a result, a combination of wave phases in a realization of Equation (4.4.1) is just as valid as any other in the ensemble.

4.4.5 Fourier Transforms

The Fourier transform of a SLAC model is simple to compute. Transformation of the position vector and frequency domains into the wavevector and delay domains respectively produces the following expression for $\tilde{H}(\tau, \vec{k})$:

$$\tilde{H}(\tau, \vec{k}) = (2\pi)^3 \sum_{i=1}^{N} V_i \exp(j\Phi_i)\, \delta(\tau - \tau_i)\, \delta(\vec{k} - \vec{k}_i) \tag{4.4.7}$$

Each discrete component in Equation (4.4.1) produces a delta function in Equation (4.4.7) at $\tau = \tau_i$ and $\vec{k} = \vec{k}_i$.

A delta function that appears in a Fourier transform or a power spectrum is called a *spectral line.* This term refers to the concentration of power about a single point in the spectral domain. Again, however, the discrete nature of Equation (4.4.7) is misleading. By allowing N to approach infinity and the amplitudes to become infinitesimally small, Equation (4.4.7) could just as easily represent a continuous function.

At this point, let us introduce a new type of spectral domain representation of a Fourier transform that eliminates spectral lines. Often, a Fourier transform is written as an integral:

$$F(\tau, \vec{k}) = \int_{-\infty}^{\tau} \int_{-\infty}^{\vec{k}} \tilde{H}(\tau', \vec{k}')\, d\vec{k}'\, d\tau' \tag{4.4.8}$$

The inverse Fourier transform may be written in terms of $F(\tau, \vec{k})$ in the form of a Riemann-Stieltjes integral:

$$\tilde{h}(f, \vec{r}) = \int_{-\infty}^{+\infty} \exp(j[2\pi f \tau + \vec{k} \cdot \vec{r}])\, dF(\tau, \vec{k}) \tag{4.4.9}$$

The Riemann-Stieltjes representation smoothes the Fourier transform, $\tilde{H}(\tau, \vec{k})$, and eliminates the need for delta functions. For the SLAC model, $F(\tau, \vec{k})$ may be written as

$$F(\tau, \vec{k}) = (2\pi)^3 \sum_{i=1}^{N} V_i \exp(j\Phi_i)\, \mathrm{u}(\tau - \tau_i)\, \mathrm{u}(\vec{k} - \vec{k}_i) \tag{4.4.10}$$

Now, spectral lines are instead represented in Equation (4.4.10) by discontinuous steps.

> *Note: The Mythical Delta Function*
>
> Mathematicians often remind engineers that the delta function is not really a function at all because it is defined by an irresolvable limit. The Riemann-Stieltjes integral representation of a Fourier transform or spectrum is used by mathematicians and physicists who wish to be rigorous and avoid the use of delta functions in their analysis. Engineers, however, unabashedly use the convenient delta function. Beware: Engineering researchers working in the field of stochastic modeling will eventually encounter the Riemann-Stieltjes integral and should not be daunted by the notation.

4.4.6 Autocorrelation Functions

Now we can analyze some of the more sophisticated statistical properties of a SLAC model, such as the autocorrelation function. Since the distribution of phase plays such a critical role in the characteristics of a SLAC model, the type of SLAC model determines the nature of the autocorrelation function. Consider Theorem 4.2 - one of the most important and useful theorems in local area channel modeling.

Theorem 4.2: U-SLAC Wide-Sense Stationarity

Statement: A SLAC model, $\tilde{h}(f, \vec{\mathrm{r}})$, is WSS-US with respect to space and frequency if and only if it is a U-SLAC model.

Proof: Start with the definition of an autocorrelation function, and substitute the inverse Fourier transforms of $\tilde{H}(\tau, \vec{\mathrm{k}})$ for the channel $\tilde{h}(f, \vec{\mathrm{r}})$.

$$C_{\tilde{h}}(f_1, f_2, \vec{\mathrm{r}}_1, \vec{\mathrm{r}}_2) = \mathrm{E}\left\{ \overbrace{\left[\frac{1}{(2\pi)^3} \iint_{-\infty}^{+\infty} \tilde{H}(\tau_1, \vec{\mathrm{k}}_1) \exp(j[\vec{\mathrm{k}}_1 \cdot \vec{\mathrm{r}}_1 + 2\pi f_1 \tau_1])\, d\vec{\mathrm{k}}_1\, d\tau_1 \right]}^{\tilde{h}(f_1, \vec{\mathrm{r}}_1)} \times \right.$$

$$\left. \overbrace{\left[\frac{1}{(2\pi)^3} \iint_{-\infty}^{+\infty} \tilde{H}^*(\tau_2, \vec{\mathrm{k}}_2) \exp(-j[\vec{\mathrm{k}}_2 \cdot \vec{\mathrm{r}}_2 + 2\pi f_2 \tau_2])\, d\vec{\mathrm{k}}_2\, d\tau_2 \right]}^{\tilde{h}^*(f_2, \vec{\mathrm{r}}_2)} \right\}$$

$$= \sum_{l=1}^{N} \sum_{m=1}^{N} V_l V_m \exp\!\left(j\left[\vec{\mathrm{k}}_l \cdot \vec{\mathrm{r}}_1 - \vec{\mathrm{k}}_m \cdot \vec{\mathrm{r}}_2 + 2\pi(f_l \tau_1 - f_m \tau_2)\right]\right) \mathrm{E}\left\{\exp(j[\Phi_l - \Phi_m])\right\} \tag{4.4.11}$$

If the above equation is a U-SLAC model, by definition the phases, Φ_l and Φ_m, are uncorrelated for $l \neq m$ and the autocorrelation function reduces to

$$C_{\tilde{h}}(f_1, f_2, \vec{\mathrm{r}}_1, \vec{\mathrm{r}}_2) = \sum_{i=1}^{N} V_i^2 \exp(j[\vec{\mathrm{k}}_i \cdot \underbrace{(\vec{\mathrm{r}}_1 - \vec{\mathrm{r}}_2)}_{\Delta\vec{\mathrm{r}}} + 2\pi\tau_i \underbrace{(f_1 - f_2)}_{\Delta f}])$$

which is WSS-US, since the autocorrelation may be written in terms of Δf and $\Delta\vec{\mathrm{r}}$ only and without space–frequency cross products. Furthermore, $\vec{\mathrm{k}}_l$ *must* equal $\vec{\mathrm{k}}_m$ and f_l *must* equal f_m in Equation (4.4.11) for the SLAC model to be WSS-US. The *only* way to force this condition for any case is with uncorrelated random phase variables.

According to Theorem 4.2, SLAC models with uncorrelated phases always lead to WSS-US channels with respect to position and frequency. The autocorrelation for all U-SLAC models may be written as

$$C_{\tilde{h}}(\Delta f, \Delta \vec{r}) = \sum_{i=1}^{N} V_i^2 \exp(j[\vec{k}_i \cdot \Delta \vec{r} + 2\pi \tau_i \Delta f]) \tag{4.4.12}$$

From this point onward, most of the models presented in the book are U-SLAC models. The WSS-US properties and the convenient form of autocorrelation for these models make them powerful tools for illustrating channel behavior.

4.4.7 Heterogeneous Scattering

At this point, it is important to define a special case of SLAC models - one that results from *heterogeneous scattering.* A SLAC model is a result of heterogeneous scattering if the following condition holds:

$$\vec{k}_l \neq \vec{k}_m \quad \text{and} \quad \tau_l \neq \tau_m \quad \text{for all } l \neq m$$

In other words, heterogeneous scattering describes a condition where no two multipath waves in Equation (4.4.1) arrive with precisely the same time delays or wavevectors.

The condition of heterogeneous scattering is mentioned as a special type of SLAC model because it often leads to some powerful theorems in stochastic channel modeling. Consider, as an example, Theorem 4.3. The theorem shows that if heterogeneous scattering can be established in a SLAC model, *then wide-sense stationarity with respect to one channel dependency automatically implies wide-sense stationarity with respect to the other.*

Theorem 4.3: WSS Heterogeneous Scattering

Statement: A SLAC model, $\tilde{h}(f, \vec{r})$, with heterogeneous scattering is WSS with respect to position, $\vec{r}$, if and only if it is WSS with respect to frequency, f.

Proof: Using the same proof in Theorem 4.2, it may be shown that a heterogeneous SLAC model is WSS with respect to position if and only if *all* phases, $\{\Phi_i\}$, are uncorrelated. Likewise, it is also true that the same model is WSS with respect to frequency if and only if all phases are uncorrelated. By transitivity, the theorem is proven.

Without the condition of heterogeneous scattering, Theorem 4.3 would not hold. If two delays, τ_i and τ_j, were equal, then correlated phases Φ_i and Φ_j will cause nonstationarity with respect to position, $\vec{r}$, but will not effect the wide-sense stationarity with respect to frequency, f. The dual problem exists for matching wavevectors.

4.4.8 SLAC Power Spectral Density

Since a U-SLAC model is WSS-US, it is possible to define a PSD using the Wiener-Khintchine theorem. The wavevector–delay spectrum for the U-SLAC model is

$$S_{\tilde{h}}(\tau, \vec{k}) = (2\pi)^3 \sum_{i=1}^{N} V_i^2 \, \delta(\tau - \tau_i) \, \delta(\vec{k} - \vec{k}_i) \tag{4.4.13}$$

Just as in the case of the Fourier transform, an integrated version of the PSD may be defined to eliminate the delta functions:

$$\begin{aligned} F_{\tilde{h}}(\tau, \vec{k}) &= \int_{-\infty}^{\tau} \int_{-\infty}^{\vec{k}} S_{\tilde{h}}(\tau', \vec{k}') \, d\tau' \, d\vec{k}' \\ &= (2\pi)^3 \sum_{i=1}^{N} V_i^2 \, \mathrm{u}(\tau - \tau_i) \, \mathrm{u}(\vec{k} - \vec{k}_i) \end{aligned} \tag{4.4.14}$$

The space–frequency autocorrelation function may then be written as a Riemann-Stieltjes integral:

$$C_{\tilde{h}}(\Delta f, \Delta \vec{r}) = \frac{1}{(2\pi)^3} \int_{-\infty}^{+\infty} \exp\left(j[2\pi\tau\Delta f + \vec{k} \cdot \Delta\vec{r}]\right) \, dF_{\tilde{h}}(\tau, \vec{k}) \tag{4.4.15}$$

Although this work will use the standard power spectrum, $S_{\tilde{h}}(\tau, \vec{k})$, whenever possible, there are certain types of analysis (usually involving the spectral lines caused by large specular components) that are best performed with the integrated power spectrum.

Diffuse Components

Often the PSD is more flexible than the summation of Equation (4.4.1) for characterizing diffuse, nonspecular power in a SLAC model. Specifically, a continuous PSD is a simple way to represent multipath waves that are diffusely spread over a range of delays and wavevectors - a case that would require an infinite number of terms in the SLAC model summation in Equation (4.4.1).

The integrated power spectrum is a particularly useful method for characterizing diffuse nonspecular power. A local area channel is *perfectly diffuse* if no discontinuities are present in the integrated power spectrum, $F_{\tilde{h}}(\tau, \vec{k})$. Mathematically, this condition may be expressed as

$$\lim_{\substack{\delta\tau \to 0 \\ |\delta\vec{k}| \to 0}} \left| F_{\tilde{h}}(\tau + \delta\tau, \vec{k} + \delta\vec{k}) - F_{\tilde{h}}(\tau - \delta\tau, \vec{k} - \delta\vec{k}) \right| = 0 \quad \text{for all } \tau \text{ and } \vec{k} \tag{4.4.16}$$

Figure 4.10 illustrates different types of delay spectra (only one dependency, delay, is shown for graphical simplicity). Notice how the discontinuities in the spectrum indicate the presence of a specular component.

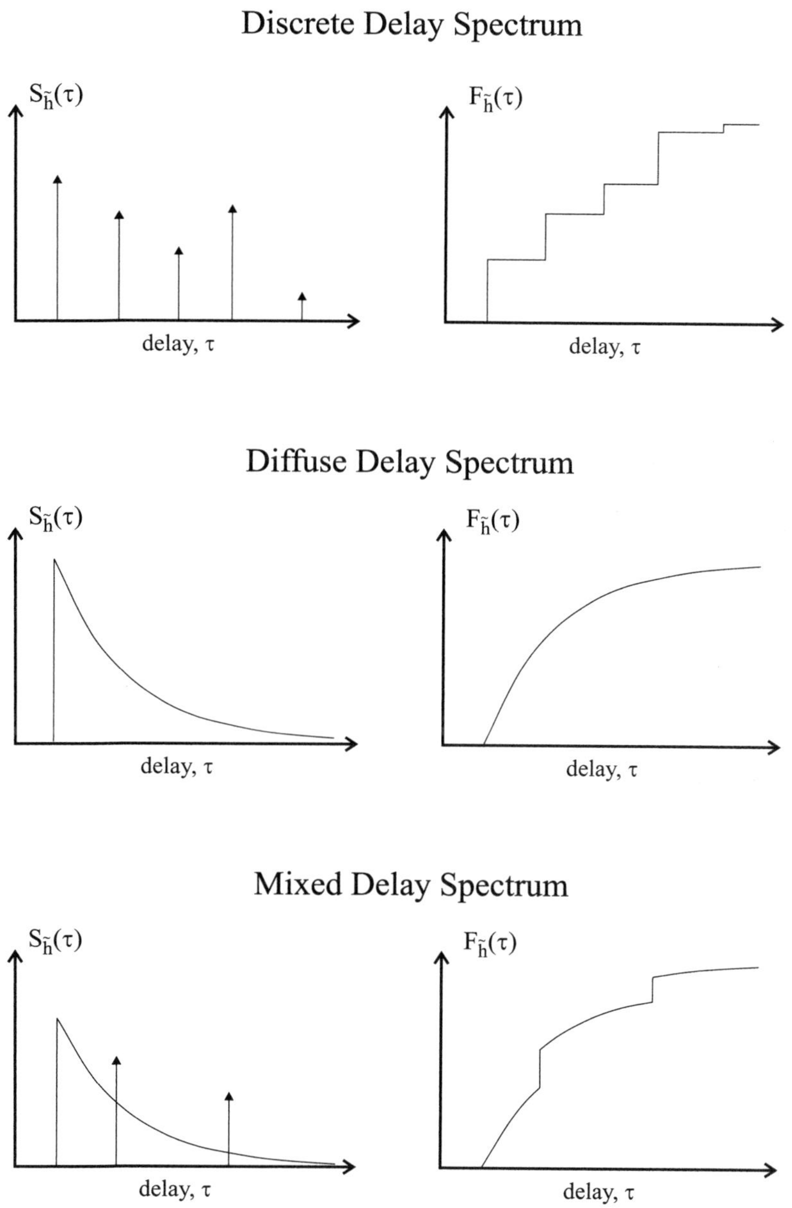

Figure 4.10 Different delay spectra using the standard power spectrum, $S_{\tilde{h}}(\tau)$, and the integrated power spectrum, $F_{\tilde{h}}(\tau)$.

Specular Components

The integrated power spectrum allows us to formalize a definition for the presence of a specular component in a local area. A specular component with magnitude V_i, delay τ_i, and wavevector $\vec{k}_i$ exists if a discontinuity is present in the integrated power spectrum, $F_{\tilde{h}}$. Mathematically, this condition may be stated as

$$\lim_{\substack{\delta\tau \to 0 \\ |\delta\vec{k}| \to 0}} \frac{1}{(2\pi)^3} \left| F_{\tilde{h}}(\tau_i + \delta\tau, \vec{k}_i + \delta\vec{k}) - F_{\tilde{h}}(\tau_i - \delta\tau, \vec{k}_i - \delta\vec{k}) \right| = V_i^2$$

Examples of specular spectral components are also illustrated in Figure 4.10.

4.5 Chapter Summary

This chapter presented an introduction to the physics behind local area modeling of the small-scale radio channel. Below is a useful outline of key points:

- Propagation in a region of free-space may be expressed as the sum of two types of plane waves:
  - ▷ *Homogeneous* plane waves have constant amplitude.
  - ▷ *Inhomogeneous* plane waves have an exponentially decaying amplitude.
- A *local area* is a region of free-space in which the bandwidth-limited propagation is approximated by the sum of only homogeneous plane waves.
- Multipath waves may be characterized into three different component groups:
  - ▷ A *specular* component is a single homogeneous plane wave.
  - ▷ A *nonspecular* component is a collection of multiple waves.
  - ▷ A *diffuse* component is a nonspecular component consisting of numerous individual waves with small amplitudes.
- A *reduced wave grouping* is a minimal collection of N specular components and a diffuse, nonspecular component. All cases of local area propagation have a unique reduced wave grouping once a rigid criterion for diffuseness is asserted.
- The *stochastic local area channel* (SLAC) model represents propagation as the sum of randomly phased homogeneous plane waves. There are several specific types of SLAC models:
  - ▷ A *U-SLAC* model has uncorrelated random phases and is WSS.
  - ▷ An *I-SLAC* model has independent (as well as uncorrelated) random phases.

▷ A SLAC model with *heterogeneous scattering* consists of homogeneous plane waves that have unique delays and wavevectors.

This chapter, when taken together with Chapter 2 and Chapter 3, represents the foundational theory and terminology behind stochastic modeling of the physical radio channel. All of the subsequent analysis in this book applies the terminology and theory presented in these chapters.

PROBLEMS

1. In the table below are equations for time-harmonic plane waves in free-space. Fill in the table with the appropriate plane wave propagation parameters. Assume standard international (SI) units of meters and seconds for all quantities. (The first line is given as an example.)

EQUATION	V_0	α	β	λ	f_c
$6\exp(-j10\pi\hat{\mathrm{x}}\cdot\vec{\mathrm{r}})$	6	0	10π	0.2 m	1.5 GHz
$3[\cos(3\hat{\mathrm{y}}\cdot\vec{\mathrm{r}}) + j\sin(3\hat{\mathrm{y}}\cdot\vec{\mathrm{r}})]\exp(-2\hat{\mathrm{x}}\cdot\vec{\mathrm{r}})$					
$\exp(20\pi[\hat{\mathrm{x}}+2\hat{\mathrm{y}}]\cdot\vec{\mathrm{r}} + j30\pi[2\hat{\mathrm{x}}-\hat{\mathrm{y}}]\cdot\vec{\mathrm{r}})$					
$19\exp(-j\pi[10\hat{\mathrm{x}}+30\hat{\mathrm{y}}-15\hat{\mathrm{z}}]\cdot\vec{\mathrm{r}})$					
$\exp(j\hat{\mathrm{k}}\cdot\vec{\mathrm{r}}+2)$					

2. Here is a nostalgic problem from basic circuit theory. Use the following circuit diagram to answer the questions below:

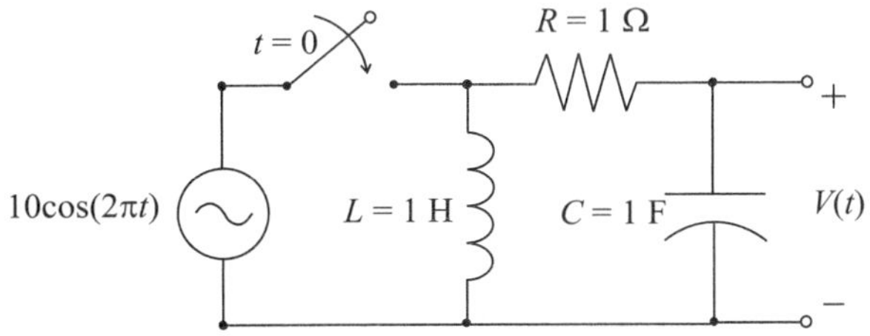

a. Assuming that the circuit has no stored energy prior to $t = 0$, calculate the voltage $V(t)$ for $t \geq 0$.

b. Sketch $V(t)$ from the answer you obtained in (a).

c. Break your answer in (a) into transient and steady-state solutions and sketch each.

3. One of the criteria for defining a local area was proximity to scatterers: close scatterers imply a smaller local area to minimize deviations due to inhomogeneous waves. Use the linear circuit analogy to explain why this is necessary.

4. Describe the most likely reduced-wave grouping that describes each of the following local area propagation scenarios:

a. Satellites communicating with one another in deep space.

b. Cellular phone operated in a cluttered city without a line-of-sight path to the transmitter.

c. Cellular phone operated in a line-of-sight channel next to a large, flat wall.

d. A low incidence satellite radio link to an plane flying over rough seas.

(d)

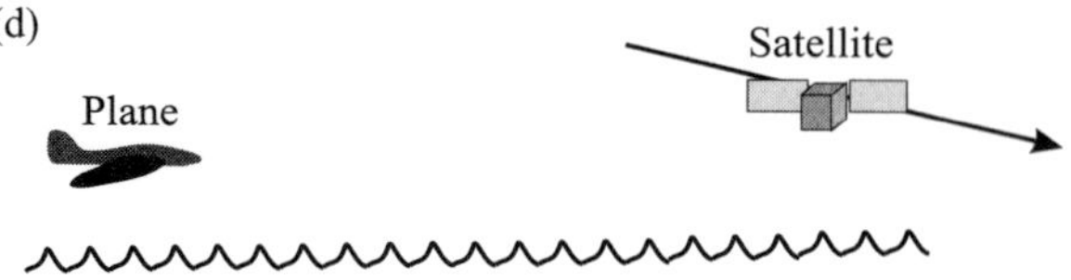

e. A ducting channel in a point-to-point radio link where one or two coherent waves are refracted back towards the line-of-sight path [Rum86].

(e)

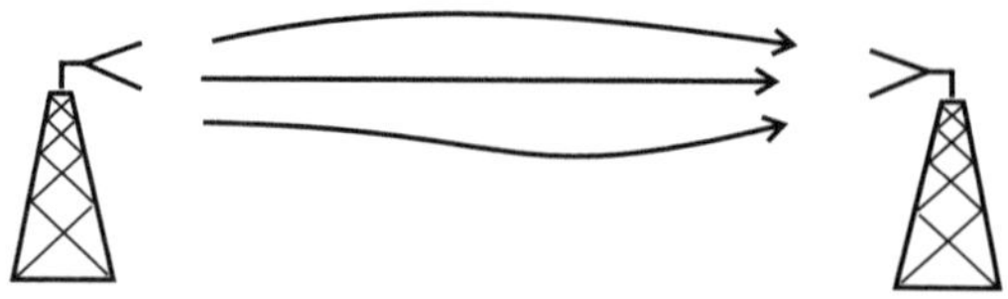

f. An indoor rectangular room penetrated by a single wave that reflects and resonates off all four flat walls.

(f)

g. A mm-wave point-to-point link operating with highly directional antennas where transmitter and receiver are on mountaintops.

(g)

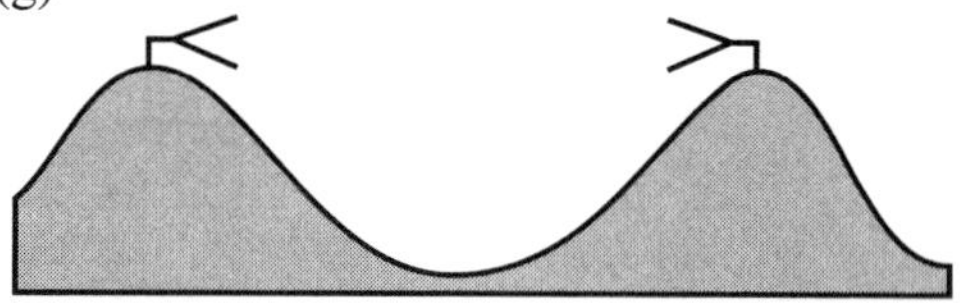

h. The same point-to-point link in (g), except the line of sight barely clears the treetops of a forest.

(h)

5. Prove the following theorem: For perfectly diffuse propagation, the channel is WSS with respect to frequency if and only if it is WSS with respect to space.

6. Discuss what physical mechanisms might lead to multipath waves that have *correlated* phases within a local area.

7. The form of the SLAC model in Equation (4.4.1) is a complex baseband channel. Write the SLAC model representation for the passband channel, $h(f, \vec{\mathrm{r}})$.

8. Prove that a random power process, $P(f, \vec{\mathrm{r}}) = |\tilde{h}(f, \vec{\mathrm{r}})|^2$, based on a SLAC model representation has the following form:

$$P(f, \vec{\mathrm{r}}) = \sum_{l=1}^{N} \sum_{m=1}^{N} V_l V_m \cos\left([\vec{\mathrm{k}}_m - \vec{\mathrm{k}}_l] \cdot \vec{\mathrm{r}} + 2\pi[\tau_m - \tau_l] f + \Phi_l - \Phi_m\right)$$

9. Use the following two-wave, narrowband SLAC model to answer the questions below:

$$\tilde{h}(\vec{\mathrm{r}}) = V_1 \exp(\Phi_1 - jk_0 \hat{\mathrm{x}} \cdot \vec{\mathrm{r}}) + V_2 \exp(\Phi_2 + jk_0 \hat{\mathrm{x}} \cdot \vec{\mathrm{r}})$$

   a. Prove that if we desire to study envelope and power statistics, the following model, with single random phase $\Phi_0 = \Phi_2 - \Phi_1$ and $\vec{\mathrm{r}} = r\hat{\mathrm{x}}$, is equivalent to the one above.

$$\tilde{h}(r) = V_1 + V_2 \exp(\Phi_0 + 2jk_0 r)$$

   b. If the initial model is a U-SLAC model, what is the PDF of Φ_0?

   c. What is the average power, $\mathrm{E}\left\{|\tilde{h}(r)|^2\right\}$, for the U-SLAC model?

   d. What is the spatial autocorrelation function, $C_{\tilde{h}}(\Delta r)$, for the U-SLAC model?

   e. Now let us assume that phases are correlated, such that the PDF of Φ_0 is not uniform over $[0, 2\pi)$, but instead is truncated to the constant ϕ_{max}:

$$f_{\Phi_0}(\phi_0) = \begin{cases} \frac{1}{\phi_{\mathrm{max}}} & 0 \leq \phi_0 < \phi_{\mathrm{max}} \\ 0 & \text{elsewhere} \end{cases}$$

   What is the average power, $\mathrm{E}\left\{|\tilde{h}(r)|^2\right\}$, for this correlated SLAC model?

   f. What is the spatial autocorrelation function, $C_{\tilde{h}}(\Delta r)$, for the model in (d)?

10. Complete the following table based on the Riemann-Stieltjes integral. In the category Type, put either *discrete*, *diffuse*, or *mixed*.

SPECTRAL DOMAIN	RIEMANN-STIELTJES	TYPE
$\tilde{H}(\tau) =$	$F(\tau) = 3\exp(-\tau)\mathrm{u}(\tau)$	
$\tilde{H}(\tau) = 3\exp(-\tau)\mathrm{u}(\tau)$	$F(\tau) =$	
$S_{\tilde{h}}(k) =$	$F_{\tilde{h}}(k) = \frac{k+2}{4}\mathrm{u}(2-\lvert k\rvert) + \mathrm{u}(k-2)$	
$S_{\tilde{h}}(\tau) =$	$F_{\tilde{h}}(\tau) = \frac{\tau^2}{1+\tau^2}\mathrm{u}(\tau)$	
$\tilde{H}(\vec{\mathrm{k}}) = 10\exp(j\vec{\mathrm{k}}\cdot\vec{\mathrm{r}}_0)\mathrm{u}(\lvert\vec{\mathrm{k}}\rvert-1)$	$F(\vec{\mathrm{k}}) =$	
$S_{\tilde{h}}(\tau) = \delta(\tau) + \tau\exp(-4\tau)$	$F_{\tilde{h}}(\tau) =$	

11. Complete the following table based on the Wiener-Khintchine relationship between autocorrelation function and PSD.

AUTOCORRELATION	POWER SPECTRAL DENSITY
$C_{\tilde{h}}(\Delta f, \Delta\vec{\mathrm{r}}) = V_0\delta(\Delta f - f_0)\delta(\Delta\vec{\mathrm{r}} - \vec{\mathrm{r}}_0)$	$S_{\tilde{h}}(\tau, \vec{\mathrm{k}}) =$
$C_{\tilde{h}}(\Delta f, \Delta r) =$	$S_{\tilde{h}}(\tau, k) = \frac{S_0\exp(-\alpha\tau)\mathrm{u}(\tau)}{\sqrt{k_0 - k^2}}$ for $\lvert k\rvert \leq k_0$
$C_{\tilde{h}}(\Delta f, \Delta\vec{\mathrm{r}}) =$	$S_{\tilde{h}}(\tau, \vec{\mathrm{k}}) = A\mathrm{sn}^2\left(\frac{\tau}{T} - 1\right)$
$C_{\tilde{h}}(\Delta f, \Delta\vec{\mathrm{r}}) = \exp\left(-\frac{\lvert\Delta\vec{\mathrm{r}}\rvert^2}{2\sigma^2} - T\lvert\Delta f\rvert\right)$	$S_{\tilde{h}}(\tau, \vec{\mathrm{k}}) =$

12. Although the modeling concepts are similar for space and frequency, there are physical limitations placed on SLAC models that are unique to each domain.

a. The principle of *causality* states that a signal cannot arrive before it is sent. If $\tau = 0$ is taken to be zero elapsed delay from transmitter to receiver, express the causality criterion mathematically using the delay-wavevector PSD, $S_{\tilde{h}}(\tau, \vec{\mathrm{k}})$.

b. In a SLAC model, the wavenumber of any multipath component must be equal to the free-space wavenumber, k_0. Express this criterion mathematically using the delay-wavevector PSD, $S_{\tilde{h}}(\tau, \vec{\mathrm{k}})$.

Chapter 5

FIRST-ORDER CHANNEL STATISTICS

The received power in fading wireless channels fluctuates between maxima and minima as a function of space, time, and frequency. A wireless engineer must accept the fact that for certain regions in space, time, or frequency, the receiver may have to operate with received signal strength below an acceptable *signal-to-noise* or *signal-to-interference* ratio. To quantify the effects of a fading channel on receiver performance, we must first quantify the distribution of received power or voltage envelope that a receiver experiences in a randomly selective channel.

This chapter develops the principles of first-order analysis for stochastic frequency-selective and spatially selective radio channels. The most important tool in a first-order analysis is the probability density function (PDF). In developing the use of a PDF for modeling random channels, the chapter discusses the following key points:

- Section 5.1: Discussion of mean received power.
- Section 5.2: Construction of envelope PDFs for I-SLAC models.
- Section 5.3: Analytical solutions to I-SLAC envelope PDFs.
- Section 5.4: Analysis of the two-wave with diffuse power (TWDP) PDF.
- Section 5.5: Summary of important concepts.

This chapter uses the definitions and concepts from the preceding chapters to develop many of the classical distributions used in wireless engineering to describe small-scale fading. Furthermore, new fading distributions that augment the classical understanding are developed and discussed. Indeed, the SLAC model defined in Chapter 4 is shown to produce an incredible variety of first-order channel behavior.

5.1 Mean Received Power

Of all first-order statistics to calculate for stochastic channel model, mean received power is perhaps the most common. Regardless of other statistical fading properties, mean received power is an intuitive measure of receiver performance, since, according to information theory, received power is related to the fundamental limitations on the amount of information that can be sent through a noisy channel [Cov91].

5.1.1 Average Versus Received Power

Consider a SLAC model with three constant-amplitude waves. Movement in frequency or space causes a steady phase rotation of each, as illustrated in Figure 5.1. Different components rotate at different speeds and directions (clockwise or counter-clockwise in Figure 5.1). When the three waves are summed from one moment to the next, the composite signal undergoes constructive-destructive interference. In this way, the total envelope of a received radio signal is rarely near its average value. The distribution of a random envelope process about its average level has a dramatic effect on radio link performance. Thus, this chapter presents a thorough discussion of the moments and the PDFs of envelope processes.

5.1.2 Stationarity

Stationarity is one of the most important and powerful attributes to determine in a stochastic model. There are many different types of stationarity (we have already discussed *wide-sense stationarity* in previous chapters). In fact, the term *stationary* may be used to describe any type of statistic that is invariant of channel dependencies. For example, a time-varying process is *mean stationary* if its mean value does not change as a function of time.

Often, statistics of an entire *order* are said to be stationary. The *order* of a statistic refers to the number of stochastic process samples used in its calculation. For example, the autocorrelation function is a *second-order* statistic because it is calculated from two samples of the stochastic process, as shown in Equation (3.1.2). One sample is taken at t_1 and the other sample is taken at t_2.

In this chapter, it is useful to establish *first-order stationarity* in the stochastic channel models. A process is first-order stationary if all of its first-order statistics - mean, variance, PDF, and so on - are invariant of the dependencies space and frequency. The U-SLAC model is a first-order stationary channel model, which makes it a convenient tool for analytical channel description. This trait of the U-SLAC model is proven in Theorem 5.1.

Theorem 5.1: First-Order Stationarity

Statement: The U-SLAC model is first-order stationary.

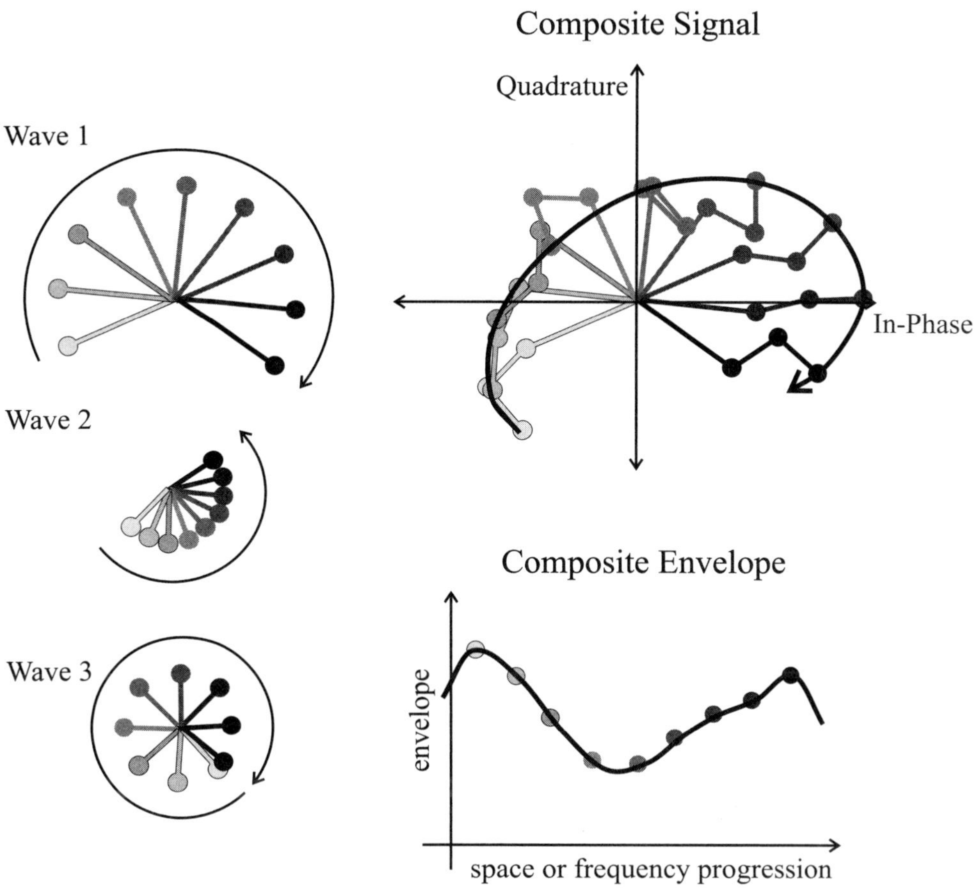

Figure 5.1 Example of three specular waves that add to form a small-scale fading channel.

Proof: Consider the U-SLAC model of Equation (4.4.1) evaluated at frequency $f = 0$ and position $\vec{r} = \vec{0}$. The resulting channel expression is shown below

$$\tilde{h}(0, \vec{0}) = \sum_{i=1}^{N} V_i \exp\left(j\left[\Phi_i - \vec{k}_i \cdot \vec{r} - 2\pi f \tau_i\right]\right)\Bigg|_{\substack{f=0\\ \vec{r}=\vec{0}}} = \sum_{i=1}^{N} V_i \exp(j\Phi_i) \tag{5.1.1}$$

Now consider the case of the channel evaluated at any arbitrary point in frequency, f, and space, $\vec{r}$. By making the following substitution,

$$\Phi_i' = \left(\Phi_i - \vec{k}_i \cdot \vec{r} - 2\pi f \tau_i\right) \bmod 2\pi$$

we could express the U-SLAC model in the following form:

$$\tilde{h}(f, \vec{r}) = \sum_{i=1}^{N} V_i \exp(j\Phi_i') \qquad (5.1.2)$$

The random phases, $\{\Phi_i'\}$, in Equation (5.1.2) are the same set of random phases, $\{\Phi_i\}$, in Equation (5.1.1) with arbitrary constants added. Otherwise, the two equations look identical. Since a uniform phase random variable remains uniformly distributed if any arbitrary constant is added to it, the expressions in Equation (5.1.1) and Equation (5.1.2) are identically distributed random variables. Adding the constant does not change the cyclical correlation properties, either. Thus, the distribution of values for $\tilde{h}(f, \vec{r})$ is identical and independent of frequency, f, and position, $\vec{r}$ - the definition of first-order stationary.

Note: Confusion of Order and Moments

The difference between the *orders* and the *moments* of a stochastic process causes endless confusion. The term *moment* refers to a family of first order statistics. For example, $\mathrm{E}\{R(f, \vec{r})\}$, $\mathrm{E}\{R(f, \vec{r})^2\}$, and $\mathrm{E}\{R(f, \vec{r})^3\}$ are the first, second, and third moments of the envelope process $R(f, \vec{r})$ - but all three are first-*order* statistics.

The most sweeping type of stationarity is *strict-sense stationary* (SSS) behavior. A stochastic process is SSS if it is stationary with respect to every order. While establishing SSS behavior in a process is highly desirable, it is extremely difficult to prove in most types of stochastic processes. The I-SLAC model is an exception, as Theorem 5.2 demonstrates.

Theorem 5.2: Strict-Sense Stationarity

Statement: The I-SLAC model is strict-sense stationary.

Proof: Consider a simple stochastic channel consisting of a single wave term:

$$\tilde{h}_1(f, \vec{r}) = V_1 \exp\left(j\left[\Phi_1 - \vec{k}_1 \cdot \vec{r} - 2\pi f \tau_1\right]\right)$$

where V_1 is a constant amplitude and Φ_1 is a random phase variable, uniformly distributed over the interval $[0, 2\pi)$. This process is SSS [Pap91, p. 301]. Since the sum of two independent SSS processes will produce an SSS process, the following process

$$\tilde{h}_2(f, \vec{r}) = \tilde{h}_1(f, \vec{r}) + V_2 \exp\left(j\left[\Phi_2 - \vec{k}_2 \cdot \vec{r} - 2\pi f \tau_2\right]\right)$$

is SSS provided the random phase variable, Φ_2, is independent of Φ_1 (as it is in an I-SLAC model). An arbitrary number of independent wave components may be added in this manner and, by induction, any I-SLAC model may be constructed and shown to be an SSS process.

It is probably evident now why we draw a distinction between SLAC, U-SLAC, and I-SLAC models: The nature of the phase distributions in the model of Equation (4.4.1) (arbitrary, uncorrelated, or independent) determines the level of stationarity. A SLAC model may be nonstationary with respect to every statistic. A U-SLAC model, however, is guaranteed to be at least WSS (Theorem 4.2 and Theorem 5.1). An I-SLAC model is guaranteed to be stationary with respect to *every* order and statistic (Theorem 5.2).

5.1.3 Mean U-SLAC Power

For a WSS U-SLAC model, it is possible to calculate a closed-form expression for the mean received power, $P(f, \vec{r})$. Inserting the SLAC model expression into the definition of mean received power produces

$$\mathrm{E}\{P(f,\vec{r})\} = \mathrm{E}\left\{|\tilde{h}(f,\vec{r})|^2\right\}$$

$$= \sum_{l=1}^{N}\sum_{m=1}^{N} V_l V_m \underbrace{\mathrm{E}\{\exp(j[\Phi_l - \Phi_m])\}}_{=\,0 \text{ for } l\neq m} \exp\left(j\left[(\vec{k}_m - \vec{k}_l)\cdot\vec{r} + 2\pi(\tau_m - \tau_l)f\right]\right)$$

$$= \sum_{m=1}^{N} V_m^2 \tag{5.1.3}$$

The expectation in the second line of Equation (5.1.3) is zero for $m \neq l$ by definition of a U-SLAC process. The end result of Equation (5.1.3) may be stated in words as follows: The mean power in a U-SLAC model is equal to the sum of the powers carried by the individual multipath waves [Rap02a]. Stated this way, Equation (5.1.3) can be considered a type of conservation-of-power law.

5.1.4 Frequency and Spatial Averaging

In the application of first-order statistics to real-life problems, it is important to speak of *averaging*. This is particularly true when attempting to measure the local area channel. In practice, it is not possible to generate an infinitum of realizations and calculate an ensemble. Rather, it is most feasible to average a measured statistic over a single realization.

This averaging may occur over each dependency of the measured channel. For example, sweeping a narrowband transmitted carrier over a range of frequencies and measuring the received voltage is a form of *frequency averaging*. Mathematically, we will define the operation of frequency averaging as

$$\left\langle\,\cdot\,\right\rangle_f = \lim_{B\to\infty}\frac{1}{B}\int_{-B/2}^{B/2}(\,\cdot\,)\,df \tag{5.1.4}$$

In this definition, a statistic is calculated by averaging the value over the range of all allowable frequencies. In practice, the integration of Equation (5.1.4) cannot be taken over an infinite range.

A similar definition exists for spatial averaging. Mathematically, we will define the operation of spatial averaging as

$$\left\langle \cdot \right\rangle_{\vec{r}} = \lim_{L\to\infty} \frac{1}{L^3} \int_{-L/2}^{L/2} (\ \cdot\ )\ d\vec{r} \tag{5.1.5}$$

where the integration in Equation (5.1.5) is a threefold integration over the x, y, and z coordinates.

5.1.5 Ergodicity

Stochastic processes with statistics that exhibit *ergodicity* are particularly useful for measurements and analysis of real-world processes. A statistic is said to be ergodic if, when measured by averaging a single realization of a stochastic ensemble, it is equal to the ensemble average. For example, a process is *mean-ergodic* if the frequency- and spatially-averaged mean of a single realization equals the ensemble-averaged mean. We express mean ergodicity for a stochastic channel as

$$\text{Condition for Mean Ergodicity:}\ \left\langle \left\langle \tilde{h}(f,\vec{r}) \right\rangle_{\vec{r}} \right\rangle_f = \mathrm{E}\left\{ \tilde{h}(f,\vec{r}) \right\}$$

Of course, this property must hold for *all* realizations of $\tilde{h}(f,\vec{r})$ in the stochastic ensemble.

Since there are two dependencies in the SLAC model - frequency and space - it is convenient to describe ergodicity with respect to each individual dependency. For example, it is possible that a statistic is ergodic with respect to frequency but not space. The converse is also possible. The basic principles governing the frequency and spatial ergodicity of mean received power are summarized in Theorem 5.3.

Theorem 5.3: Power Ergodicity

Statement:

1. A U-SLAC model is power-ergodic with respect to frequency if its scattering is heterogeneous with respect to frequency.
2. A U-SLAC model is power-ergodic with respect to space if its scattering is heterogeneous with respect to space.
3. All U-SLAC models with heterogeneous scattering are power-ergodic with respect to both space and frequency.

Proof:

1. The frequency averaged power of any SLAC model is given by inserting Equation (4.4.1) into the frequency-averaging operation defined in Equation (5.1.4):

$$\begin{aligned}\langle P(f,\vec{\mathrm{r}})\rangle_f &= \left\langle |\tilde{h}(f,\vec{\mathrm{r}})|^2 \right\rangle_f \\ &= \sum_{l=1}^{N}\sum_{m=1}^{N} V_l V_m \exp\left(j\left[\Phi_l - \Phi_m + (\vec{\mathrm{k}}_m - \vec{\mathrm{k}}_l)\cdot\vec{\mathrm{r}}\right]\right) \\ &\quad \times \underbrace{\langle \exp\left(j2\pi[\tau_m - \tau_l]f\right)\rangle_f}_{=\,0 \text{ for } \tau_m \neq \tau_l} \end{aligned} \tag{5.1.6}$$

If all delays, $\{\tau_i\}$, in the SLAC model are heterogeneous (dissimilar), then the result in Equation (5.1.6) reduces to

$$\langle P(f,\vec{\mathrm{r}})\rangle_f = \sum_{m=1}^{N} V_m^2 \tag{5.1.7}$$

which is the ensemble average for power in a U-SLAC model.

2. The spatially averaged power of any SLAC model is given by inserting Equation (4.4.1) into the space-averaging operation defined in Equation (5.1.5):

$$\begin{aligned}\langle P(f,\vec{\mathrm{r}})\rangle_{\vec{\mathrm{r}}} &= \left\langle |\tilde{h}(f,\vec{\mathrm{r}})|^2 \right\rangle_{\vec{\mathrm{r}}} \\ &= \sum_{l=1}^{N}\sum_{m=1}^{N} V_l V_m \exp\left(j\left[\Phi_l - \Phi_m + 2\pi(\tau_m - \tau_l)f\right]\right) \\ &\quad \times \underbrace{\left\langle \exp\left(j[\vec{\mathrm{k}}_m - \vec{\mathrm{k}}_l]\cdot\vec{\mathrm{r}}\right)\right\rangle_{\vec{\mathrm{r}}}}_{=\,0 \text{ for } \vec{\mathrm{k}}_m \neq \vec{\mathrm{k}}_l} \end{aligned} \tag{5.1.8}$$

If all wavevectors, $\{\vec{\mathrm{k}}_i\}$, in the SLAC model are heterogeneous (dissimilar), then the result in Equation (5.1.8) reduces to

$$\langle P(f,\vec{\mathrm{r}})\rangle_{\vec{\mathrm{r}}} = \sum_{m=1}^{N} V_m^2 \tag{5.1.9}$$

which is also the ensemble average for power in a U-SLAC model.

3. This result follows logically from 1 and 2.

An immediate consequence of Theorem 5.3 is the equivalence of averaging received power with respect to frequency or space. If a SLAC model has heterogeneous scattering, then the following result holds:

$$\langle P(f,\vec{\mathrm{r}})\rangle_f = \langle P(f,\vec{\mathrm{r}})\rangle_{\vec{\mathrm{r}}} \tag{5.1.10}$$

In other words, received power averaged as a function of frequency produces the same result as received power averaged as a function of space. Both are equal to the ensemble-averaged power of the U-SLAC model that describes the location.

This equivalence is particularly powerful when *measuring* local area power in a real-life radio channel. The experimentalist has the freedom to calculate mean power in a channel measurement by using

(a) a wideband transmitted signal and a fixed-antenna receiver [Rap02a].

(b) a narrowband transmitted signal and a receiver antenna moving in space [Dur98b].

For case (a), the mean power is calculated from the received power averaged over the frequencies in the wideband signal. For case (b), the mean power is calculated from the received power averaged at the different receiver positions in space. The two techniques are equivalent.

> *Note: Importance of Heterogeneous Scattering*
>
> By now it is clear why Chapter 4 introduced the concept of *heterogeneous scattering*. Heterogeneous scattering is a crucial property for determining ergodicity. Ergodicity is a crucial property for determining the "measurability" of a real channel. Hence, the property of heterogeneous scattering determines how statistics may be calculated from a real-world channel measurement.

5.2 Envelope Probability Density Functions

Mean received power is just one aspect of first-order stochastic channel behavior. To understand the complete first-order behavior, it is necessary to calculate the power or envelope PDF. This section describes the generation of PDFs for a variety of I-SLAC models.

5.2.1 Notes and Concepts

The PDFs developed in this section are strictly for I-SLAC models. In order to derive a deterministic expression for the distribution of received power or voltage envelope, it is necessary to have a uniquely defined joint PDF on the phases of a SLAC model. A U-SLAC model, which only asserts uncorrelated phases, is not restrictive enough for this criterion. The I-SLAC model, with its independent, uniformly distributed phases, does have a specific joint PDF describing phases.

Furthermore, the envelope PDFs of this section are based on the *reduced wave grouping* (see Section 4.3.4). As will become apparent later, envelope PDFs involving a diffuse, nonspecular component depends only on the mean-squared power of that component - not the fine multipath wave structure within the component. Thus, when constructing a PDF from an arbitrarily complicated I-SLAC model, the reduced wave grouping may be used without loss of generality.

Finally, all of the PDFs presented in this chapter are with respect to received voltage envelope. If a distribution with respect to received power is desired, the envelope PDF may be converted to a power PDF. Since envelope, R, is simply equal to the square root of power, P, the following relationships hold:

$$\text{Converting envelope PDF to power PDF:} \quad f_P(\mathrm{p}) = \frac{1}{2\sqrt{\mathrm{p}}} f_R(\sqrt{\mathrm{p}}) \tag{5.2.1}$$

$$\text{Converting power PDF to envelope PDF:} \quad f_R(\rho) = 2\rho f_P(\rho^2) \tag{5.2.2}$$

Most wireless journal papers, as a convention, report envelope PDFs.

5.2.2 Characteristic Functions

Since the reduced wave grouping of Equation (4.3.4) is a sum of independent random variables, the envelope PDF may be found by using a characteristic function approach. Recall from probability theory that if a real-valued random variable, Z, is the sum of a set of independent random variables, $\{W_i\}$:

$$Z = \sum_i W_i$$

then the characteristic function of Z - the Fourier transform of the PDF - is equal to the product of the characteristic functions of each individual W_i random variable [Pap91].

The same characteristic function technique may be used for the independent, *complex* random variable terms that constitute an I-SLAC model. The diffuse, nonspecular term and the N specular terms of Equation (4.3.4) are independent complex variables, each having a characteristic function, $\Phi_{XY}(v)$. The product of these individual characteristic functions may then be transformed into a PDF describing the envelope of Equation (4.3.4).

The PDF of a scalar random variable and its characteristic function are Fourier transform pairs. Since fading PDFs are functions of *envelope*, and since complex *voltages* must be summed as phasors, the transforms for an envelope PDF, $f_R(\rho)$, and its characteristic function, $\Phi_{XY}(v)$, are modified to the following expressions, as shown in Appendix 5.A:

$$\Phi_{XY}(v) = \int_0^\infty f_R(\rho) \mathrm{J}_0(v\rho)\, d\rho \tag{5.2.3}$$

$$f_R(\rho) = \rho \int_0^\infty \Phi_{XY}(v) \mathrm{J}_0(v\rho)\, v\, dv \tag{5.2.4}$$

To solve the problem of constructing envelope PDFs for a reduced wave grouping, we must only find the characteristic functions of specular components and diffuse components.

> *Note: The Fourier-Bessel Transform*
>
> Equation (5.2.3) and Equation (5.2.4) define a transform pair in the form of a *Fourier-Bessel* transform. Also called the *Hankel* transform, the formal relationship is defined for the following pair:
>
> $$\frac{f_R(\rho)}{\rho} \longleftrightarrow \Phi_{XY}(\upsilon)$$

5.2.3 Specular Characteristic Function

A specular voltage term of the form $V_0 \exp(j\Phi_0)$ has constant amplitude, V_0, and uniformly random phase, Φ_0. The envelope of this term is always V_0, regardless of the phase. Therefore, the envelope PDF of a specular component is best described by a delta function:

$$f_R(\rho) = \delta(\rho - V_0) \tag{5.2.5}$$

Plugging this function into the transform in Equation (5.2.3) produces the specular characteristic function:

$$\Phi_{XY}(\upsilon) = \mathrm{J}_0(V_0\upsilon) \tag{5.2.6}$$

Thus, the characteristic function of a specular wave component is a zero-order Bessel function that depends only on the amplitude of the specular component [Ben48].

5.2.4 Diffuse, Nonspecular Characteristic Function

Diffuse, nonspecular received voltage may be written as the sum of numerous small-valued wave components.

$$\begin{aligned}
\tilde{V}_{\text{dif}} &= \sum_{i=1}^{N} V_i \exp(j\Phi_i) \\
&= \sum_{i=1}^{N} V_i \cos\Phi_i \quad + \quad j\sum_{i=1}^{N} V_i \sin\Phi_i \\
&= X + jY
\end{aligned} \tag{5.2.7}$$

which may be grouped into in-phase, X, and quadrature, Y, random variables. In the limit of large N, the real and imaginary components of $\tilde{V}_{\text{dif}}$ follow the central limit theorem, each tending to zero-mean, identically distributed Gaussian distributions. Provided all the individual amplitudes, $\{V_i\}$, remain small relative to the total power, the distributions of X and Y will be uncorrelated. Thus, the joint distribution may be written as

$$f_{XY}(x, y) = \frac{1}{2\pi\sigma^2} \exp\left(-\frac{x^2 + y^2}{2\sigma^2}\right) \tag{5.2.8}$$

where σ^2 is the variance of X and Y.

The probability that the random envelope, R, of the diffuse, nonspecular voltage is below a level ρ is defined as the *cumulative density function* (CDF). The envelope CDF, $F_R(\rho)$, may be calculated from the joint distribution, $f_{XY}(x, y)$, by performing the following integration:

$$F_R(\rho) = \Pr\left[R < \rho\right] = \int_0^{2\pi}\int_0^{\rho} f_{XY}(\rho'\cos\phi, \rho'\sin\phi)\rho' d\rho' d\phi \tag{5.2.9}$$

Equation (5.2.9) has a simple geometrical interpretation. If we view the in-phase component, X, and quadrature component, Y, as coordinates on a Cartesian plane, then the envelope, R, is simply the Pythagorian distance to the origin, $R = \sqrt{X^2 + Y^2}$. Thus, the probability that the envelope will be less than a threshold value ρ is equal to the integration of the joint PDF over a circular region centered at the origin with radius ρ. Equation (5.2.9) performs this integration.

The PDF is simply the derivative of the CDF. Differentiation of Equation (5.2.9) produces

$$f_R(\rho) = \frac{dF_R(\rho)}{d\rho} = \rho \int_0^{2\pi} f_{XY}(\rho\cos\phi, \rho\sin\phi) d\phi \tag{5.2.10}$$

To obtain the PDF of the diffuse, nonspecular voltage component, we insert the joint Gaussian PDF of Equation (5.2.8) into Equation (5.2.10). This produces the following PDF:

$$\begin{aligned} f_R(\rho) &= \frac{\rho}{2\pi\sigma^2}\int_0^{2\pi} \exp\left(-\frac{(\rho\cos\phi)^2 + (\rho\sin\phi)^2}{2\sigma^2}\right) d\phi \\ &= \frac{\rho}{\sigma^2}\exp\left(-\frac{\rho^2}{2\sigma^2}\right) \\ &= \frac{2\rho}{P_{\text{dif}}}\exp\left(-\frac{\rho^2}{P_{\text{dif}}}\right) \end{aligned} \tag{5.2.11}$$

The substitution $P_{\text{dif}} = 2\sigma^2$ is made to give the PDF more physical meaning. The value P_{dif} is the mean power of the nonspecular voltage component,

$$P_{\text{dif}} = \mathrm{E}\left\{\left|\tilde{V}_{\text{dif}}\right|^2\right\}$$

where $\tilde{V}_{\text{dif}}$ is the diffuse component. The power, P_{dif}, is less nebulous than the value σ.

The final step simply involves plugging the diffuse, nonspecular envelope PDF into Equation (5.2.3) to produce the characteristic function

$$\Phi_{XY}(v) = \exp\left(-\frac{v^2 P_{\text{dif}}}{4}\right) \tag{5.2.12}$$

Thus, the characteristic function of a diffuse, nonspecular voltage component is a Gaussian-shaped function that depends only on the average power of the voltage component, P_{dif} [Ric44], [Ric45]. The distribution of amplitudes, $\{V_i\}$, is not important - only the mean-squared power affects the shape of this characteristic function.

5.2.5 The I-SLAC PDF Generator

Now that the characteristic functions for specular components and diffuse, nonspecular components have been determined, it is possible to construct an envelope PDF for the reduced wave grouping by chaining the characteristic functions together and inverting the product using Equation (5.2.4). The general form for the I-SLAC envelope PDF of the reduced wave grouping, Equation (4.3.4), is given by

$$f_R(\rho) = \rho \int_0^\infty \mathrm{J}_0(v\rho) \exp\left(\frac{-v^2 P_{\text{dif}}}{4}\right) \left[\prod_{i=1}^{N} \mathrm{J}_0(V_i v)\right] v\, dv \qquad (5.2.13)$$

$$\text{for} \quad R = \left|\tilde{V}_{\text{dif}} + \sum_{i=1}^{N} V_i \exp(j\Phi_i)\right| \quad \text{and} \quad \mathrm{E}\left\{|\tilde{V}_{\text{dif}}|^2\right\} = P_{\text{dif}}$$

which is valid for $\rho \geq 0$. Equation (5.2.13) allows for *any* combination of single, constant-amplitude voltage waves and a diffuse group of voltage waves. As the next section demonstrates, Equation (5.2.13) has a number of closed-form solutions that result from standard definite integrals [Abr70], [Gra94].

5.3 Closed-Form PDF Solutions

There are five different types of reduced wave groupings that produce closed-form solutions for the envelope PDF generator in Equation (5.2.13). This section describes the five PDFs and discusses their use in wireless communications. Table 5.1 summarizes these cases.

5.3.1 The One-Wave PDF

A trivial case of fading is the *one-wave PDF*, in which only one constant-amplitude wave is present in a local area. Its characteristic function, however, is a building block for other canonical fading PDFs. Integrating Equation (5.2.13) for $N = 1$ and $P_{\text{dif}} = 0$ produces a result of 0 for all values of r except $r = V_1$, which is infinite [Abr70, p. 485]. Thus, the PDF is represented by

$$f_R(\rho) = \delta(\rho - V_1) \qquad (5.3.1)$$

The one-wave PDF, as illustrated in the IQ plot of Figure 5.2, results in no envelope fading.

Table 5.1 Summary of Envelope PDFs in Different Fading Environments

PDF	Expression for Envelope PDF $f_R(\rho)$	Characteristic Function $\Phi_{XY}(v)$	$\mathrm{E}\{R\}$ (volts)	$\mathrm{E}\{R^2\}$ (volts^2)
One-Wave No Fading	$\delta(\rho - V_1)$	$\mathrm{J}_0(V_1 v)$	V_1	V_1^2
Two-Wave Simple Fading	$\frac{2\rho}{\pi\sqrt{4V_1^2V_2^2-(V_1^2+V_2^2-\rho^2)^2}}$ $\lvert V_1 - V_2\rvert \leq \rho \leq V_1 + V_2$	$\mathrm{J}_0(V_1 v)\mathrm{J}_0(V_2 v)$	$\frac{2(V_1+V_2)}{\pi}\mathrm{E}\left[\frac{4V_1V_2}{(V_1+V_2)^2}\right]$	$V_1^2+V_2^2$
Three-Wave† Max. Discrete Fading	$\frac{\rho^{\frac{1}{2}}}{\pi^2\sqrt{V_1V_2V_3}}K\left(\frac{\Delta_\rho^2}{V_1V_2V_3\rho}\right),\ \Delta_\rho^2 < V_1V_2V_3\rho^*$ $\frac{\rho}{\pi^2\Delta_\rho}K\left(\frac{V_1V_2V_3\rho}{\Delta_\rho^2}\right),\ \Delta_\rho^2 < V_1V_2V_3\rho^*$	$\mathrm{J}_0(V_1 v)\mathrm{J}_0(V_2 v)\mathrm{J}_0(V_3 v)$	no closed-form solution	$V_1^2+V_2^2+V_3^2$
Rayleigh Numerous Multipath	$\frac{2\rho}{P_{\text{dif}}}\exp\left(\frac{-\rho^2}{P_{\text{dif}}}\right)\mathrm{u}(\rho)$	$\exp\left(\frac{-v^2 P_{\text{dif}}}{4}\right)$	$\sqrt{\frac{\pi P_{\text{dif}}}{4}}$	P_{dif}
Rician Dominant Component	$\frac{2\rho}{P_{\text{dif}}}\exp\left(\frac{-\rho^2-V_1^2}{P_{\text{dif}}}\right)\mathrm{I}_0\left(\frac{2V_1\rho}{P_{\text{dif}}}\right)\mathrm{u}(\rho)$	$\mathrm{J}_0(V_1 v)\exp\left(\frac{-v^2 P_{\text{dif}}}{4}\right)$	no closed-form solution	$P_{\text{dif}}+V_1^2$

* also provided that $2\max(V_1, V_2, V_3)-V_1-V_2-V_3 \leq \rho \leq V_1+V_2+V_3$; otherwise $f_R(\rho) = 0$.

† $16\Delta_\rho^2 = \left[(\rho+V_1)^2-(V_2-V_3)^2\right]\left[(V_2+V_3)^2-(\rho-V_1)^2\right]$

$\mathrm{K}(x)$ – Complete First Kind Elliptical Integral

$\mathrm{J}_n(x)$ – Bessel Function

$\mathrm{E}(x)$ – Complete Second Kind Elliptical Integral

$\mathrm{I}_n(x)$ – Modified Bessel Function

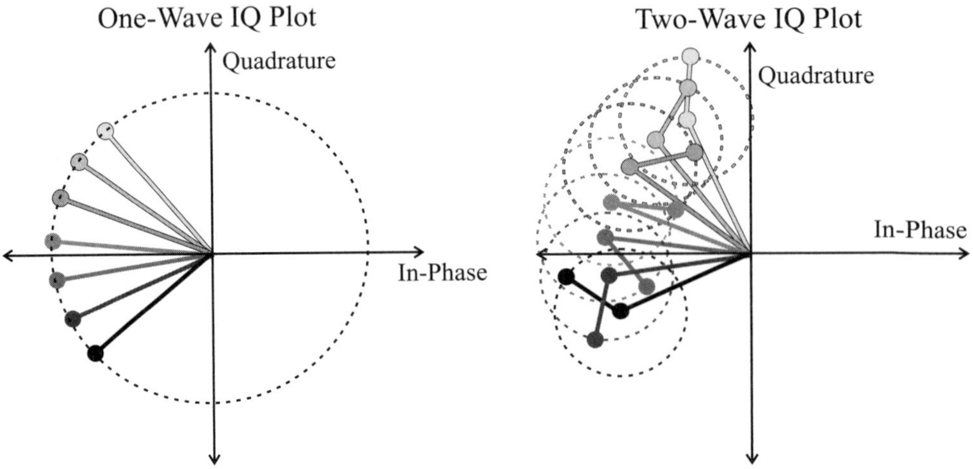

Figure 5.2 IQ plots of a single specular wave with no envelope fading (left) and two specular waves with constructive-destructive interference (right).

5.3.2 The Two-Wave PDF

The *two-wave PDF* represents the envelope fading caused by the interference of only two constant-amplitude waves in a local area, corresponding to $N = 2$ and $P_{\text{dif}} = 0$ in Equation (5.2.13). Integration of Equation (5.2.13) under these conditions is a well-understood result that produces the following PDF [Gra94, p. 718]:

$$f_R(\rho) = \frac{2\rho}{\pi\sqrt{4V_1^2V_2^2 - (V_1^2 + V_2^2 - \rho^2)^2}}, \qquad |V_1 - V_2| \leq \rho \leq V_1 + V_2 \tag{5.3.2}$$

Equation (5.2.13) evaluates to zero for $\rho < |V_1 - V_2|$ and $\rho > V_1 + V_2$, leading to the limits placed on ρ in Equation (5.3.2). Figure 5.2 shows an IQ sketch of two-wave envelope fading.

Figure 5.3 plots several examples of the two-wave PDF and CDF using a convenient parameter, Δ, which we have defined to relate the amplitudes of V_1 and V_2 to one another [Dur99c]. The Δ-parameter ranges between 0 and 1 and is defined by

$$\Delta = \frac{\text{Peak Specular Power}}{\text{Average Specular Power}} - 1 = \frac{2V_1V_2}{V_1^2 + V_2^2} \tag{5.3.3}$$

As shown in Figure 5.3, when the magnitudes of two multipath waves are equal, $\Delta = 1$. In the absence of a second component (V_1 or $V_2 = 0$), $\Delta = 0$. For dissimilar voltage values, the two-wave PDF exhibits two prominent spikes, which mark the interval over which the PDF is nonzero. For the limiting case of $\Delta = 1$ ($V_1 = V_2$), the lower spike disappears and the PDF permits envelope values of zero, which correspond to complete destructive cancellation.

Example 5.1: Mean of a Two-Wave PDF

Problem: Calculate the mean of a fading envelope that follows a two-wave distribution with voltages V_1 and V_2.

Solution: The mean of the two-wave distribution follows from the basic integral:

$$\mathrm{E}\{R\} = \int_0^\infty \rho f_R(\rho)\, d\rho$$

After inserting Equation (5.3.2) for $f_R(\rho)$, we write

$$\mathrm{E}\{R\} = \int_{|V_1-V_2|}^{V_1+V_2} \frac{2\rho^2\, d\rho}{\pi\sqrt{4V_1^2V_2^2 - (V_1^2+V_2^2-\rho^2)^2}}$$

This problem becomes much easier after a clever change of integration variables, $\rho = \sqrt{V_1^2 + V_2^2 + 2V_1V_2\sin(2\theta)}$. This substitution produces the following simplification:

$$\begin{aligned}
\mathrm{E}\{R\} &= \frac{2}{\pi}\int_{-\frac{\pi}{4}}^{\frac{\pi}{4}} \sqrt{V_1^2+V_2^2+2V_1V_2\sin(2\theta)}\, d\theta \\
&= \frac{2(V_1+V_2)}{\pi} \underbrace{\int_0^{\frac{\pi}{2}} \sqrt{1+\frac{4V_1V_2}{(V_1+V_2)^2}\sin^2(\theta)}\, d\theta}_{\mathrm{E}\left(\frac{4V_1V_2}{(V_1+V_2)^2}\right)}
\end{aligned}$$

Thus, the final answer may be expressed in terms of a complete elliptic integral of the second kind (see Appendix A.5). Note that, as in all other cases, it is much easier to calculate the power mean from the two-wave distribution than the envelope mean. Following Equation (5.1.3), the average two-wave power is simply $V_1^2 + V_2^2$.

5.3.3 The Three-Wave PDF

Two- and three-wave models often are used to describe fading in microwave digital radio communications [Rum86]. The *three-wave PDF* in Table 5.1 is the solution of Equation (5.2.13) for $N = 3$ and $P_{\text{dif}} = 0$, a result formulated by Nicholson in 1920

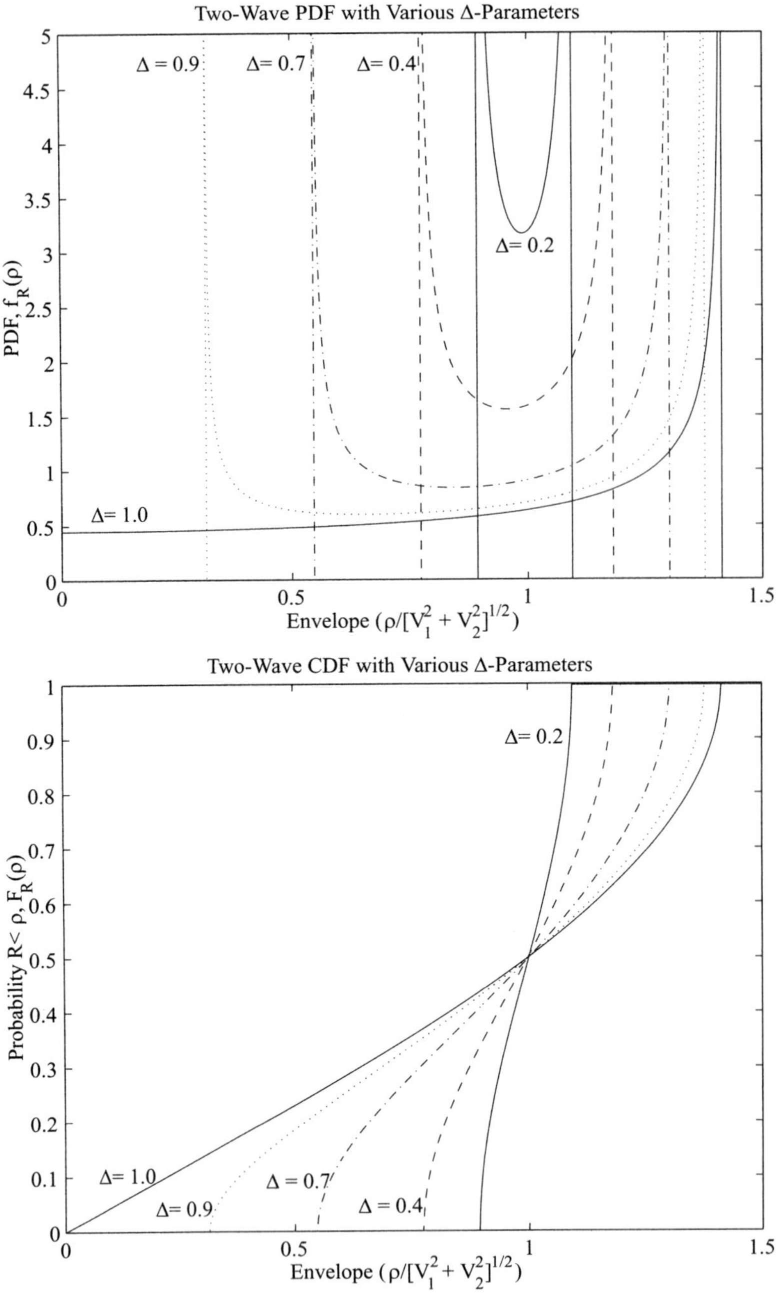

Figure 5.3 Two-wave PDF and CDF with varying Δ [Dur02].

[Nic20] and given below:

$$f_R(\rho) = \begin{cases} 0, & \rho < \rho_{\text{min}} \text{ or } \rho > \rho_{\text{max}} \\ \frac{\rho^{\frac{1}{2}}}{\pi^2\sqrt{V_1V_2V_3}}\mathrm{K}\left[\frac{\Delta_\rho^2}{V_1V_2V_3\rho}\right], & \Delta_\rho^2 < V_1V_2V_3\rho \\ \frac{r}{\pi^2\Delta_\rho}\mathrm{K}\left[\frac{V_1V_2V_3\rho}{\Delta_\rho^2}\right], & \Delta_\rho^2 > V_1V_2V_3\rho \end{cases} \tag{5.3.4}$$

The function $\mathrm{K}(\cdot)$ is an elliptic integral of the first kind. Note that Δ_ρ in Equation (5.3.4) is a function of ρ. Here we have used the subscript ρ, which is appended to Nicholson's notation to avoid any confusion between this parameter and the Δ-parameter used to describe the two-wave PDF. The values ρ_{min} and ρ_{max} define the interval over which the integration of Equation (5.2.13) and subsequently the PDF is nonzero. They are given by

$$\rho_{\text{max}} = V_1 + V_2 + V_3 \qquad \rho_{\text{min}} = \max[2\max(V_1, V_2, V_3) - V_1 - V_2 - V_3, 0] \tag{5.3.5}$$

The expressions in Equation (5.3.5) have an appealing geometric interpretation: The three-wave PDF is 0 for all ρ such that four line segments of lengths ρ, V_1, V_2, and V_3 are incapable of forming a quadrilateral [Nic20]. An IQ sketch of three-wave envelope fading was the example shown in Figure 5.1.

As one might expect, the behavior of the three-wave PDF is varied and complicated. Figure 5.4 plots just a few examples of the PDF and corresponding CDF. A comparison of the three-wave CDFs of Figure 5.4 and the two-wave CDFs of Figure 5.3 provides insight into the difference between specular and nonspecular power. Unlike the two-wave case, the different plots of the three-wave CDF are much more similar to one another. This similarity is due to the central limit theorem: The addition of another constant-amplitude wave to two-wave propagation makes the *total* multipath power more nonspecular. Despite the complex shape of the CDF, the general cases of three-wave propagation begin to approach the CDF for purely nonspecular power. Thus, even if it were possible to analytically calculate a "four-wave PDF," its usefulness would be limited, since the most general cases will appear to be even more Rayleigh-distributed than the three-wave PDF.

5.3.4 The Rayleigh PDF

The Rayleigh PDF assumes that *all* multipath power is nonspecular and occurs from the integration of Equation (5.2.13) under the condition $N = 0$ and nonzero P_{dif}. This definite integral is a standard result and produces the following PDF [Gra94, p. 738]:

$$f_R(\rho) = \frac{2\rho}{P_{\text{dif}}} \exp\left(\frac{-\rho^2}{P_{\text{dif}}}\right), \quad \rho \geq 0 \tag{5.3.6}$$

This result was derived as Equation (5.2.11). Unlike the purely specular wave PDFs, the Rayleigh PDF is nonzero over the entire range of $0 \leq \rho < \infty$. The Rayleigh

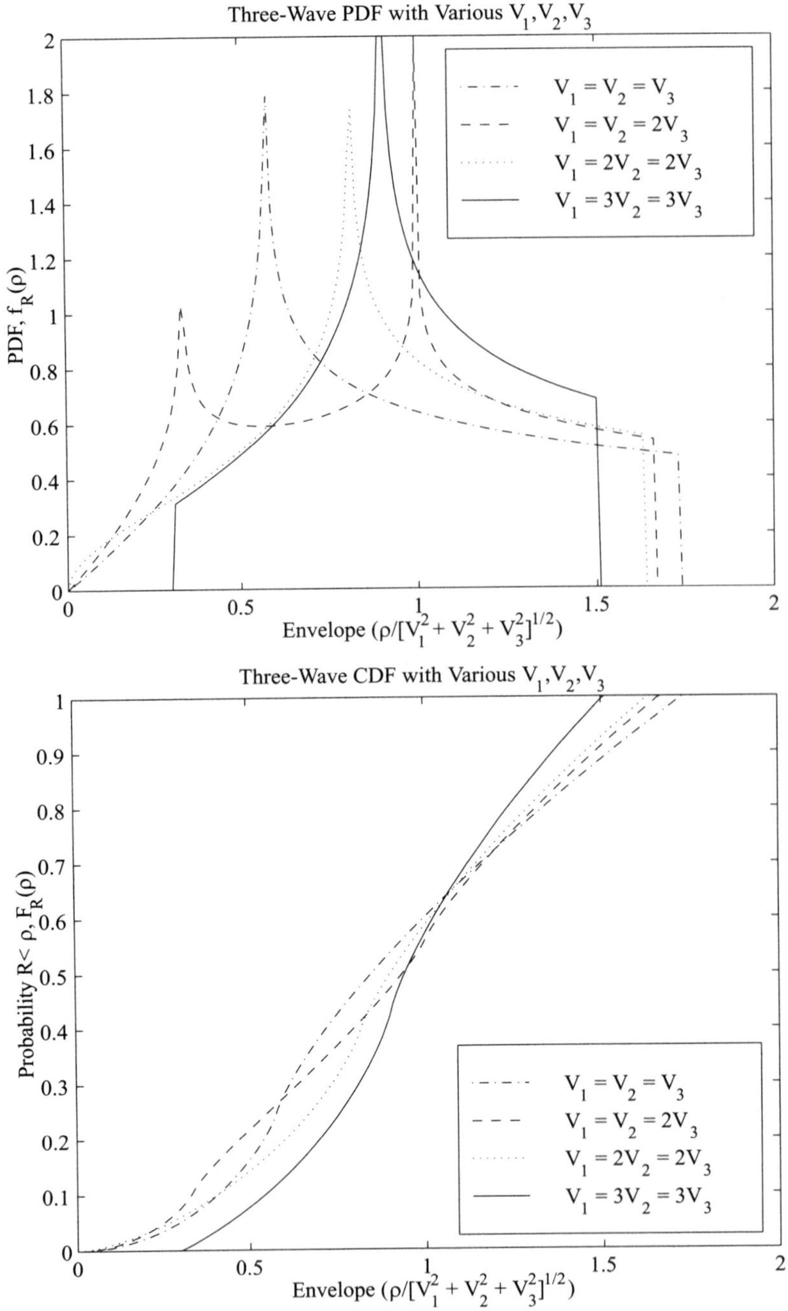

Figure 5.4 Three-wave CDF and PDF for four cases [Dur02].

PDF has been used extensively to describe narrowband local area fading for mobile radio receivers [Par92], [Reu74], [Jak74]. An IQ sketch of the Rayleigh PDF is shown in Figure 5.5.

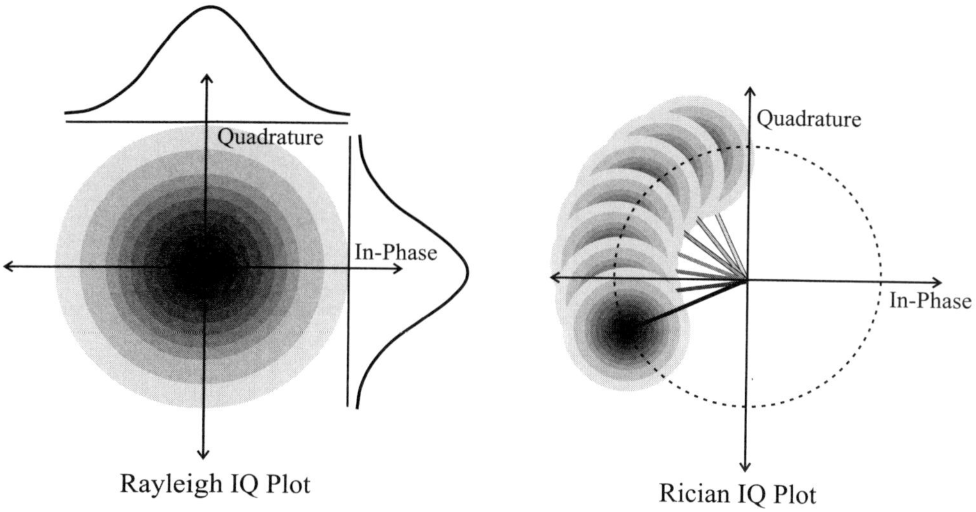

Figure 5.5 IQ coordinates with identical, independent Gaussian distributions produce a Rayleigh distribution (left), and adding a specular component produces a Rician distribution (right).

The Rayleigh distribution is the most popular distribution for calculating *fade margins* in radio links for wireless local area networks, cellular phones, and other mobile radio applications. A fade margin is the difference, usually in dB, between the average local area power and the minimum power level for reliable communications. For example, even on the fringe of a coverage area, a wireless engineer plans for an average local area power that is 14 dB to 18 dB higher than the minimum power required for maintaining an acceptable signal-to-noise+interference ratio. Without this extra margin, the channel will fade and create numerous *outages* (i.e., losses of data or service). Example 5.2 illustrates one type of outage calculation.

Example 5.2: Prediction of Link Outage

Problem: An indoor wireless link sends packet data through a Rayleigh fading radio channel. If the link suffers a signal strength fade of more than 10 dB with respect to the average power, packets begin to drop and data is lost. Given this information, about what percentage of data packets are dropped?

Solution: If the average power of a Rayleigh fading radio link is P_{dif}, then a 10 dB fade corresponds to a power of $0.1P_{\text{dif}}$ or, equivalently, a drop of $0.3162\sqrt{P_{\text{dif}}}$ in envelope. To calculate the probability of a Rayleigh channel dropping below this threshold, we setup and evaluate the following integral:

$$\begin{aligned}
\Pr[0 \le R < 0.3162\sqrt{P_{\text{dif}}}] &= \int_0^{0.3162\sqrt{P_{\text{dif}}}} \frac{2\rho}{P_{\text{dif}}} \exp\left(\frac{-\rho^2}{P_{\text{dif}}}\right) d\rho \\
&= -\exp\left(\frac{-\rho^2}{P_{\text{dif}}}\right)\Bigg|_0^{\rho=0.3162\sqrt{P_{\text{dif}}}} \\
&= 0.0952
\end{aligned}$$

If we assume that fades are slow with respect to data packet length, we can estimate that 9.5% of the packets will be dropped.

5.3.5 The Rician PDF

The Rician PDF describes the fading of nonspecular power in the presence of a dominant, nonfluctuating multipath component [Reu74], [Ric48]. The analytical expression for the Rician distribution results from the integration of Equation (5.2.13) under the condition $N = 1$ and nonzero P_{dif}. After applying a well-understood definite integral relationship [Gra94, p. 739], the resulting PDF is

$$f_R(\rho) = \frac{2\rho}{P_{\text{dif}}} \exp\left(\frac{-\rho^2 - V_1^2}{P_{\text{dif}}}\right) \mathrm{I}_0\left(\frac{2\rho V_1}{P_{\text{dif}}}\right), \quad \rho \ge 0 \tag{5.3.7}$$

where $\mathrm{I}_0(\cdot)$ is a zero-order modified Bessel function. An IQ sketch of the Rician PDF is shown in Figure 5.5.

Figure 5.6 shows several different kinds of Rician PDFs and CDFs. The plots are labeled using a Rician K factor, which is the ratio of the power of the dominant multipath component to the power of the remaining nonspecular multipath:

$$K = \frac{\text{Specular Power}}{\text{Nonspecular Power}} = \frac{V_1^2}{P_{\text{dif}}} \tag{5.3.8}$$

In the literature, the parameter K is often given as a dB value, which is $10\log_{10}$ of the quantity in Equation (5.3.8). Notice from Figure 5.6 that $K = -\infty$ dB corresponds to the Rayleigh PDF and the complete disappearance of the specular power.

> *Note: A Useful Approximation*
>
> As the Rician K-factor becomes large ($K \gg 1$), it is possible to approximate the Rician distribution with a Gaussian PDF of the following form:
>
> $$f_R(\rho) = \frac{1}{\sqrt{\pi P_{\text{dif}}}} \exp\left(-\frac{(\rho - V_1)^2}{P_{\text{dif}}}\right)$$

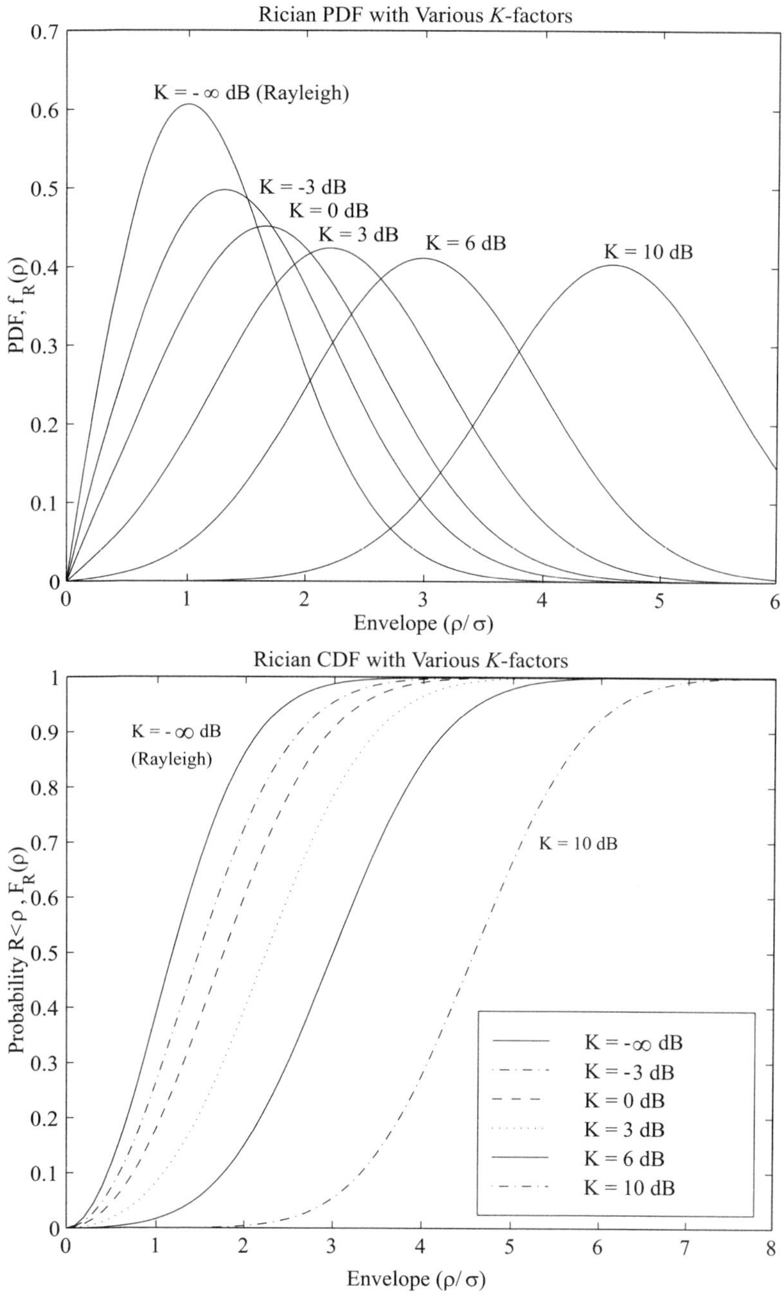

Figure 5.6 Rician PDF and CDF as the dominant multipath component increases ($\sigma = \sqrt{P_{\mathrm{dif}}/2}$) [Dur02].

> *Note: Rice or Nakagami*
>
> The Rician distribution is also called the *Rice-Nakagami* distribution in the literature to recognize the result that was independently formulated by outstanding Japanese researcher M. Nakagami. The term *Rician* is used in this work not to diminish Nakagami's contribution, but to avoid confusion with another popular PDF in radio channel modeling that bears his name: the *Nakagami-m* distribution [Nak60]. This distribution was originally formulated for characterizing temporal fading measurements from upper-atmosphere propagation but has been applied liberally to the small-scale fading of terrestrial wireless systems as well [Cou98a], [Yac00].

5.4 Two-Wave with Diffuse Power PDF

If Equation (5.2.13) is evaluated with $N = 2$ and nonzero P_{dif}, then the two-wave with diffuse power (TWDP) PDF results [Esp73], [Dur02]. Such a distribution, while difficult to model analytically, provides the greatest wealth of fading behavior for an I-SLAC model.

5.4.1 Approximate Representation

We will use parameters similar to the physical Rician K-parameter of Equation (5.3.8) and the two-wave Δ-parameter of Equation (5.3.3) to classify the shape of the TWDP PDF:

$$K = \frac{V_1^2 + V_2^2}{P_{\text{dif}}} \qquad \Delta = \frac{2V_1V_2}{V_1^2 + V_2^2} \tag{5.4.1}$$

There is no exact closed-form equation for TWDP fading, but this section presents a family of closed-form PDFs that closely approximate the behavior of the exact TWDP PDF.

An IQ sketch of TWDP fading is shown in Figure 5.7. One common approximation to the TWDP PDF is presented in [Dur02]:

$$f_R(\rho) = \frac{2\rho}{P_{\text{dif}}} \exp\left(\frac{-\rho^2}{P_{\text{dif}}} - K\right) \sum_{i=1}^{M} a_i \mathrm{D}\left(\frac{\rho}{\sqrt{P_{\text{dif}}/2}}; K, \Delta\cos\frac{\pi(i-1)}{2M-1}\right) \tag{5.4.2}$$

where

$$\mathrm{D}\left(x; K, \alpha\right) = \frac{1}{2}\exp(\alpha K)\mathrm{I}_0\left(x\sqrt{2K(1-\alpha)}\right) + \frac{1}{2}\exp(-\alpha K)\mathrm{I}_0\left(x\sqrt{2K(1+\alpha)}\right).$$

The value M in the summation is the *order* of the approximate TWDP PDF. By increasing the order in Equation (5.4.2), the approximate PDF becomes a more accurate representation of the true TWDP PDF. However, using the first few orders ($M = 1$ through 5) yields accurate representations *over the most useful range of K and Δ parameters.* Table 5.2 records the exact $\{a_i\}$ coefficients for the first five orders of Equation (5.4.2).

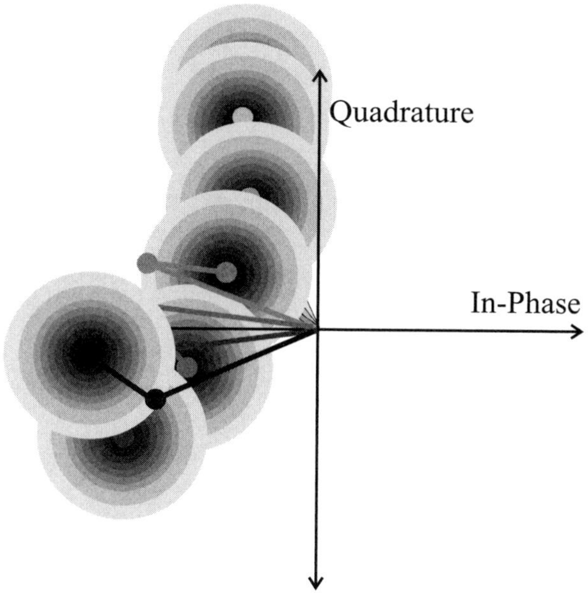

Figure 5.7 A diffuse, Rayleigh component added to two randomly phased specular waves to produce a TWDP distribution.

Table 5.2 Exact Coefficients for the First Five Orders of the Approximate TWDP Fading PDF

ORDER	a_1	a_2	a_3	a_4	a_5
1	1				
2	$\frac{1}{4}$	$\frac{3}{4}$			
3	$\frac{19}{144}$	$\frac{25}{48}$	$\frac{25}{72}$		
4	$\frac{751}{8640}$	$\frac{3577}{8640}$	$\frac{49}{320}$	$\frac{2989}{8640}$	
5	$\frac{2857}{44800}$	$\frac{15741}{44800}$	$\frac{27}{1120}$	$\frac{1209}{2800}$	$\frac{2889}{22400}$

The product of the parameters K and Δ determines which order of Equation (5.4.2) should be used when representing TWDP fading. As the product of these two parameters increases, a higher order approximation is needed to model the TWDP PDF accurately. As a general rule of thumb, the minimum order is

$$\text{Order } (M) = \left\lceil \frac{1}{2} K\Delta \right\rceil \tag{5.4.3}$$

where $\lceil \cdot \rceil$ is the ceiling function (round up). Equation (5.4.3) is based on a graphical comparison between the approximate analytical functions and the true, numerical solution of the TWDP PDF. The approximate PDF will deviate from the exact TWDP PDF only if the specular power is much larger than the nonspecular power (large K value) *and* if the amplitudes of the specular voltage components are relatively equal in magnitude (Δ approaches 1).

Example 5.3: Order-2 Approximate TWDP PDF

Problem: Using Table 5.2 and Equation (5.4.2), calculate the order-2 approximate TWDP PDF.

Solution: Plugging the coefficients a_1 and a_2 into Equation (5.4.2) produces

$$f_R(\rho) = \frac{2\rho}{P_{\text{dif}}} \exp\left(\frac{-\rho^2}{P_{\text{dif}}} - K\right)\left[\frac{1}{4}\text{D}\left(\frac{\rho}{\sqrt{P_{\text{dif}}/2}}; K, \Delta\right) + \frac{3}{4}\text{D}\left(\frac{\rho}{\sqrt{P_{\text{dif}}/2}}; K, \frac{\Delta}{2}\right)\right]$$

which, in this form, is not much more complicated than a Rician PDF.

Despite being an approximate result, the family of PDFs in Equation (5.4.2) have a number of extraordinary characteristics that are independent of order, M, and parameters, K and Δ:

- *They are mathematically exact PDFs.* They integrate to 1 over the range $0 \leq \rho < \infty$.
- *They are accurate over their upper and lower tails.* These regions are important for modeling noise-limited or interference-limited mobile communication systems [Cou98b].
- *They all exactly preserve the second moment of the true PDF.* The second moment is the most important moment to preserve, since it physically represents the average local area power [Rap02a].
- *They can be entirely described with three physically intuitive parameters.* The physical parameters P_{dif}, K, and Δ - as defined in this book - have straightforward physical definitions.
- *They exhibit the proper limiting behavior.* All of the PDFs contain, as a special case of $\Delta = 0$, the exact Rician PDF and, as a special case of $K = 0$, the exact Rayleigh PDF.

Accurate analytical representation of these PDFs reveals interesting behavior in fading channels that goes unnoticed using Rician PDFs, which are capable of modeling the envelope fading of diffuse power in the presence of only *one* specular component.

It should be noted that there are many interesting ways to approximate the TWDP PDF and other nonanalytic forms of Equation (5.2.13) (see the work by Esposita and Wilson in [Esp73] and Abdi et al. in [Abd00]).

5.4.2 Graphical Analysis

Figures 5.8 through 5.11 plot a series of PDFs and CDFs for TWDP fading. As shown by Figure 5.8, there is little difference between the Rician PDF and the TWDP PDF when K is less than 3 dB. The difference gradually becomes more pronounced as K increases, particularly when the specular power is divided equally between the two discrete components ($\Delta = 1$). The $K = 10$ dB graph of Figure 5.11 illustrates these distortions most dramatically. In fact, as the product of parameters K and Δ becomes large, the graph of the PDF becomes bimodal, exhibiting two maxima.

5.4.3 Rayleigh and Rician Approximations

For the limiting parameter cases of Table 5.3, the exact TWDP PDF contains the Rayleigh, Rician, one-wave, and two-wave PDFs. This demonstrates the generality of the exact and approximate TWDP PDFs. It also shows the utility of the three-wave PDF, since it is the only analytical expression in Table 5.1 that is not a general case of the TWDP PDF.

Table 5.3 The TWDP PDF Contains the Rayleigh, Rician, One-Wave, and Two-Wave PDFs as Special Cases

Parameter Value		Type of Fading
$K = 0$	–	Rayleigh
$K > 0$	$\Delta = 0$	Rician
$K \to \infty$	$\Delta = 0$	One-Wave
$K \to \infty$	$\Delta > 0$	Two-Wave

Since the Rician and Rayleigh PDFs are special cases of the TWDP PDF, it is useful to know the range of parameters over which TWDP fading may be approximated by these simpler distributions. An inspection of the graphs of Figure 5.8 through Figure 5.11 reveals the range of K and Δ over which a Rician PDF approximates a TWDP PDF. In general, the TWDP PDF resembles a Rician PDF in shape for $K\Delta < 2$. Under this condition, the smallest of the two specular components may be grouped with the nonspecular power so that only one large specular component remains. After computing a Rician K-factor for this new grouping, the resulting Rician PDF will approximately describe the envelope of the TWDP fading.

TWDP fading may be further approximated by a Rayleigh PDF if, in addition to the above-mentioned criterion, the power of the largest specular component is less than the power of the smaller specular component *plus* the average nonspecular power:

$$\begin{aligned}\max(V_1^2, V_2^2) &< \min(V_1^2, V_2^2) + P_{\text{dif}} \\ \frac{1}{2} + \frac{1}{2}\sqrt{1-\Delta^2} &< \frac{1}{2} - \frac{1}{2}\sqrt{1-\Delta^2} + \frac{1}{K+1}\end{aligned} \qquad (5.4.4)$$

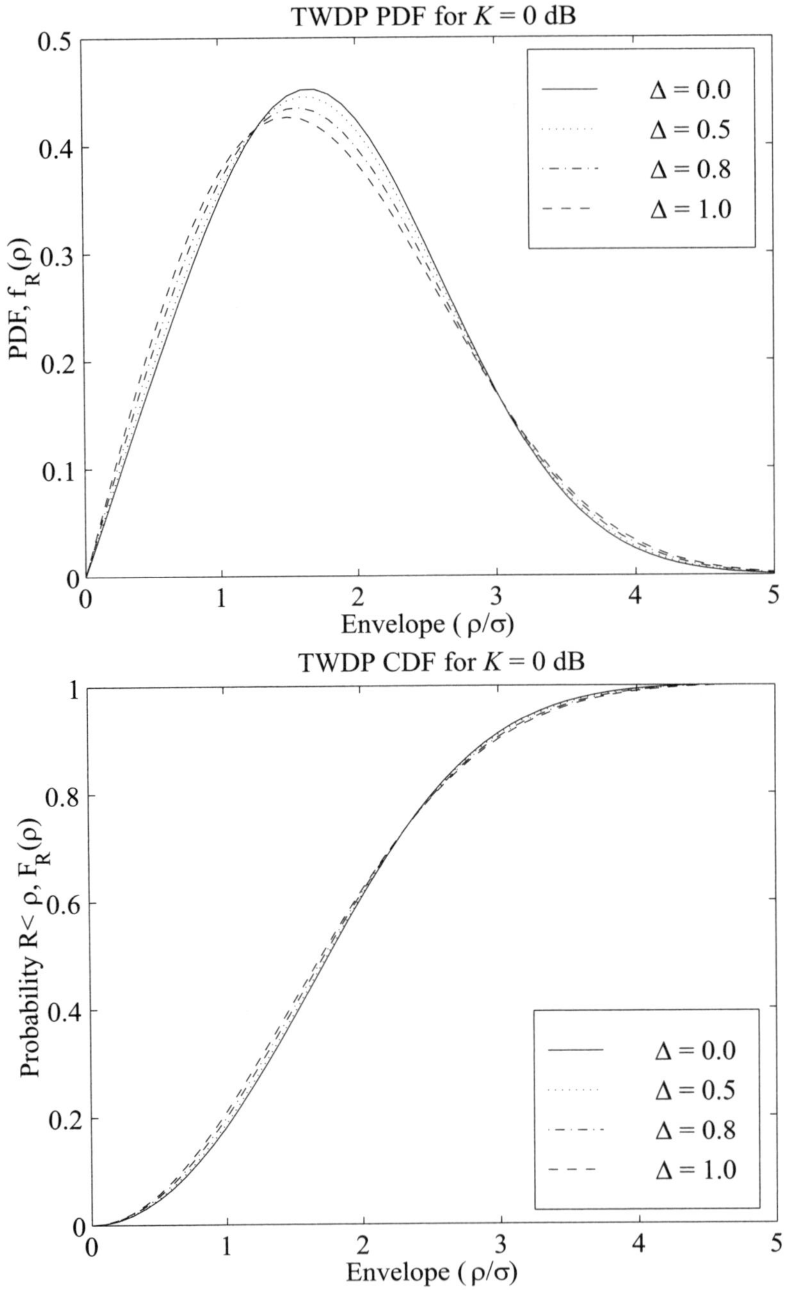

Figure 5.8 TWDP PDF and CDF for $K = 0$ dB [Dur02] ($\sigma = \sqrt{P_{\text{dif}}/2}$).

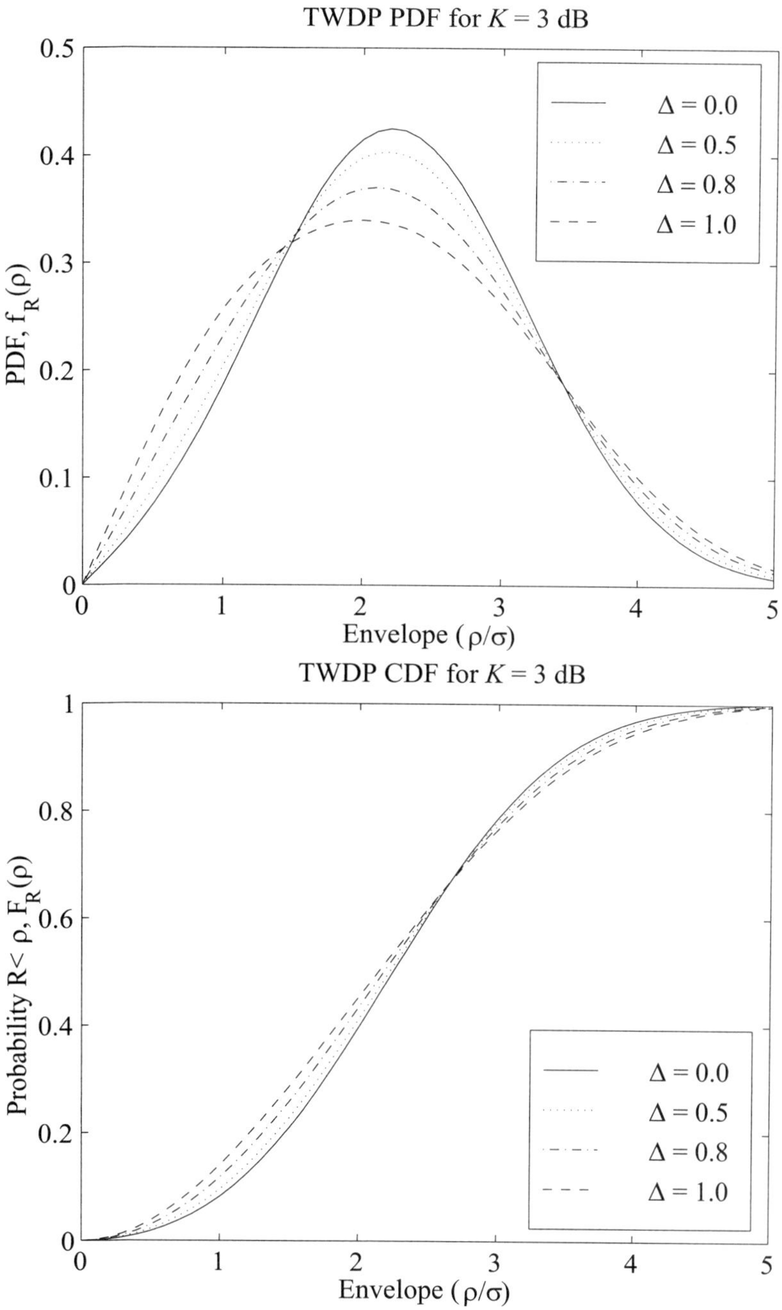

Figure 5.9 TWDP PDF and CDF for $K = 3$ dB [Dur02] ($\sigma = \sqrt{P_{\text{dif}}/2}$).

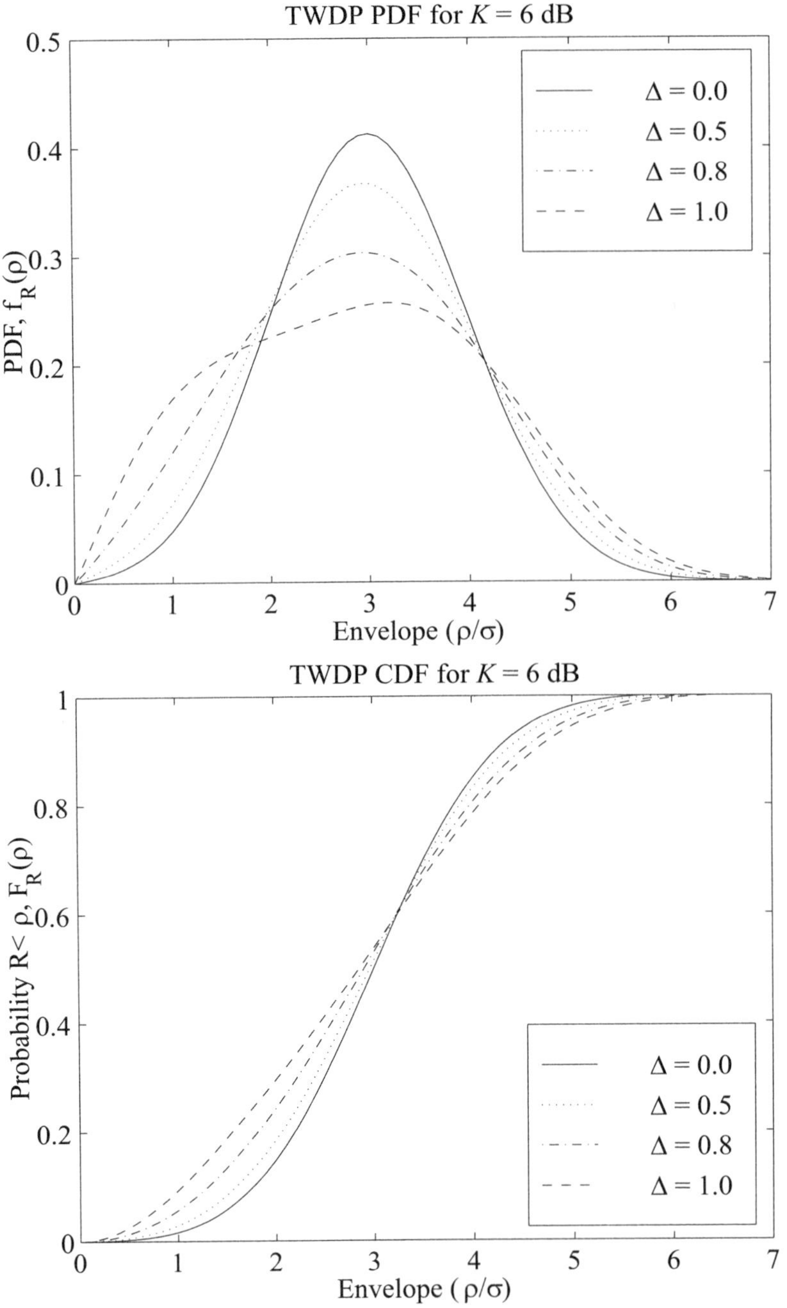

Figure 5.10 TWDP PDF and CDF for $K = 6$ dB [Dur02] ($\sigma = \sqrt{P_{\text{dif}}/2}$).

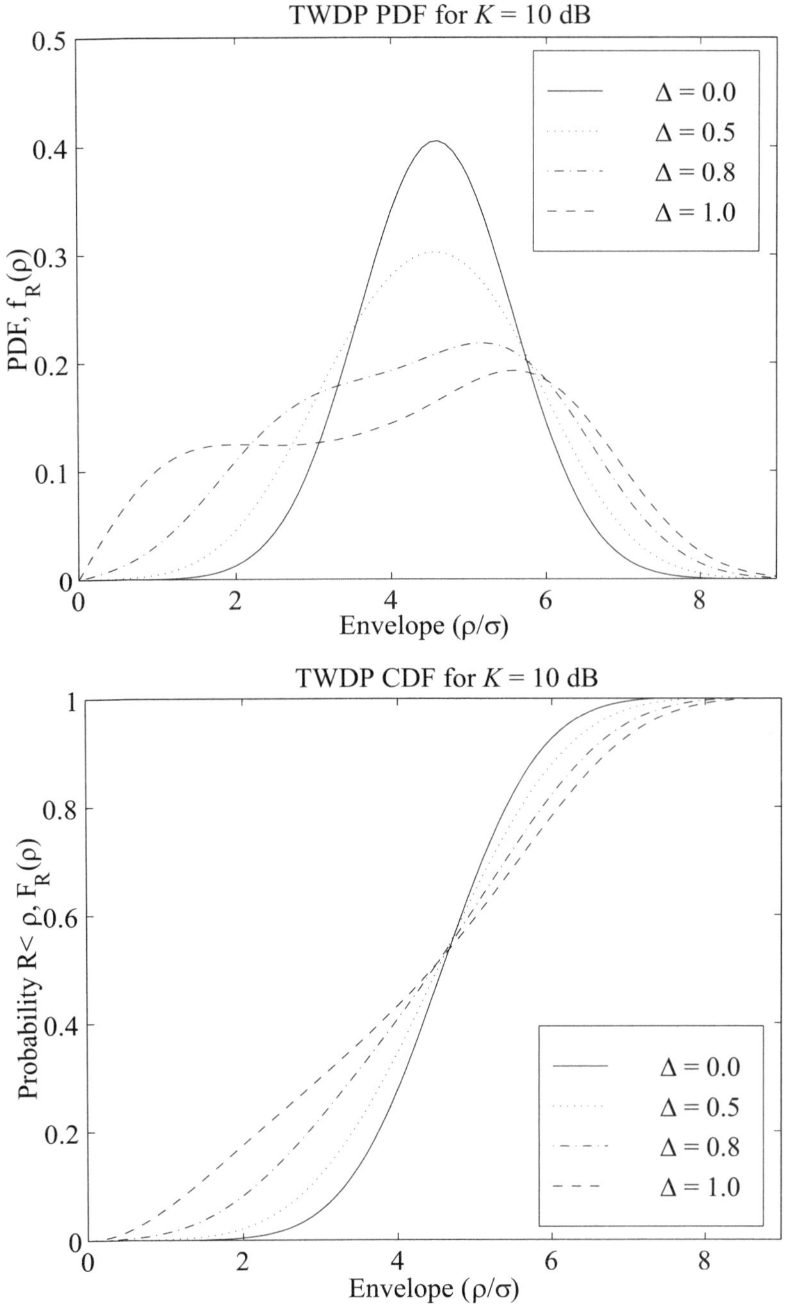

Figure 5.11 TWDP PDF and CDF for $K = 10$ dB [Dur02] ($\sigma = \sqrt{P_{\mathrm{dif}}/2}$).

This condition derives from Figure 5.6, which shows that Rician PDFs resemble the shapes of Rayleigh PDFs (after scaling) for a Rician K-factor less than 0 dB. Under this condition, the entire sum of voltage components may be treated together as diffuse, nonspecular power, despite the presence of two specular components.

The Rician and Rayleigh approximation conditions, therefore, are best summarized in terms of the TWDP K and Δ parameters by the following:

$$\text{Rician Condition:} \quad K < \frac{2}{\Delta} \tag{5.4.5}$$

$$\text{Rayleigh Condition:} \quad K < \min\left(\frac{2}{\Delta}, \frac{1}{\sqrt{1-\Delta^2}} - 1\right) \tag{5.4.6}$$

These conditions show the parameter range over which a TWDP PDF may be approximated by an analytically simpler Rician or Rayleigh PDF. If these conditions are not met, then the only recourse is to use Equation (5.4.2) or some other evaluation of Equation (5.2.13) for $N = 2$ and nonzero P_{dif}.

Table 5.4 shows three examples of TWDP fading and determines the simplest approximate PDF that describes the voltage envelope of each. Case A in Table 5.4 satisfies the Rayleigh condition of Equation (5.4.6). Case B, on the other hand, satisfies only the Rician condition of Equation (5.4.5). Case C satisfies neither condition and may not be approximated by a Rayleigh or Rician PDF. Note how the subtle changes in voltage amplitudes between the three cases drastically affects the overall shape and calculation of the PDF, emphasizing the need for careful and accurate representation of TWDP PDFs. See Example 5.4 for another example of finding the optimum PDF representation.

Table 5.4 Three Examples of TWDP Fading That May Simplify to Rayleigh or Rician PDFs

	EXAMPLE TWDP VOLTAGE VALUES					
	1st Specular	2nd Specular	Diffuse RMS	Parameters		Simplest
Case	Voltage (V_1)	Voltage (V_2)	Voltage ($\sqrt{P_{\text{dif}}}$)	K	Δ	PDF
A	2 μV	2 μV	3 μV	0.89	1.0	Rayleigh
B	4 μV	2 μV	3 μV	2.22	0.8	Rician
C	4 μV	4 μV	3 μV	3.56	1.0	TWDP

Example 5.4: PDF Grouping

Problem: It is known that an I-SLAC model is composed of three multipath waves with voltage amplitudes 4 μV, 3 μV, and 2 μV and a diffuse, nonspecular component with $P_{\text{dif}} = (1\mu\text{V})^2$. Find the simplest analytical representation, if any, of this envelope PDF.

Solution: Perform the following steps to ascertain the best PDF:

1. The initial grouping of voltages is $[V_1 = 4,\ V_2 = 3,\ V_3 = 2,\ P_{\mathrm{dif}} = 1]$ (units dropped for simplicity).
2. There is diffuse power ($P_{\mathrm{dif}} \neq 0$) and there are more than two specular components ($N = 3$), so all but the two largest specular components must be grouped with the nonspecular component. The new grouping is $[V_1 = 4,\ V_2 = 3,\ P_{\mathrm{dif}} = 5]$.
3. The TWDP factors for this distribution are $K = 5$ and $\Delta = 0.96$. This TWDP distribution is too complicated to simplify to a Rayleigh distribution ($K > 1$) or a Rician distribution ($K\Delta > 1$), but can be approximated accurately by Equation (5.4.2) ($K\Delta < 10$).
4. The value of $\left\lceil \frac{K\Delta}{2} \right\rceil$ is 3, so an order-3 approximation of Equation (5.4.2) should be used to represent the PDF.

5.4.4 TWDP PDF Applications

The TWDP PDF and its approximations are important for characterizing fading in a variety of propagation scenarios. Small-scale fading is characterized by the TWDP PDF whenever the received signal contains two strong, specular multipath waves. While this may occur for typical narrowband receiver operation, directional antennas and wideband signals increase the likelihood of TWDP small-scale fading.

The use of directive antennas or arrays at a receiver, for example, amplifies several of the strongest multipath waves that arrive in one particular direction while attenuating the remaining waves [God97], [Win98]. This effectively increases the ratio of specular to nonspecular received power, turning a Rayleigh or Rician fading channel into a TWDP fading channel.

Wideband signal fading will likely exhibit TWDP fading characteristics for similar reasons. A wideband receiver has the ability to reject multipath components that arrive with largely different propagation time delays [Rap02a], [Bra91]. This property of a wideband receiver separates specular multipath components from other nonspecular multipath waves. Under these circumstances, the ratio of specular to nonspecular received power increases for a given propagation delay and a TWDP fading channel may result.

5.4.5 Closing Remarks on TWDP Fading

Beyond the TWDP PDF, a three-wave with diffuse power (3WDP) PDF is the next logical step. The value of such an analytically difficult PDF, however, is questionable. Much like the previously discussed four-wave PDF, the central-limit theorem would begin to dominate the behavior of an I-SLAC model, making it difficult to distinguish between the different cases of a 3WDP PDF. For example, a 3WDP PDF may be approximated by the TWDP PDF if the smallest of the three specular voltage components is grouped with the nonspecular power. This approximation would fail only if the nonspecular power were small compared to the third smallest specular component - yet such a situation implies that the nonspecular power is so small that it could be ignored: A 3WDP PDF could then be approximated by the three-wave PDF. Therefore, it is safe to say that the analytical expressions of Equation (5.4.2) and Table 5.1 provide a near-complete description of the possible envelope fading of complex voltages in an I-SLAC model.

5.5 Chapter Summary

In a randomly varying small-scale channel, the distribution of received signal power or envelope dramatically affects the performance of a receiver. These fluctuations are best described using a PDF, which characterizes all of the first-order statistics of a channel. The following key points summarize the first-order analysis described in this chapter:

- Mean received power is one of the most fundamental first-order statistics in channel modeling and measurement.
  - ▷ For a U-SLAC model, the mean power of the channel is equal to the sum of the powers carried by individual multipath waves.
  - ▷ An I-SLAC model is strict-sense stationary.
  - ▷ A U-SLAC model is ergodic if its scattering is heterogeneous.
  - ▷ For a U-SLAC model with heterogeneous scattering, spatial averaging and frequency averaging produce identical results for mean received power.
- The canonical PDF generator of Equation (5.2.13) describes the distribution of received envelope voltages for I-SLAC models.
  - ▷ The generator is based on the reduced wave grouping and uses a characteristic function approach.
  - ▷ There are five closed-form solutions to the PDF generator.
  - ▷ The simplest solutions have become popular in wireless engineering.
- The two-wave with diffuse power (TWDP) PDF models the most general type of fading behavior.
  - ▷ This PDF has no closed-form solution.
  - ▷ There are several techniques for approximating the TWDP PDF.
  - ▷ The TWDP PDF contains Rayleigh, Rician, one-wave, and two-wave PDFs as special cases.
  - ▷ TWDP behavior can deviate substantially from Rayleigh or Rician PDFs.

Envelope PDF calculation completely characterizes the first-order power statistics of a random radio channel. While useful, first-order statistics do not provide any information as to how processes develop as a function of frequency and space. To understand the space-varying characteristics of random channels, we will further develop our analysis of the 3D spatial channel and introduce the concept of multipath *angle spectrum*. This is the subject of Chapter 6.

PROBLEMS

1. The following statistic, χ, is used to describe a frequency-varying I-SLAC model, $\tilde{h}(f)$:
$$\chi(f_1, f_2, f_3) = \mathrm{E}\left\{\tilde{h}(f_1)\tilde{h}^*(f_2)\tilde{h}(f_3)\right\}$$
Demonstrate how it is possible to simplify the number of dependencies in this statistic.

2. You decide to measure local area power using two techniques. First, you measure the spatial average of a narrowband channel, $\tilde{h}(\vec{r})$. Then, you measure the frequency average of a fixed channel, $\tilde{h}(f)$. Explain a possible physical interpretation for the multipath waves if the two averages do not agree.

3. Why are the following functions invalid for use as an envelope PDF?

   a. $f_R(\rho) = \frac{1}{\sigma\sqrt{2\pi}} \exp\left(-\frac{\rho^2}{2\sigma^2}\right)$
   b. $f_R(\rho) = \mathrm{u}(\rho - 1) - \mathrm{u}(\rho - 3)$
   c. $f_R(\rho) = \mathrm{sn}(\rho - 3)\mathrm{u}(\rho)$
   d. $f_R(\rho) = \mathrm{sn}^2(\rho)\mathrm{u}(\rho)$

4. Transform the following envelope PDFs into power PDFs based on the relationship $p = \rho^2$:

   a. Weibull PDF: $f_R(\rho) = a\exp(-ap)\,\mathrm{u}(p), \quad a > 0$
   b. Half-Gaussian PDF: $f_R(\rho) = \frac{1}{\sigma}\sqrt{\frac{2}{\pi}} \exp\left(-\frac{\rho^2}{2\sigma^2}\right)\mathrm{u}(\rho)$
   c. Two-Wave PDF: $f_R(\rho) = \frac{2\rho}{\pi\sqrt{4V_1^2V_2^2 - (V_1^2+V_2^2-\rho^2)^2}}$, for $|V_1 - V_2| \le \rho \le V_1 + V_2$
   d. Rician PDF: $f_R(\rho) = \frac{2\rho}{P_{\mathrm{dif}}} \exp\left(\frac{-\rho^2 - V_1^2}{P_{\mathrm{dif}}}\right) \mathrm{I}_0\left(\frac{2V_1\rho}{P_{\mathrm{dif}}}\right)\mathrm{u}(\rho)$
   e. Half Sinc-Squared PDF: $f_R(\rho) = 2V_0\mathrm{sn}^2(V_0\rho)\mathrm{u}(\rho)$

5. Transform the following power PDFs into envelope PDFs based on the relationship $p = \rho^2$:

   a. Exponential: $f_P(p) = \frac{1}{P_0} \exp\left(-\frac{p}{P_0}\right)\mathrm{u}(p)$
   b. Half-Gaussian PDF: $f_P(p) = \frac{1}{\sigma}\sqrt{\frac{2}{\pi}} \exp\left(-\frac{p^2}{2\sigma^2}\right)\mathrm{u}(p)$
   c. Triangle PDF: $f_P(p) = 2(1 - p)[\mathrm{u}(p) - \mathrm{u}(p - 1)]$
   d. Half Sinc-Squared PDF: $f_P(p) = 2P_0\mathrm{sn}^2(P_0 p)\mathrm{u}(p)$

6. Recall the transmission line problem from previous engineering courses. A lossless transmission line of length $L = l\lambda$ and real impedance, Z_0, terminates in a complex load with impedance, $\tilde{Z}_L$. This configuration is illustrated below:

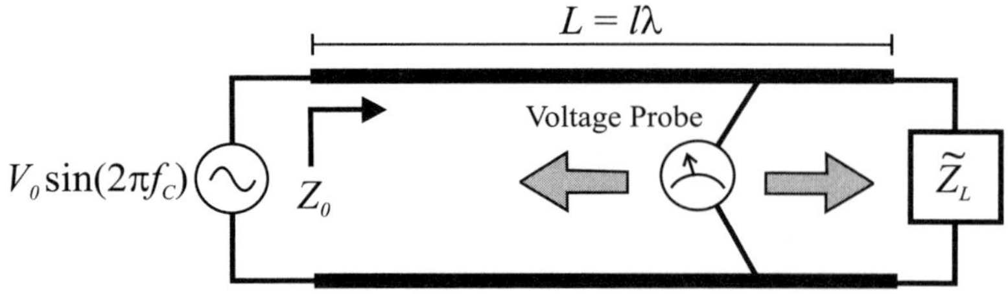

The transmission line is very long, and a voltage probe takes measurements at random positions along the length of the line. Find an expression for the envelope PDF measured by the probe in terms of l, Z_0, and $\tilde{Z}_L$.

7. Write a computer program to numerically compute any PDF from the I-SLAC PDF generator of Equation (5.2.13). Use this program to graph the PDFs for the following cases:

a. $N = 4$, $V_1 = V_2 = V_3 = V_4 = 1\text{V}$, $P_{\text{dif}} = 1\text{V}^2$

b. $N = 4$, $V_1 = V_2 = V_3 = V_4 = 1\text{V}$, $P_{\text{dif}} = 0$

c. $N = 4$, $V_1 = V_2 = 2V_3 = 2V_4 = 1\text{V}$, $P_{\text{dif}} = 0$

d. $N = 3$, $V_1 = V_2 = 2V_3 = 1\text{V}$, $P_{\text{dif}} = 1\text{V}^2$

e. $N = 3$, $V_1 = 2V_2 = 2V_3 = 1\text{V}$, $P_{\text{dif}} = 1\text{V}^2$

8. Consider a local area propagation scenario where three equal-amplitude specular waves ($V_1 = V_2 = V_3 = 1\text{V}$) are received in the presence of other diffuse multipath waves with total power P_0. We may can calculate this case by either grouping one of the specular voltages with the diffuse power (a TWDP approximation) or using the full 3WDP representation:

TWDP:	$V_1 = V_2 = 1\text{V}$	3WDP:	$V_1 = V_2 = V_3 = 1\text{V}$
	$P_{\text{dif}} = P_0 + 1\text{V}^2$		$P_{\text{dif}} = P_0$

Evaluate and graph both the exact 3WDP and approximate TWDP envelope PDFs for different values of P_0. At which value of P_0 does this approximation fail?

9. Use the Rayleigh PDF to calculate the following information about a Rayleigh fading channel with average power P_{dif}:

a. What is the mean of the Rayleigh fading envelope?

b. What is the most likely value of the envelope?

c. What is the *median* of the envelope? (The median is the voltage, ρ_m, at which $\Pr[R > \rho_m] = \Pr[R < \rho_m] = 0.5$.)

10. Analytically write and solve the generating integral for the Rician PDF. Hint: See Appendix A.4 for formulas that help evaluate this integral.

11. Prove that the following mathematical relationship holds for any positive values of A_i:

$$\int_0^\infty \int_0^\infty x^3 \nu \prod_{i=1}^{N} \mathrm{J}_0(A_i \nu)\mathrm{J}_0(\nu x)\, d\nu\, dx = \sum_{i=1}^{N} A_i^2$$

Hint: Try applying some theorems learned in this chapter before computing any integrals.

12. Prove that the Rician distribution may be approximated as a Gaussian distribution for $K \gg 1$. (See Table A.3 in Appendix A.)

13. Which of the following descriptions of wave groupings will produce envelopes that never fade to zero?

a. $\{V_i\} = \{4, 3, 2, 1\}$

b. $\{V_i\} = \{4, 2, 1\}$

c. $\{V_i\} = \{1, 3, 4, 10, 1\}$

d. $\{V_i\} = \{200, 1, 1\}$ and $P_{\text{dif}} = 0.1\text{V}^2$

14. Compute $\mathrm{E}\{P\}$ for the groups of waves in the previous problem.

15. Assume that N specular waves with equal amplitude are received by an antenna. Test how the exact I-SLAC PDF compares with an approximation of the PDF using the Rayleigh distribution of equal power. Graph the cases for $N = 3, 4, 5, 7, 10, 15$, and 20.

5.A Envelope Characteristic Functions

In the study of PDFs, it is convenient to define a characteristic function, which is the Fourier transform of the PDF [Pap91]. The standard mathematical definitions for finding a characteristic function, $\Phi_X(v)$, from a PDF, $f_X(x)$, and vice versa are given below:

$$\Phi_X(v) = \int_{-\infty}^{+\infty} f_X(x) \exp(-jvx)\, dx \tag{5.A.1}$$

$$f_X(x) = \frac{1}{2\pi} \int_{-\infty}^{+\infty} \Phi_X(v) \exp(jvx)\, dv \tag{5.A.2}$$

Characteristic functions are useful for studying the addition of independent random variables. If random variables A, B, and C satisfy the relationship $C = A + B$ and A and B are independent, then their characteristic functions satisfy the relationship $\Phi_C(v) = \Phi_A(v)\Phi_B(v)$ [Sta94].

Characteristic functions are also useful for studying the superposition of two independent random voltages, such as those in the SLAC model. Since voltage, $\tilde{V}$, is complex-valued, its characteristic function must be a *double* Fourier transform over the joint PDF of the random in-phase, X, and quadrature, Y, voltage components ($\tilde{V} = X + jY$). This transformation is demonstrated below:

$$\Phi_{XY}(v_x, v_y) = \int_{-\infty}^{+\infty}\int_{-\infty}^{+\infty} f_{XY}(x, y) \exp(-jv_x x) \exp(-jv_y y)\, dx\, dy \tag{5.A.3}$$

$$f_{XY}(x, y) = \frac{1}{4\pi^2} \int_{-\infty}^{+\infty}\int_{-\infty}^{+\infty} \Phi_{XY}(v_x, v_y) \exp(jv_x x) \exp(jv_y y)\, dv_x\, dv_y \tag{5.A.4}$$

Starting with the basic envelope PDF, $f_R(\rho)$, as a function of only envelope, ρ, it is possible to extend this PDF into a joint PDF using the relationship

$$f_{R\Phi}(\rho, \phi) = \frac{1}{2\pi\rho} f_R(\rho) \tag{5.A.5}$$

Equation (5.A.5) is a joint PDF, albeit in terms of envelope, ρ, and phase, ϕ, variables instead of in-phase, x, and quadrature, y, variables. Equation (5.A.5) assumes that the net phase, ϕ, is uniformly distributed, independent of ρ - consistent with the I-SLAC model.

Rather than convert Equation (5.A.5) into an XY joint PDF, it is more convenient to make a change of variables in the transform definition of Equation (5.A.3). With the polar-coordinate substitutions $x = -\rho\cos\phi$, $y = -\rho\sin\phi$, and $dx\, dy =$

$\rho\, d\rho\, d\phi$, Equation (5.A.3) becomes

$$\Phi_{XY}(\upsilon_x, \upsilon_y) = \frac{1}{2\pi} \int_0^{\infty} \int_0^{2\pi} f_R(\rho) \exp(j\upsilon_x \rho \cos\phi) \exp(j\upsilon_y \rho \sin\phi)\, d\phi\, d\rho \tag{5.A.6}$$

Equation (5.A.6) may be grouped:

$$\Phi_{XY}(\upsilon) = \int_0^{\infty} f_R(\rho) \left[\frac{1}{2\pi} \int_0^{2\pi} \exp\left[-j\upsilon\rho\cos(\phi + \phi_0)\right] d\phi \right] d\rho \tag{5.A.7}$$

where $\upsilon = \sqrt{\upsilon_x^2 + \upsilon_y^2}$ and $\tan(\phi_0) = \upsilon_y / \upsilon_x$. The angle ϕ_0 is unimportant, since the integration of ϕ_0 is over the entire period of the cosine function in Equation (5.A.7). Thus, the characteristic function is solely dependent on the variable υ.

The bracketed term in Equation (5.A.7) is a standard definite integral that evaluates to a zero-order Bessel function of the first kind [Gra94]. The final expression for the transformation from envelope PDF to characteristic function is Equation (5.2.3). Using a similar set of reductions, the reverse transformation from characteristic function to envelope PDF becomes Equation (5.2.4). The only assumption made in these transformations is the statistical independence and uniform distribution of the complex voltage phase.

Chapter 6

THE ANGLE SPECTRUM

The concept of a delay spectrum is intuitive: Multipath power arrives with different propagation delays, causing a smear of received power as a function of time delay. A Doppler spectrum is equally intuitive: Moving objects in the channel introduce nonzero Doppler frequencies to selected multipath components.

The concept of a wavenumber spectrum is not so intuitive, as scientists and engineers do not often work in the spectral domain of space. Even in the field of wireless channel modeling, most engineers characterize multipath in space using an *angle* spectrum rather than a wavenumber spectrum. The angle spectrum describes the angle-of-arrival of incoming multipath power. This geometrically intuitive angle spectrum provides an alternative for analyzing spatial channels, which are already complicated by the multidimensional nature of space.

The bulk of this chapter develops the concepts of stochastic local area channel modeling using an angle spectrum. An outline of the contents appears below:

- Section 6.1: Switching between vector and scalar space.
- Section 6.2: Presents definitions for angle spectrum.
- Section 6.3: Introduces the concept of *shape factors*.
- Section 6.4: Calculates shape factors for example channel models.
- Section 6.5: Summary of important concepts.

This chapter emphasizes the use of an azimuthal angle spectrum and multipath *shape factors*, a simplified method for characterizing the complexities of spatial propagation.

6.1 Vector and Scalar Space

Engineers must decide whether they require a wireless channel with simple, scalar position dependency or a full three-dimensional representation. This section presents the basic relationships between these two representations, demonstrating how to convert between the two.

6.1.1 Scalar Collapse of Position Vectors

A function that depends on a three-dimensional position vector may be treated as a function of a scalar position variable by fixing the angular orientation of the spatial dependence. Thus, a full spatial channel, $\tilde{h}(\vec{r})$, may be expressed as a function of scalar position, r, by sacrificing two degrees of freedom in the representation. This reduction is useful for applications requiring only one degree of spatial freedom, such as a linear array of receiver antennas or constant-velocity motion in space.

The first step in converting a channel of vector space, $\tilde{h}(\vec{r})$, to a channel of scalar space, $\tilde{h}(r)$, is to break down the position variable $\vec{r}$ into its constituent Cartesian coordinates:

$$\vec{r} = x\hat{x} + y\hat{y} + z\hat{z}$$

where $\hat{x}$, $\hat{y}$, and $\hat{z}$ are the orthonormal Cartesian basis vectors for three-dimensional free-space. We desire to fix the orientation of the movement in space as a function of a scalar value, r, constraining movement on a line that points towards a direction in azimuth, θ, and an elevation angle above the horizon, φ. In terms of these spherical coordinates (r, θ, and φ), the Cartesian coordinates become

$$\begin{aligned} x &= r\cos\varphi\cos\theta \\ y &= r\cos\varphi\sin\theta \\ z &= r\sin\varphi \end{aligned}$$

Figure 6.1 illustrates these relationships between Cartesian and spherical coordinates. Therefore, we can write

$$\tilde{h}(\vec{r}) = \tilde{h}(r; \theta, \varphi) \tag{6.1.1}$$

where θ and φ are fixed constants in all subsequent analysis. To avoid cumbersome notation, the scalar-distance channel is written as $\tilde{h}(r)$, implying that θ and φ have been set to some arbitrary direction. This vector-to-scalar collapse may be applied to any function that has three dimensional space as one of its dependencies (space–time channels, received signals, and autocorrelation functions, for example).

Example 6.1: Plane Wave Channel

Problem: A static, narrowband channel of the form $\tilde{h}(\vec{r}) = V_0 \exp(-jk_0\vec{r}\cdot\hat{x})$ is a function of three-dimensional space. If azimuth, θ, and elevation, φ, angles are known, calculate the scalar representation, $\tilde{h}(r)$.

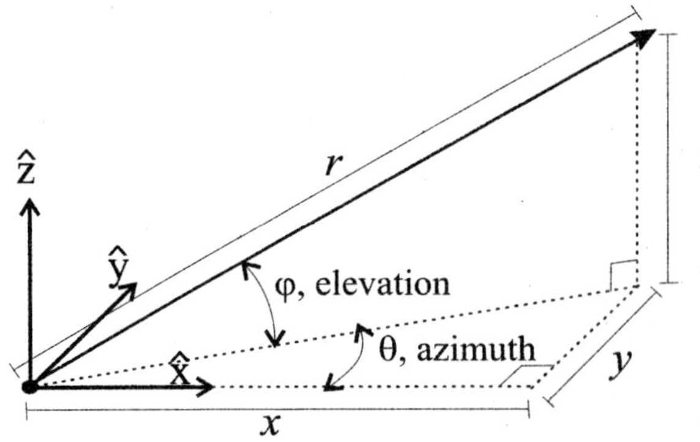

Figure 6.1 Relationship between (x, y, z) and (r, θ, φ) coordinates.

Solution: The vector, $\vec{r}$, may be expressed as $r\hat{r}$, where r is the scalar position and $\hat{r}$ is a unit vector given by

$$\hat{r} = \cos\varphi\cos\theta\,\hat{x} + \cos\varphi\sin\theta\,\hat{y} + \sin\varphi\,\hat{z}$$

Substitution of $\vec{r} = r\hat{r}$ into the given function of $\tilde{h}(\vec{r})$ produces the scalar representation:

$$\tilde{h}(r) = V_0 \exp(-jk_0 r \cos\varphi\cos\theta)$$

Note: Volume Integration in Spherical Coordinates

Integration of a function over all space in spherical coordinates is related to its Cartesian-coordinate integration as follows:

$$\iiint_{-\infty}^{+\infty} f(x, y, z)\, dx\, dy\, dz = \int_{-\frac{\pi}{2}}^{\frac{\pi}{2}} \int_{0}^{2\pi} \int_{0}^{\infty} f(r, \theta, \varphi)\, r^2 \cos(\varphi)\, dr\, d\theta\, d\varphi$$

6.1.2 Scalar Collapse of Wavevectors

A scalar distance dependence implies a scalar Fourier transform dependence as well. For a channel, $\tilde{h}(r)$, the Fourier transform is given by $\tilde{H}(k)$, where k is a wavenumber instead of a wavevector. We then use the standard scalar definition to define the Fourier pairs, $\tilde{h}(r) \longleftrightarrow \tilde{H}(k)$, which are simpler than the relationship between the Fourier transform pairs defined for functions of three-dimensional position.

The relationship between scalar wavenumber Fourier transforms, $\tilde{H}(k)$, and wavevector Fourier transforms, $\tilde{H}(\vec{k})$, is not as simple as the relationship between scalar position and vector position channels. For example, it is tempting to write a wavevector as a function of spherical coordinates:

$$\vec{k} = k\left[\cos\varphi\cos\theta\,\hat{x} + \cos\varphi\sin\theta\,\hat{y} + \sin\varphi\,\hat{z}\right] \tag{6.1.2}$$

and then treat the orientation of azimuth angle θ and elevation angle φ as fixed. This is incorrect, because the scalar Fourier transform is *not* given by

$$\times \quad \textbf{Incorrect:} \quad \tilde{H}(k) = \tilde{H}(k;\theta,\varphi)$$

since this is inconsistent with the position-wavenumber transform definitions discussed in Chapter 2.

The appropriate definition is derived from the vector Fourier relationship after making the substitution $\vec{r} = r\hat{r}$, where $\hat{r}$ is a constant unit vector that points in the direction of scalar position displacement:

$$\tilde{h}(r) = \frac{1}{(2\pi)^3}\int_{-\infty}^{+\infty} \tilde{H}(\vec{k})\exp(jr\vec{k}\cdot\hat{r})\,d\vec{k} \tag{6.1.3}$$

Equation (6.1.3) is then substituted into the *forward* Fourier transform of scalar position (see Table 2.2) to produce

$$\begin{aligned}\tilde{H}(k) &= \frac{1}{(2\pi)^3}\int_{-\infty}^{+\infty}\int_{-\infty}^{+\infty} \tilde{H}(\vec{k})\exp\left(-jr\left[k - \vec{k}\cdot\hat{r}\right]\right)dr\,d\vec{k} \\ &= \frac{1}{(2\pi)^2}\int_{-\infty}^{+\infty} \tilde{H}(\vec{k})\,\delta(k - \vec{k}\cdot\hat{r})\,d\vec{k} \end{aligned} \tag{6.1.4}$$

Equation (6.1.4) is the correct way to convert a wavevector Fourier transform to a wavenumber Fourier transform.

Geometrically, the integration over the delta function of Equation (6.1.4) collects all of the components from the three-dimensional Fourier transform, $\tilde{H}(\vec{k})$, which satisfy

$$k = \vec{k}\cdot\hat{r} \tag{6.1.5}$$

Equation (6.1.5) represents a collection of wavevectors, $\vec{k}$, whose tips lie in a geometrical plane that is perpendicular to the direction of scalar position displacement, $\hat{r}$, and a distance, k, from the origin. Thus, for a given k, all of the values of $\tilde{H}(\vec{k})$ associated with this plane integrate and collapse to a single Fourier transform value in $\tilde{H}(k)$.

The relationship of Equation (6.1.4) also holds for wavevector spectrum, $S_{\tilde{h}}(\vec{k})$, and wavenumber spectrum, $S_{\tilde{h}}(k)$. An example of the vector-to-scalar mapping in the wavevector domain follows as Example 6.2.

Example 6.2: Spherical Shell Spectrum

Problem: For a channel wavevector spectrum that takes the form of a "spherical shell," the spectrum is zero everywhere except for values of $\vec{k}$ that lie on a sphere of radius k_0; the value of the power spectrum on this sphere is constant. Mathematically, this wavevector spectrum takes the form

$$S_{\tilde{h}}(\vec{k}) = \frac{2\pi S_0}{k_0}\delta(|\vec{k}| - k_0)$$

Calculate the *wavenumber* spectrum for any arbitrary orientation in space.

Solution: Start by inserting the three-dimensional wavevector spectrum into Equation (6.1.4) using Cartesian coordinates (k_x, k_y, k_z). Recognize that the wavevector spectrum is *isotropic* (identical regardless of orientation), so that the wavenumber spectrum is identical for all possible values of $\hat{r}$. We set $\hat{r} = \hat{z}$ for convenience:

$$\begin{aligned} S_{\tilde{h}}(k) &= \frac{S_0}{2\pi k_0} \iiint_{-\infty}^{+\infty} \delta\left(\sqrt{k_x^2 + k_y^2 + k_z^2} - k_0\right) \delta(k - k_z)\, dk_x\, dk_y\, dk_z \\ &= \iint_{-\infty}^{+\infty} \delta\left(\sqrt{k_x^2 + k_y^2 + k^2} - k_0\right) dk_x\, dk_y \end{aligned}$$

This double integration is best performed in spherical coordinates (k', θ'). After substituting $k'^2 = k_x^2 + k_y^2$ and $dk_x\, dk_y = k'\, dk'\, d\theta'$, the final wavenumber spectrum results:

$$\begin{aligned} S_{\tilde{h}}(k) &= \frac{S_0}{2\pi k_0} \int_0^{+\infty} \int_0^{2\pi} k' \delta\left(\sqrt{k'^2 + k^2} - k_0\right) d\theta'\, dk' \\ &= S_0 \mathrm{u}(k_0 - |k|) \end{aligned}$$

The wavenumber spectrum is a constant value, S_0, over the range $-k_0 \leq k \leq +k_0$ and is zero elsewhere.

6.2 Angle Spectrum Concepts

The *angle spectrum* is one of the most important concepts in spatial channel modeling, since it provides a simpler, more intuitive way to characterize a wavevector spectrum. This section rigorously defines the angle spectrum and discusses its relationship to other spectra used in channel modeling.

6.2.1 Definition of the Angle Spectrum

Recall from Equation (4.4.13) in Section 4.4.8 that for the U-SLAC model, the wavevector spectrum for a static, narrowband channel may be written as

$$S_{\tilde{h}}(\vec{k}) = (2\pi)^3 \sum_{i=1}^{N} P_i \delta(\vec{k} - \vec{k}_i) \tag{6.2.1}$$

where P_i is the power of the ith multipath component ($P_i = V_i^2$, units of *Volts*^{2}). In fact, Equation (6.2.1) *can only* represent the power spectrum of a U-SLAC model - power spectra are only defined for WSS processes, and a SLAC model is WSS if and only if it is a U-SLAC model (Theorem 4.2).

By definition, if Equation (6.2.1) represents a stochastic local area channel model, then all values of $\vec{k}_i$ must have a magnitude equal to the free-space wavenumber. For power spectrum analysis, we denote the free-space wavenumber as k_0, since this represents the magnitude of the maximum possible wavenumber for which a wavenumber spectrum has nonzero components. Recall from Chapter 4 that the value of k_0 is equal to $\frac{2\pi}{\lambda}$, where λ is the wavelength of carrier-frequency radiation.

Now the vectors $\vec{k}_i$ and $\vec{k}$ in Equation (6.2.1) may be expressed in terms of spherical coordinates involving azimuth and elevation angles. The relationships between the spherical coordinates and the vectors are given below:

$$\begin{aligned}
\vec{k}_i &= k_0 \left(\cos\varphi_i \cos\theta_i\, \hat{x} + \cos\varphi_i \sin\theta_i\, \hat{y} + \sin\varphi_i\, \hat{z}\right) \\
\vec{k} &= |\vec{k}| \left(\cos\varphi \cos\theta\, \hat{x} + \cos\varphi \sin\theta\, \hat{y} + \sin\varphi\, \hat{z}\right)
\end{aligned}$$

The azimuth angle, θ_i, and the elevation angle, φ_i, may be thought of as the *angle-of-arrival* coordinates of the ith multipath. With these two expressions for $\vec{k}_i$ and $\vec{k}$, it is possible to rewrite the delta function term in Equation (6.2.1) using the following identity:

$$\delta(\vec{k} - \vec{k}_i) = \frac{\delta(|\vec{k}| - k_0)\delta(\varphi - \varphi_i)\delta(\theta - \theta_i)}{k_0^2 \cos\varphi_i} \tag{6.2.2}$$

Thus, the original wavevector spectrum may now be written as

$$S_{\tilde{h}}(\vec{k}) = \frac{(2\pi)^3 \delta(|\vec{k}| - k_0)}{k_0^2} \underbrace{\sum_{i=1}^{N} \frac{P_i \delta(\varphi - \varphi_i)\delta(\theta - \theta_i)}{\cos\varphi_i}}_{p(\theta,\varphi)} \tag{6.2.3}$$

The underbraced term on the right-hand side of Equation (6.2.3) is the *angle spectrum* of the multipath power.

There is a key principle in channel modeling that results from Equation (6.2.3): *Any U-SLAC wavevector spectrum may be completely described in terms of an angle spectrum,* $p(\theta, \varphi)$. To understand the advantages of an angle spectrum representation, consider the simplified form of Equation (6.2.3):

$$S_{\tilde{h}}(\vec{\mathrm{k}}) = \frac{(2\pi)^3 \delta(|\vec{\mathrm{k}}| - k_0)}{k_0^2} p(\theta, \varphi) \qquad (6.2.4)$$

There are several key advantages to using an angle spectrum to represent spatial propagation:

- *The angle spectrum is physically intuitive.* The units of $p(\theta, \varphi)$ are *power per steradian.* The angles θ and φ represent the respective azimuth and elevation angles-of-arrivals of radio waves that impinge upon the receiver antenna. Compared to a wavevector spectrum, the angle spectrum provides a closer physical representation to how engineers visualize radio wave propagation.
- *The full angle spectrum has only two dependencies.* The wavevector spectrum depends on a three-dimensional wavevector, $\vec{\mathrm{k}}$. However, a U-SLAC model will only have spectral components for $|\vec{\mathrm{k}}| = k_0$, representing a sphere in $\vec{\mathrm{k}}$-space. By casting the dependencies in terms of azimuth and elevation angles-of-arrival, the spatial propagation may be completely described with one fewer dependency - a convenient simplification.
- *Diffuse propagation is best represented with an angle spectrum.* Recall that whenever a discrete sum of impulse functions were used to represent a spectrum in Chapter 4, we had to mention that a diffuse continuum of waves was possible if the summation contained an infinite number of terms and the amplitudes of each wave became infinitesimally small. We no longer require such a disclaimer. The angle spectrum representation, $p(\theta, \varphi)$, may contain impulses (specular components) and intervals of finite or continuous power (diffuse, nonspecular components) as shown in Figure 6.2.

Also note that an angle spectrum may be used to describe the original propagation model in Equation (4.4.1) with random *voltage amplitudes* as well as random phases. In other words, Equation (6.2.4) is also valid for certain types of stochastic macro area channel (SMAC) analysis, although most of our applications are concerned with SLAC analysis.

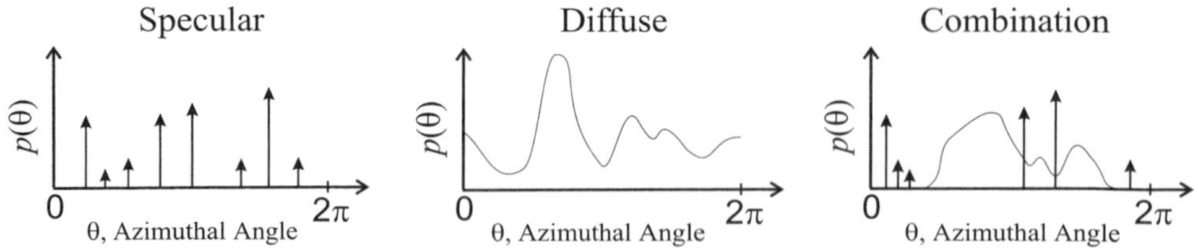

Figure 6.2 Multipath power in an angles spectrum can be specular, diffuse, or a combination.

6.2.2 Mapping Angles to Wavenumbers

Recall from Section 6.1.2 that characterization of small-scale fading along a particular direction in space required a one-dimensional *wavenumber spectrum*, $S_{\tilde{h}}(k)$. Given a direction of characterization in space, $\hat{\mathrm{r}}$, the wavenumber spectrum may be calculated from the wavevector spectrum:

$$S_{\tilde{h}}(k) = \frac{1}{(2\pi)^2} \int_{-\infty}^{+\infty} S_{\tilde{h}}(\vec{\mathrm{k}})\delta(k - \vec{\mathrm{k}} \cdot \hat{\mathrm{r}})\, d\vec{\mathrm{k}} \qquad (6.2.5)$$

which is the PSD version of Equation (6.1.4). By substituting Equation (6.2.4) and simplifying, the wavenumber spectrum may be calculated directly from the angle spectrum:

$$S_{\tilde{h}}(k) = 2\pi \int_{0}^{2\pi} \int_{-\frac{\pi}{2}}^{+\frac{\pi}{2}} p(\theta, \varphi)\, \delta(k - k_0 \hat{\mathrm{k}} \cdot \hat{\mathrm{r}}) \cos\varphi \, d\varphi \, d\theta \qquad (6.2.6)$$

where

$$\hat{\mathrm{k}} = \cos\varphi \cos\theta \, \hat{\mathrm{x}} + \cos\varphi \sin\theta \, \hat{\mathrm{y}} + \sin\varphi \, \hat{\mathrm{z}}$$

Thus, Equation (6.2.6) is a useful relationship for calculating a wavenumber spectrum from an angle spectrum. As Figure 6.3 illustrates, the mapping depends on the orientation, $\hat{\mathrm{r}}$, in Equation (6.2.6).

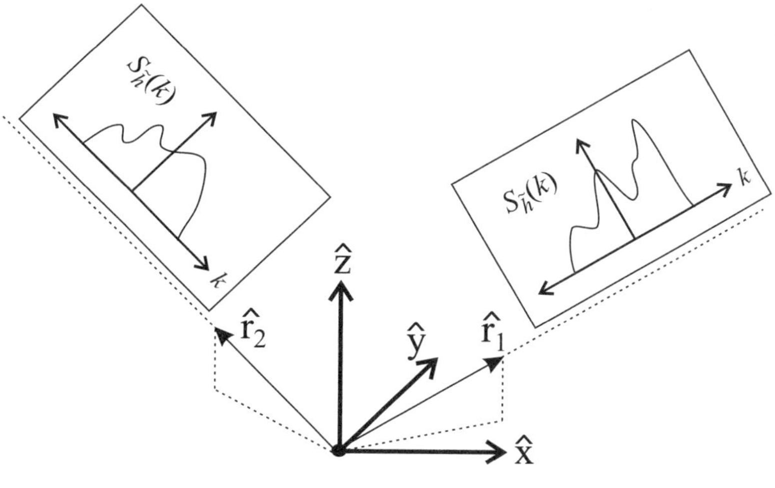

Figure 6.3 Different wavenumber spectra, $S_{\tilde{h}}(k)$, depend on the observation orientation in space ($\hat{\mathrm{r}}_1$ or $\hat{\mathrm{r}}_2$).

6.2.3 From-the-Horizon Propagation

An *azimuthal* channel model is of particular interest in wireless communications. Such a channel model consists of multipath that propagates from the horizon with zero elevation angle. The angle spectrum for this type of channel may be written as

$$p(\theta, \varphi) = p(\theta)\,\delta(\varphi) \tag{6.2.7}$$

where $p(\theta)$ is the azimuthal angle spectrum, with units of *power per radian*. Besides being a useful simplification, the azimuthal channel model of Equation (6.2.7) is an excellent approximation to most terrestrial propagation scenarios [Cla68].

Given an azimuthal angle spectrum, $p(\theta)$, it is possible to calculate a wavenumber spectrum by substituting Equation (6.2.7) into Equation (6.2.6) and integrating out the φ variable:

$$S_{\tilde{h}}(k) = 2\pi \int_{0}^{2\pi} p(\theta)\,\delta(k - k_0 \cos[\theta - \theta_R])\,d\theta \tag{6.2.8}$$

where θ_R is the azimuthal orientation of observation. The resulting integration produces one of the most famous formulas in wireless communications, the Gans mapping for angle spectra [Gan72]:

$$S_{\tilde{h}}(k) = \frac{2\pi}{\sqrt{k_0^2 - k^2}}\left[p\left(\theta_R + \cos^{-1}\frac{k}{k_0}\right) + p\left(\theta_R - \cos^{-1}\frac{k}{k_0}\right)\right], \quad |k| \le k_0 \tag{6.2.9}$$

The Gans mapping assumes that the orientation vector, $\hat{\mathrm{r}}$, of Equation (6.2.6) also points towards the horizon and may be written in the form

$$\hat{\mathrm{r}} = \cos\theta_R\,\hat{\mathrm{x}} + \sin\theta_R\,\hat{\mathrm{y}} + 0\,\hat{\mathrm{z}}$$

This assumption is not very restrictive. Most motion by a mobile receiver, for example, is going to be towards the horizon with virtually no vertical $\hat{\mathrm{z}}$-component. In terrestrial systems, receivers with multiple antennas also tend to distribute their array or diversity antennas in the same azimuthal plane, eliminating the need to model the $\hat{\mathrm{z}}$ dimension.

> *Note: Off-Azimuth Observation*
>
> Although it is not common to study the case of off-azimuth movement ($\varphi_R \neq 0$) for propagation from the horizon, it is fairly easy to modify Equation (6.2.9) to do so. First, calculate the wavenumber spectrum $S_{\tilde{h}}(k)$ from Equation (6.2.9) as if observation was in the horizon. Then, simply modify the spectrum by $S'_{\tilde{h}}(k) = S_{\tilde{h}}(k)\cos(\varphi_R)$. Keep in mind that this simple modification is only possible when the *propagation* is from the horizon, as in Equation (6.2.7).

The mapping of Equation (6.2.9) has a straightforward physical interpretation from the geometry of propagation shown in Figure 6.4. A multipath wave arrives from the horizon at angle θ, and the direction of azimuthal motion that we

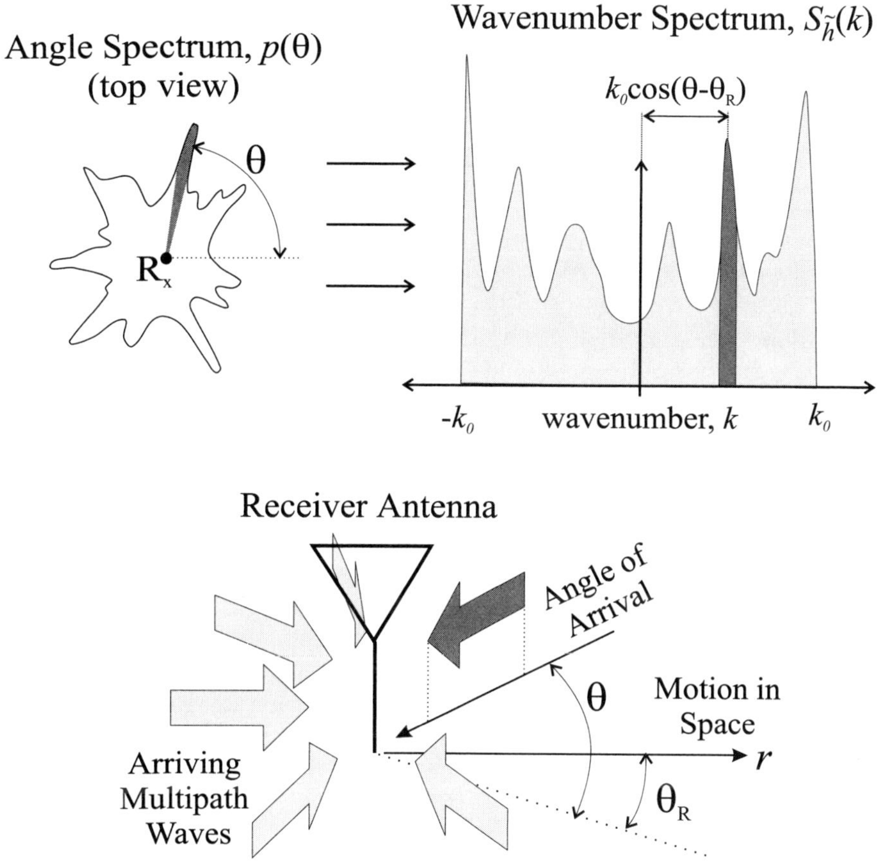

Figure 6.4 Multipath power is mapped from the angle spectrum, $p(\theta)$, to the wavenumber spectrum, $S_{\tilde{h}}(k)$, as a function of its angle-of-arrival [Dur00a].

wish to map is θ_R. The phase progression of this multipath wave is the free-space wavenumber, k_0. However, to a receiver moving along the θ_R direction, the actual wavenumber, k, appears to be foreshortened by a factor $\cos(\theta - \theta_R)$. Thus,

$$k = k_0 \cos(\theta - \theta_R) \quad \text{and} \quad \frac{dk}{d\theta} = -k_0 \sin(\theta - \theta_R) \tag{6.2.10}$$

We can arrive at Equation (6.2.9) equivalently by equating the wavevector spectrum power to the angle spectrum power: $S(k)|dk| = 2\pi p(\theta)|d\theta|$. The mapping of Equation (6.2.9) provides a useful bridge between spatial selectivity and multipath angle-of-arrival characteristics. Example 6.3 illustrates the well-known mapping from omnidirectional azimuth spectrum to a wavenumber spectrum.

Example 6.3: Omnidirectional Mapping

Problem: In a cluttered multipath environment it is common to approximate the angle spectrum of incoming multipath power as a uniform distribution:

$$p(\theta) = \frac{P_T}{2\pi} \tag{6.2.11}$$

where P_T - the total average power - is a constant. Calculate the wavenumber spread for this angle spectrum.

Solution: Using Equation (6.2.9), the wavenumber spectrum for this propagation scenario is

$$S(k) = \frac{2P_T}{\sqrt{k_0^2 - k^2}}, \quad |k| \leq k_0 \tag{6.2.12}$$

which is independent of orientation, θ_R. The wavenumber spread is

$$\sigma_k^2 = \frac{k_0^2}{2} \tag{6.2.13}$$

Note: Top View of Multipath Propagation

In our discussion of the angle spectrum, we will often use a top view sketch of multipath propagation, such as the one shown in Figure 6.4. This plot, with receiver antenna as the center point, represents an overhead view of multipath power angle-of-arrival, much like a polar plot of an azimuth spectrum.

6.2.4 Summary of Angle Spectrum Concepts

To summarize the concepts of this section, Figure 6.5 shows a section of the space transform map, similar to Figure 3.3 in Section 3.3. This mapping diagram has been modified to include angle spectra and wavevector spectra, emphasizing how each relates to one another. An angle spectrum is simply an alternate way to represent a wavevector spectrum. Any local area channel model that has a wavevector spectrum can, by definition, be characterized by an angle spectrum. An angle spectrum is a simpler and more intuitive way to characterize spatial selectivity.

Wavenumber spectra may be calculated from the angle spectrum just as they are calculated from the wavevector spectrum. A direction in space, $\hat{\mathrm{r}}$, must be assumed before a wavenumber spectrum can be calculated. The wavenumber spectrum is the Fourier transform of a one-dimensional spatial autocorrelation. This scalar spatial autocorrelation function may be calculated by fixing the orientation of the full spatial autocorrelation function, $C_{\tilde{h}}(\Delta\vec{\mathrm{r}})\,|_{\Delta\vec{\mathrm{r}}=\Delta r\hat{\mathrm{r}}}$. The wavenumber, wavevector, and angle spectrum relationships in the bottom part of Figure 6.5 are summarized here:

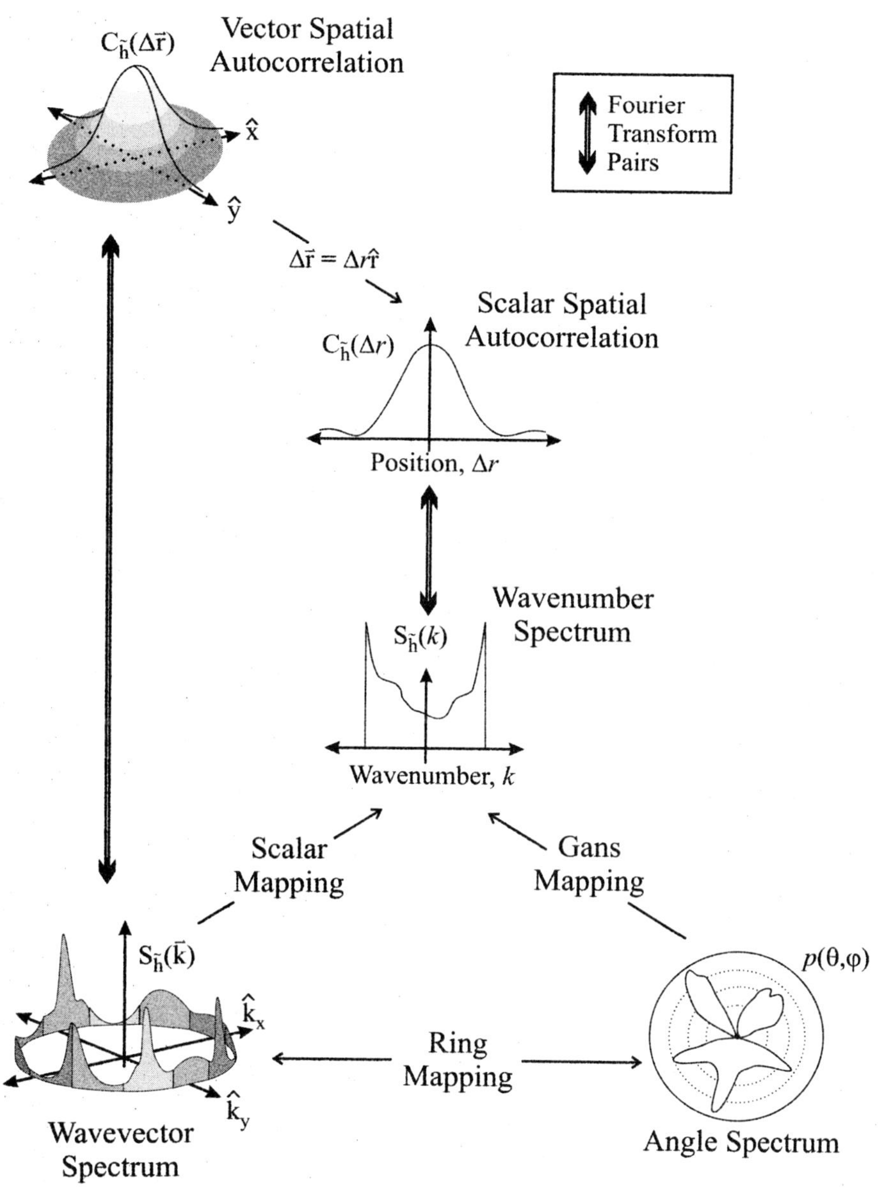

Figure 6.5 Autocorrelation and spectrum relationships for the space-varying channel.

- **Ring Mapping:** between wavevector and angle spectrum, Equation (6.2.3)
- **Scalar Mapping:** wavevector to wavenumber spectrum, Equation (6.2.5)
- **Gans Mapping:** angle to wavenumber spectrum, Equation (6.2.6) or (6.2.9)

This set of relationships allows engineers to use the conceptually simple angle spectrum to characterize multipath power.

The only drawback in using an angle spectrum is that angle-of-arrival is not the natural domain for relating spectral properties to spatial selectivity in a channel. Therefore, the angle spectrum must be converted to a wavevector spectrum or wavenumber spectrum whenever spatial autocorrelation functions, spectral spreads, or duality results are calculated. Remaining analysis in this chapter will mostly discuss the popular from-the-horizon channel model in detail. Indeed, for this channel model, it is possible to overcome the key drawback of an angle spectrum by using *shape factors* to relate the geometry of propagation to the properties of spatial selectivity.

6.3 Multipath Shape Factors

This section defines and describes multipath *shape factors* - geometrical characteristics of an azimuth angle spectrum that also describe small-scale spatial selectivity in a channel.

6.3.1 Definition of Shape Factors

This section presents the three multipath shape factors that characterize small-scale fading statistics in space. The shape factors are derived from the angular distribution of multipath power, $p(\theta)$, which is a general representation of from-the-horizon propagation in a local area. This representation of $p(\theta)$ includes antenna gains and polarization mismatch effects [Dur98a]. Shape factors are based on the complex Fourier coefficients of $p(\theta)$:

$$F_n = \int_0^{2\pi} p(\theta)\exp(jn\theta)d\theta \tag{6.3.1}$$

where F_n is the nth complex Fourier coefficient.

Angular Spread

The shape factor *angular spread*, Λ, is a measure of how multipath concentrates about a single azimuthal direction [Dur98a]. We define angular spread to be

$$\Lambda = \sqrt{1 - \frac{|F_1|^2}{F_0^2}} \tag{6.3.2}$$

where F_0 and F_1 are defined by Equation (6.3.1). There are several advantages to defining angular spread in this manner. First, since angular spread is normalized by F_0 (the total amount of local average received power), it is invariant under changes in transmitted power. Second, Λ is invariant under any series of rotational or reflective transformations of $p(\theta)$. Finally, this definition is intuitive; angular spread ranges from 0 to 1, with 0 denoting the extreme case of a single multipath component from a single direction and 1 denoting no clear bias in the angular distribution of received power. Angular spread values for several cases of multipath azimuth spectra are shown in Figure 6.6.

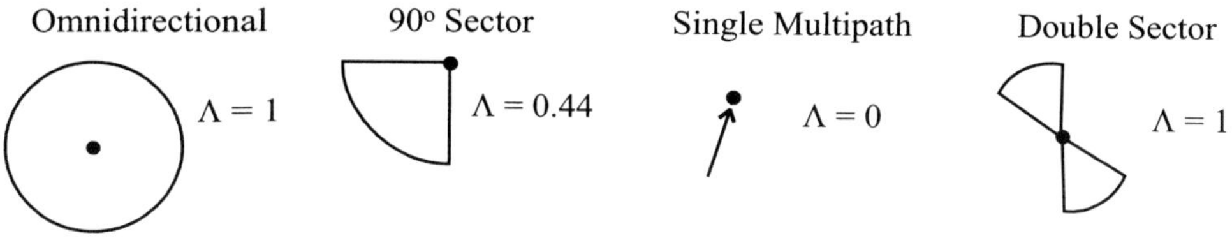

Figure 6.6 Examples of multipath angular spread values, Λ.

It should be noted that other definitions exist in the literature for angular spread. These definitions involve either beamwidth or the second centered moment of θ and are often ill-suited for general application to periodic functions such as $p(\theta)$.

Angular Constriction

The shape factor *angular constriction*, γ, is a measure of how multipath concentrates about *two* azimuthal directions. We define angular constriction to be

$$\gamma = \frac{|F_0 F_2 - F_1^2|}{F_0^2 - |F_1|^2} \tag{6.3.3}$$

where F_0, F_1, and F_2 are defined by Equation (6.3.1). Much like the definition of angular spread, the measure for angular constriction is invariant under changes in transmitted power or any series of rotational or reflective transformations of $p(\theta)$. The possible values of angular constriction, γ, range from 0 to 1, with 0 denoting no clear bias in two arrival directions and 1 denoting the extreme case of exactly two multipath components arriving from different directions. Values of angular constriction for several multipath azimuth spectra are shown in Figure 6.7.

> *Note: Memory Aids*
>
> The use of Λ and γ to denote angular spread and constriction respectively is not a gratuitous or pretentious use of Greek letters for notation - they were originally used as memory aids. The Λ denotes a spread, since its alphabetical character resembles a sector of multipath in azimuth. The γ resembles constriction, like a channel "tied into a knot."

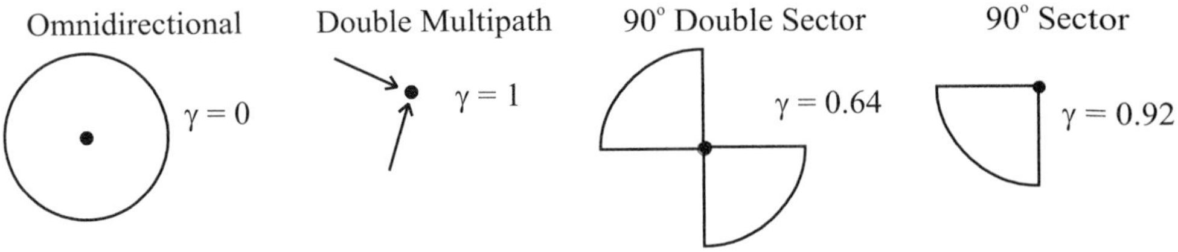

Figure 6.7 Examples of multipath angular constriction values, γ.

Azimuthal Direction of Maximum Fading

A third shape factor, which may be thought of as an orientation parameter, is the *azimuthal direction of maximum fading*, $\theta_{\max}$. We define this parameter to be

$$\theta_{\max} = \frac{1}{2}\arg\left\{F_0F_2 - F_1^2\right\} \tag{6.3.4}$$

The physical meaning of the parameter is presented in the next section.

6.3.2 Basic Wavenumber Spread Relationship

Shape factors have a particularly useful application when describing small-scale fading: the multipath angle-of-arrival dependence of wavenumber spread may be cast exclusively in terms of the three shape factors angular spread, angular constriction, and direction of maximum fading. It has been shown that the wavenumber spread for the complex voltage of a receiver traveling along the azimuthal direction θ_R is

$$\sigma_k^2 = \frac{2\pi^2\Lambda^2}{\lambda^2}\left(1 + \gamma\cos\left[2(\theta_R - \theta_{\max})\right]\right) \tag{6.3.5}$$

where λ is the wavelength of the carrier frequency [Dur00b]. The value σ_k^2 describes the spatial selectivity of a channel in a local area for a receiver moving in the θ_R direction. Equation (6.3.5) is valid for any channel in which multipath waves arrive at the receiver from the horizon - a common assumption when describing mobile radio propagation. The next section discusses the unique aspects of small-scale fading behavior described by each shape factor.

> *Note: A Great Simplification*
>
> At first glance, it is difficult to appreciate how much simpler Equation (6.3.5) makes the calculation of wavenumber spreads. Consider the following stepwise procedure for calculating σ_k without shape factors: (1) choose a direction, θ_R, (2) calculate the wavenumber spectrum using the Gans mapping, (3) calculate the moments of the wavenumber spectrum, $\bar{k}$ and $\bar{k^2}$, (4) use the moments to find σ_k, and (5) for information in another spatial orientation, θ_R, repeat this procedure. Equation (6.3.5), however, uses just three shape factor constants to bypass all this procedure for *every possible orientation*, θ_R.

6.3.3 Comparison to Omnidirectional Propagation

Applying the three shape factors, Λ, γ, and $\theta_{\max}$, to the classical omnidirectional propagation model, we find that there is not a bias in either one or two directions of angle-of-arrival, leading to maximum angular spread ($\Lambda = 1$) and minimum angular constriction ($\gamma = 0$). The statistics of omnidirectional propagation are *isotropic*, exhibiting no dependence on the azimuthal direction of receiver travel, θ_R.

If the rate variance relationship of Equation (6.3.5) is normalized against their values for omnidirectional propagation, then they reduce to the following form:

$$\begin{aligned}\sigma^2(\theta_R) &= \frac{\sigma_k^2(\theta_R)}{\sigma_{k_{\text{omni}}}^2} \\ &= \Lambda^2\left(1 + \gamma\cos\left[2(\theta_R - \theta_{\max})\right]\right) \qquad (6.3.6)\end{aligned}$$

where σ^2 is a normalized wavenumber spread. Equation (6.3.6) provides a convenient way to analyze the effects of the shape factors on the second-order statistics of small-scale fading.

First, notice that angular spread, Λ, describes the *average* fading rate within a local area. The average spatial fading rate is proportionally higher in local areas with high angular spreads. A convenient way of viewing this effect is to consider the fading rate variance taken along two perpendicular directions within the same local area. From Equation (6.3.6), the average of the two fading rate variances, regardless of the orientation of the measurement, is always given by

$$\frac{1}{2}\left[\sigma^2(\theta_R) \;+\; \sigma^2(\theta_R + \pi/2)\right] = \Lambda^2 \qquad (6.3.7)$$

Equation (6.3.7) shows that the typical fading rates within a local area decrease with respect to omnidirectional propagation as multipath power becomes more and more concentrated about a single azimuthal direction.

Second, notice that angular constriction, γ, does not affect the average fading rate within a local area, but describes the variability of fading rates taken along different azimuthal directions, θ_R. From Equation (6.3.6), fading rate variance σ_r^2 will change as a function of θ_R, but will always fall within the following range:

$$\sqrt{1-\gamma} \leq \frac{\sigma(\theta_R)}{\Lambda} \leq \sqrt{1+\gamma} \qquad (6.3.8)$$

The upper limit of Equation (6.3.8) corresponds to a receiver traveling in the azimuthal direction of maximum fading ($\theta_R = \theta_{\max}$), while the lower limit corresponds to travel in a perpendicular direction ($\theta_R = \theta_{\max} + \pi/2$). Equation (6.3.8) clearly shows that the variability of fading rates within the same local area increases as the channel becomes more and more constricted. The physical effects of all shape factors are illustrated in Figure 6.8.

It is interesting to note that the propagation mechanisms of a channel are not uniquely described by the three shape factors Λ, γ, and θ_{max}. An infinitum of propagation mechanisms exist, which may have the same set of shape factors and, by extension, lead to channels that exhibit nearly the same end-to-end performance. In fact, Equation (6.3.6) provides rigorous mathematical criteria for a multipath channel that may be treated as "pseudo-omnidirectional":

$$|F_1|, |F_2| \ll F_0 \tag{6.3.9}$$

Under the condition of Equation (6.3.9), angular spread becomes approximately 1 and angular constriction becomes approximately 0. Thus, the second-order statistics of the channel behave almost identically to the classical omnidirectional channel.

6.4 Illustrative Examples

This section presents four different analytical examples of multipath propagation channels, summarized in Figure 6.9, that provide insight into the shape factor definitions and how they describe spatial fading rate variances [Dur99b]. Each example is accompanied by a graph of the angular spread and angular constriction and a graph showing the average and limiting cases of fading rate variance with respect to omnidirectional propagation.

6.4.1 Two-Wave Channel Model

Consider the simplest small-scale fading situation where two constant-amplitude multipath components, with individual powers defined by P_1 and P_2, arrive at a mobile receiver separated by an azimuthal angle α. Figure 6.9 illustrates this angular distribution of power, which is mathematically defined as

$$p(\theta) = P_1\, \delta(\theta - \theta_o) + P_2\, \delta(\theta - \theta_o - \alpha) \tag{6.4.1}$$

where θ_o is an arbitrary offset angle and $\delta(\cdot)$ is an impulse function. By applying Equation (6.3.2), Equation (6.3.3), and Equation (6.3.4), the expressions for Λ, γ, and θ_{max} for this distribution are

$$\Lambda = \Delta \sin\frac{\alpha}{2}, \quad \gamma = 1, \quad \theta_{\text{max}} = \theta_0 + \frac{\alpha + \pi}{2} \quad \left(\text{Note: } \Delta = \frac{2\sqrt{P_1 P_2}}{P_1 + P_2}\right) \tag{6.4.2}$$

The angular constriction, γ, is always 1 because the two-wave model represents perfect clustering about two directions. The limiting case of two multipath components arriving from the same direction ($\alpha = 0$) results in an angular spread, Λ, of 0. An angular spread of 1 results only when two multipath of identical powers ($P_1 = P_2$) are separated by $\alpha = 180°$. Figure 6.10 shows how the fading behavior changes as multipath separation angle, α, increases for the case of two equal-powered waves. Thus, increasing α changes a channel with low spatial selectivity into a channel with high spatial selectivity that exhibits a strong dependence on the azimuthal direction of receiver motion.

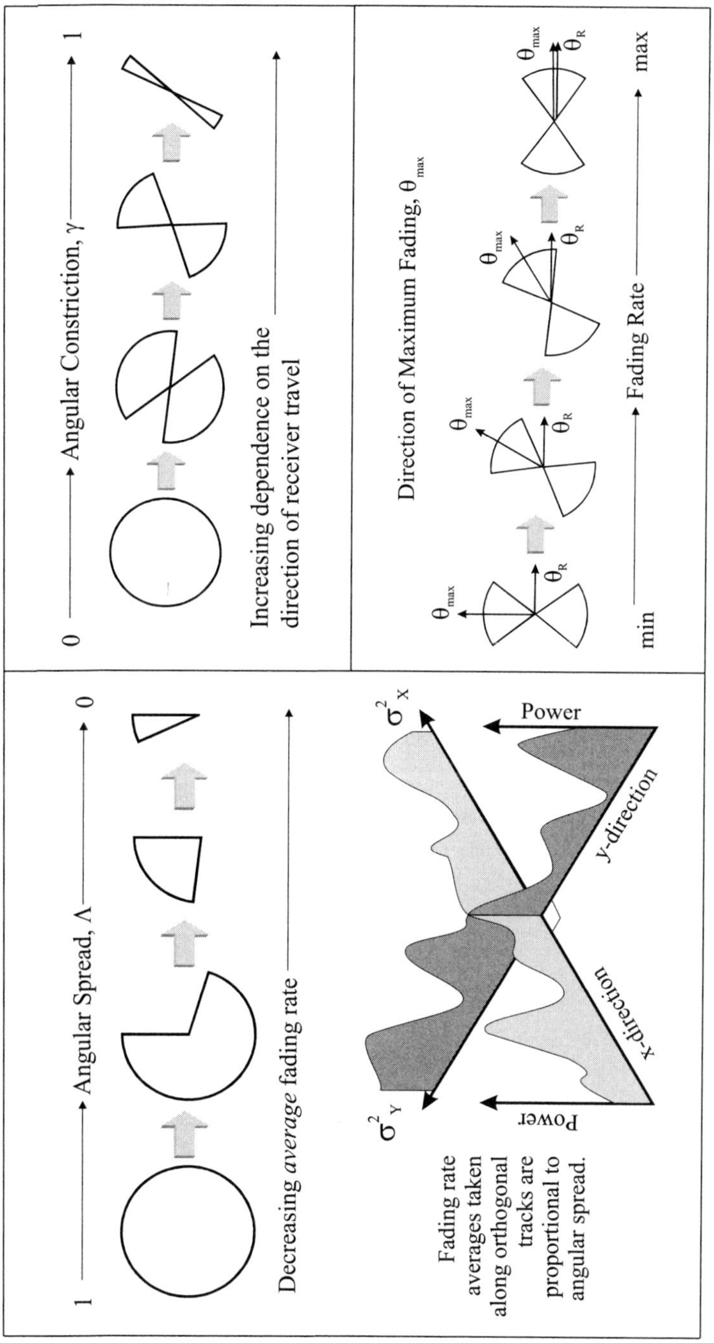

Figure 6.8 A graphical summary of shape factor behavior.

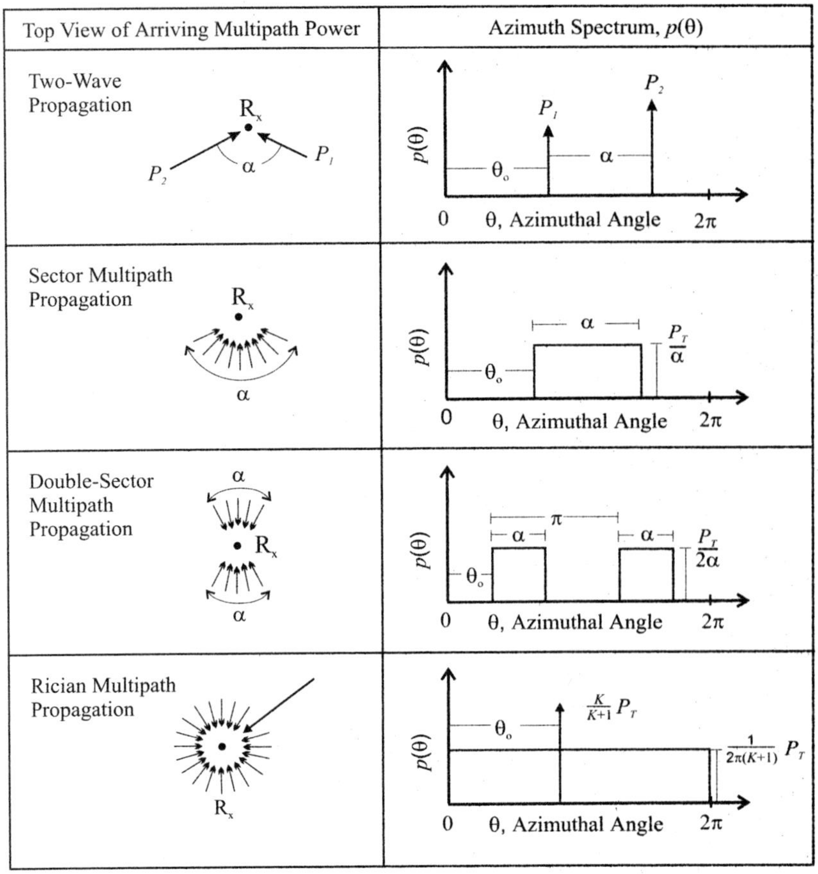

Figure 6.9 Overview of multipath azimuth spectrum studied in this section [Dur00b].

Note: Together at Last

The shape factors allow the characterization of second-order statistics for spatially selective channels. Recall from Chapter 5 that a PDF can be found to calculate first-order statistics. The two-wave channel model, for example, follows a two-wave distribution with parameter Δ. Between shape factors and reduced-wave grouping PDFs, it is possible to calculate most of the important statistics for spatial fading.

6.4.2 Sector Channel Model

Consider another theoretical situation where multipath power is arriving continuously and uniformly over a range of azimuth angles. This model has been used to describe propagation for directional receiver antennas with a distinct azimuthal

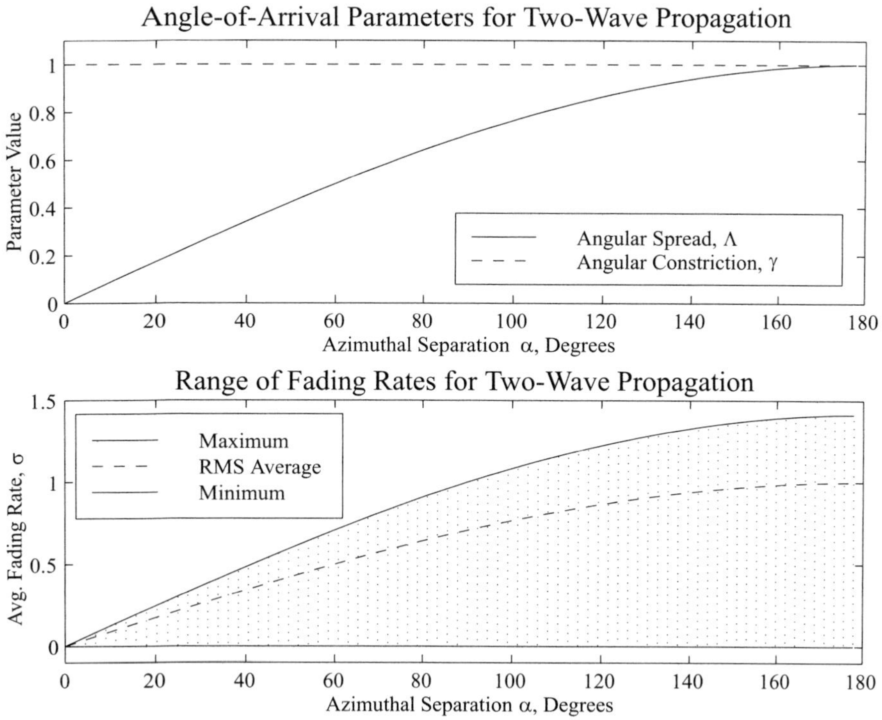

Figure 6.10 Two-wave propagation model. Graphs are for the special case of $P_1 = P_2$ [Dur00b].

beam [Gan72]. The function $p(\theta)$ will be defined by

$$p(\theta) = \begin{cases} \frac{P_T}{\alpha} & : \ \theta_o \leq \theta \leq \theta_o + \alpha \\ 0 & : \ \text{elsewhere} \end{cases} \tag{6.4.3}$$

The angle α indicates the width of the sector (in radians) of arriving multipath power, and the angle θ_o is an arbitrary offset angle, as illustrated by Figure 6.9. By applying Equation (6.3.2), Equation (6.3.3), and Equation (6.3.4), the expressions for Λ, γ, and $\theta_{\max}$ for this distribution are

$$\Lambda = \sqrt{1 - \text{sn}^2\left(\frac{\alpha}{2\pi}\right)}, \qquad \gamma = \frac{\text{sn}^2\left(\frac{\alpha}{2\pi}\right) - \text{sn}\left(\frac{\alpha}{\pi}\right)}{1 - \text{sn}^2\left(\frac{\alpha}{2\pi}\right)}, \qquad \theta_{\max} = \theta_0 + \frac{\alpha + \pi}{2} \tag{6.4.4}$$

where $\text{sn}(\cdot)$ is the sinc function. The limiting cases of these parameters and Equation (6.3.5) provide deeper understanding of angular spread and constriction.

Figure 6.11 graphs the spatial channel parameters, Λ and γ, as a function of sector width, α. The limiting case of a single multipath arriving from precisely one direction corresponds to $\alpha = 0$, which results in the minimum angular spread of

$\Lambda = 0$. The other limiting case of uniform illumination in all directions corresponds to $\alpha = 360°$ (omnidirectional Clarke model), which results in the maximum angular spread of $\Lambda = 1$. The angular constriction, γ, follows an opposite trend. It is at a maximum ($\gamma = 1$) when $\alpha = 0$ and at a minimum ($\gamma = 0$) when $\alpha = 360°$. The graph in Figure 6.11 shows that as the multipath angles-of-arrival are condensed into a smaller and smaller sector, the directional dependence of fading rates within the same local area increases. Overall, however, fading rates decrease with decreasing sector size α.

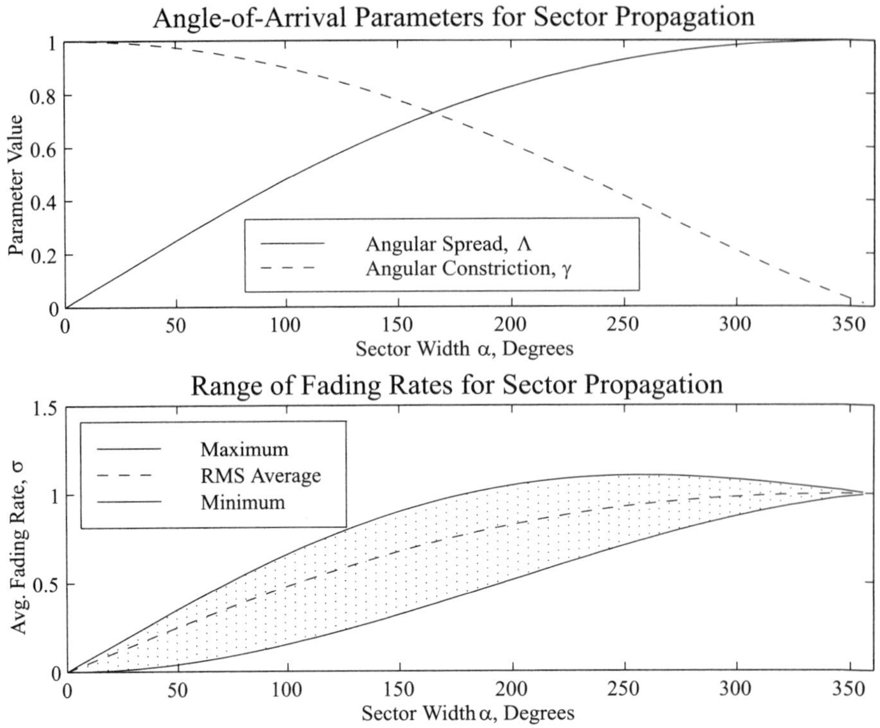

Figure 6.11 Multipath sector propagation model [Dur00b].

6.4.3 Double-Sector Channel Model

Another example of angular constriction may be studied using the double-sector multipath model. Diffuse multipath propagation over two equal and opposite sectors of azimuthal angles characterize the incoming power. The equation that describes this angular distribution of power is

$$p(\theta) = \begin{cases} \frac{P_T}{2\alpha} & : \quad \theta_o \leq \theta \leq \theta_o + \alpha, \quad \theta_o + \pi \leq \theta \leq \theta_o + \alpha + \pi \\ 0 & : \quad \text{elsewhere} \end{cases} \tag{6.4.5}$$

The angle α is the sector width, and the angle θ_o is an arbitrary offset angle. By applying Equation (6.3.2), Equation (6.3.3), and Equation (6.3.4), the expressions for Λ, γ, and $\theta_{\max}$ for this distribution are

$$\Lambda = 1, \qquad \gamma = \frac{\sin\alpha}{\alpha}, \qquad \theta_{\max} = \theta_0 + \frac{\alpha}{2} \tag{6.4.6}$$

Note that the value of angular spread, Λ, is always 1. Regardless of the value of α, an equal amount of power arrives from opposite directions, producing no clear bias in the direction of multipath arrival.

The limiting case of $\alpha = 180°$ (omnidirectional propagation) results in an angular constriction of $\gamma = 0$. As α decreases, the angular distribution of power becomes more and more constricted. In the limit of $\alpha = 0$, the value of angular constriction reaches its maximum, $\gamma = 1$. This case corresponds to the above-mentioned instance of two-wave propagation. Figure 6.12 shows how the fading behavior changes as sector width α increases, making the fading rate more and more isotropic while the RMS average remains constant.

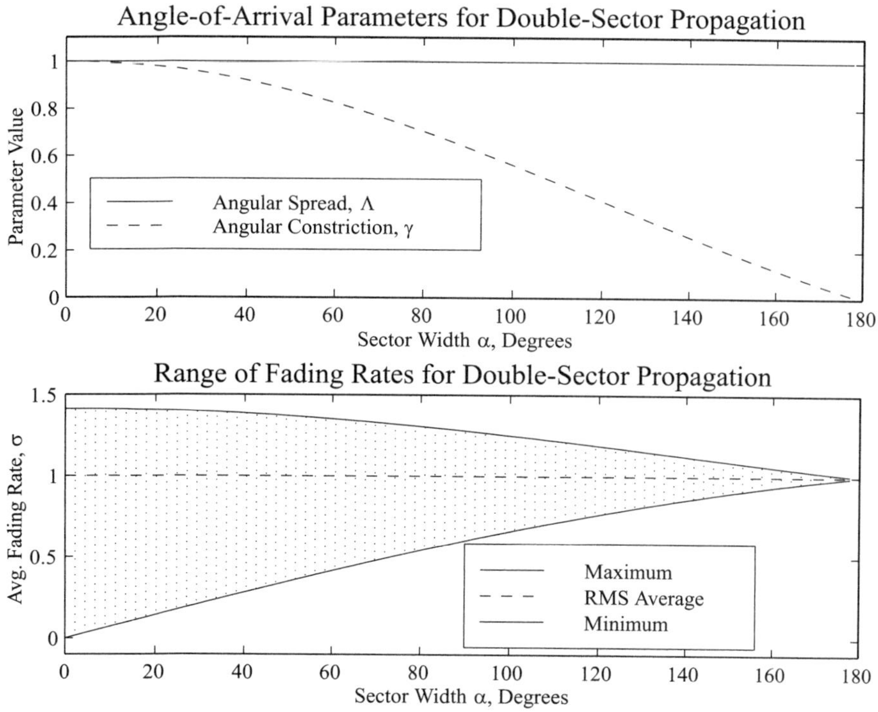

Figure 6.12 Multipath double-sector propagation model [Dur00b].

6.4.4 Rician Channel Model

A Rician channel model results from the addition of a single plane wave and numerous diffusely scattered waves [Ric48]. If the power of the scattered waves is assumed to be evenly distributed in azimuth, then the channel may be modeled by the following $p(\theta)$:

$$p(\theta) = \frac{P_T}{2\pi(K+1)}[1 + 2\pi K\delta(\theta - \theta_0)] \tag{6.4.7}$$

where K is the ratio of specular to diffuse nonspecular power, often referred to as the Rician K-factor. By applying Equation (6.3.2), Equation (6.3.3), and Equation (6.3.4), the expressions for Λ, γ, and $\theta_{\max}$ for this distribution are

$$\Lambda = \frac{\sqrt{2K+1}}{K+1}, \qquad \gamma = \frac{K}{2K+1}, \qquad \theta_{\max} = \theta_0 \tag{6.4.8}$$

Figure 6.13 depicts the spatial channel parameters, Λ and γ, as a function of K-factor. For very small K-factors, the channel appears to be omnidirectional ($\Lambda = 1$ and $\gamma = 0$). As the K-factor increases, the angular spread of the Rician channel decreases and the angular constriction increases. This indicates that the overall fading rate in the Rician channel decreases and that the differences between the minimum and maximum fading rate variances within the same local area but different directions increases.

> *Note: Don't Forget the First-Order Stats*
>
> Of course, using the principles of Chapter 5, it is possible to characterize the local area envelope and power fading statistics for all four examples in this section, since the specular and diffuse components are so easily spotted in the azimuth spectrum representation of multipath. The four examples follow two-wave, Rayleigh, Rayleigh, and Rician envelope PDFs (in that order).

6.5 Chapter Summary

This chapter presented the basic definitions and concepts behind modeling the angular distribution of multipath power in space - analysis that is usually performed using an *angle spectrum*. Below is a summary of the key points in this chapter:

- Vector channels, autocorrelation functions, and power spectra capture the 3D spatial dependencies of a radio link. There are simple formulas that convert these vector representations to scalar representations.
- Wavevector, wavenumber, and angle spectrum are all related in the following way:
  - ▷ Any wavevector spectrum for a U-SLAC model may be written as an angle spectrum.

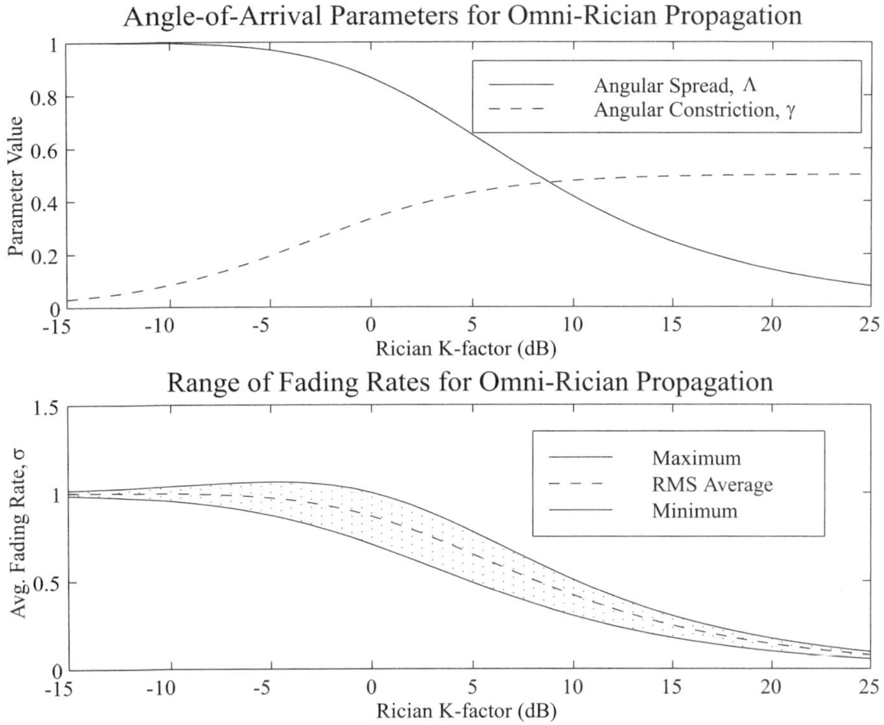

Figure 6.13 Rician propagation model [Dur00b].

- ▷ Angle spectra are simpler and more physically intuitive than wavevector spectra.
- ▷ A wavenumber spectrum for a particular orientation in space may be derived from a wavevector spectrum (scalar mapping) or an angle spectrum (Gans mapping).

■ A shape factor is a geometrical characteristic of the angle spectrum that relates to the wavenumber spread of an azimuth propagation model.

- ▷ *Angular spread*, Λ, describes concentration of multipath power in one azimuthal direction and is related to the average wavenumber spread in a local area.
- ▷ *Angular constriction*, γ, describes concentration of multipath power in two azimuthal directions and is related to how much a wavenumber spread depends on orientation.
- ▷ *Direction of maximum fading*, $\theta_{\max}$, describes the azimuthal orientation for which the wavenumber spread is maximized.

- Shape factor theory may be applied to any number of channel models, including
  - a simple two-wave model
  - a diffuse azimuthal sector of arriving multipath power
  - a diffuse double sector of arriving multipath power
  - a Rician channel with a specular component and diffuse, omnidirectional power

This chapter - much like the ones before it - has focused on the theory and definitions used to characterize the wireless radio channel. Subsequent topics will deal more with applications in wireless communications that involve the measurement and performance of a channel.

PROBLEMS

1. A U-SLAC model is described by multipath waves that arrive uniformly from all azimuthal directions with angle of elevation φ_0. Based on this scenario, answer the following questions:

a. Write an equation for the PSD's (angle and wavevector) that describe the conical propagation.

b. Calculate the wavenumber spectrum along an orientation in space with $\varphi = 0°$.

c. What is the wavenumber spread for the spectrum in (b)?

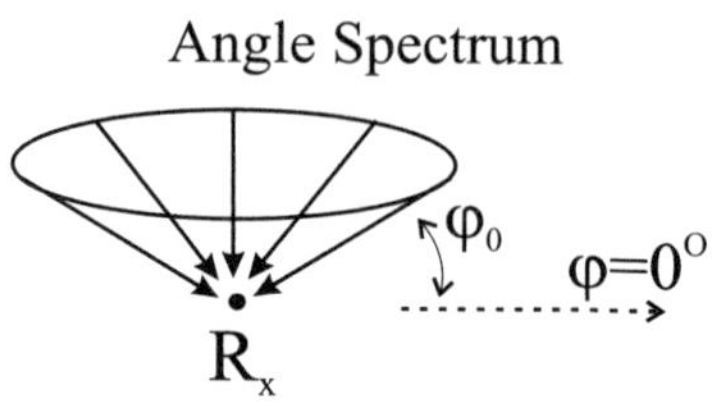

2. A U-SLAC model is described by multipath waves that arrive with a smooth, donut-shaped angle spectrum, $p(\theta, \varphi) = A\cos^2(\varphi)$. Based on this scenario, answer the following questions:

a. Write an equation for the wavevector spectrum.

b. Calculate the wavenumber spectrum along an arbitrary orientation in space, φ.

c. What is the wavenumber spread along a direction in space with $\varphi = 0°$?

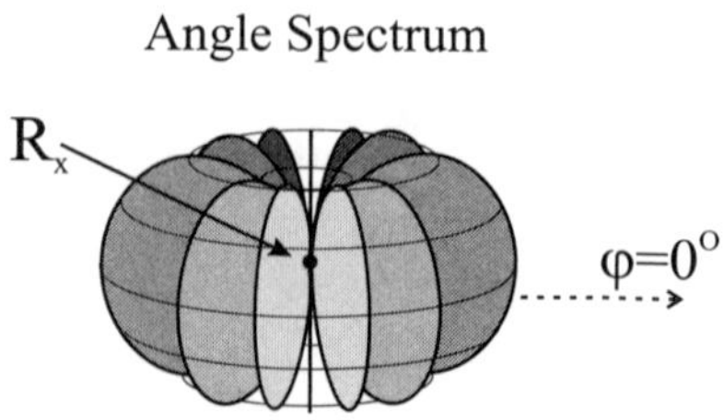

3. At first glance, we may wonder if the definition for angle spread Λ in Equation (6.3.2) might produce a negative argument for the square root operator. Prove, using the Schwartz inequality, that the argument in Equation (6.3.2) is always nonnegative.

$$\text{Schwartz Inequality:} \quad \left| \int_b^a \tilde{f}(x)\, dx \right| \le \int_b^a |\tilde{f}(x)|\, dx$$

4. In a rectangular room, a specular incoming multipath may resonate off the flat walls to create an azimuth spectrum with four equal-powered multipath, separated by azimuth angle α. Answer the following questions based on this azimuth spectrum:

a. Write a mathematical expression for $p(\theta)$.

b. Derive equations for the shape factors Λ, γ, and $\theta_{\max}$ as a function of α.

c. Evaluate the shape factors for the limiting cases of $\alpha = 0, \frac{\pi}{2}$, and π radians.

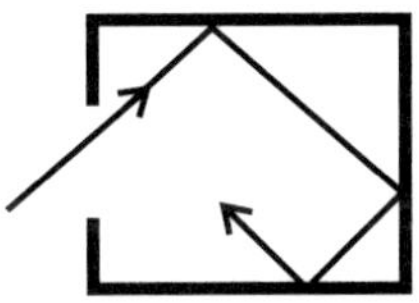

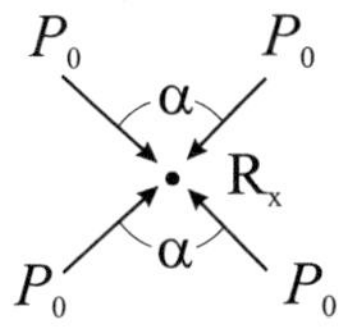

5. Consider an azimuth spectrum divided into two hemicircles of diffusely distributed multipath power. One hemicircle has power βP_0, while the other has power $(1-\beta)P_0$. Answer the following questions based on this azimuth spectrum:

a. Derive equations for the shape factors Λ, γ, and $\theta_{\max}$ as a function of β.

b. Derive equations for the level-crossing distance, N_r, and average fade length, $\bar{r}$, as a function of β and orientation θ_R.

c. Evaluate the level-crossing rates for the limiting cases of $\beta = 0, 0.5$, and 1 for $\theta_R = 0$ and $\theta_R = \frac{\pi}{2}$.

d. Describe an environment or propagation scenario where this model might be useful.

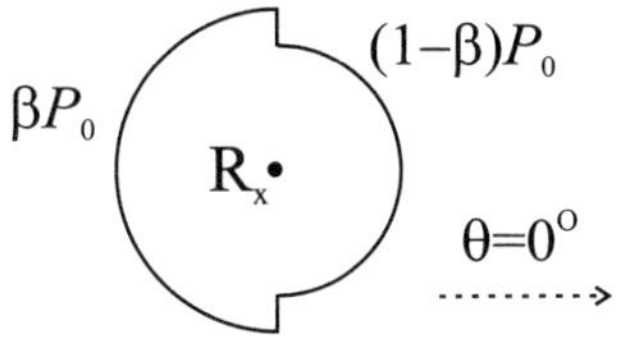

where $0 \le \beta \le 1$.

6. Consider an azimuth spectrum consisting of three equal-powered specular multipaths, evenly spaced in azimuth with respect to each other (i.e., each is 120° apart). Answer the following questions based on this azimuth spectrum:

a. Guess the shape factors Λ, γ, and $\theta_{\max}$ for this angle spectrum by visual inspection.

b. Write the mathematical expression for $p(\theta)$ and solve for the shape factors formally.

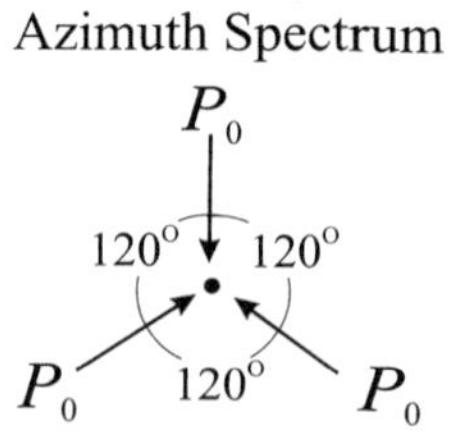

7. For a specified location, local area propagation is described by an azimuth spectrum that resembles the double-sector propagation model of Section 6.4.3, but with a large specular component arriving in the center of one sector. Based on this configuration, answer the following questions:

a. What is the K-factor of the Rician PDF describing this fading?

b. What are the multipath shape factors as a function of P_{dif}, P_1, θ_0, and α?

c. You are told that a channel has the following parameters: [$\gamma = 0.75$, $\theta_{\max} = 50°$, $K = 3\text{dB}$]. Construct an equivalent spatial channel model based on the azimuth spectrum of this problem.

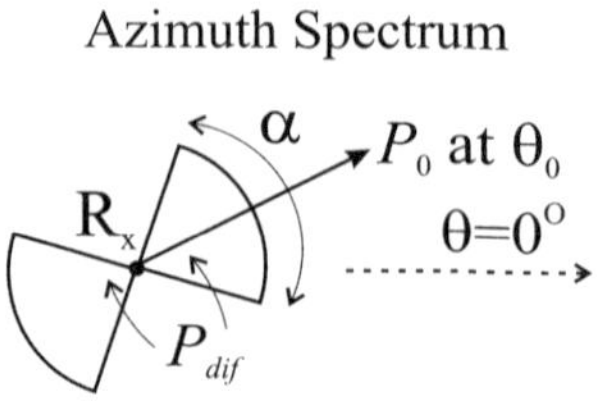

8. For a specified location, local area propagation is described by an azimuth spectrum that resembles the sector propagation model with a large specular component arriving in the center of one sector. Based on this configuration, answer the following questions:

a. What is the K-factor of the Rician PDF describing this fading?

b. What are the multipath shape factors as a function of P_{dif}, P_1, θ_0, and α?

c. You are told that a channel has the following parameters: [$\Lambda = 0.25$, $\theta_{\max} = 135°$, $K = 6\text{dB}$]. Construct an equivalent spatial channel model based on the azimuth spectrum of this problem.

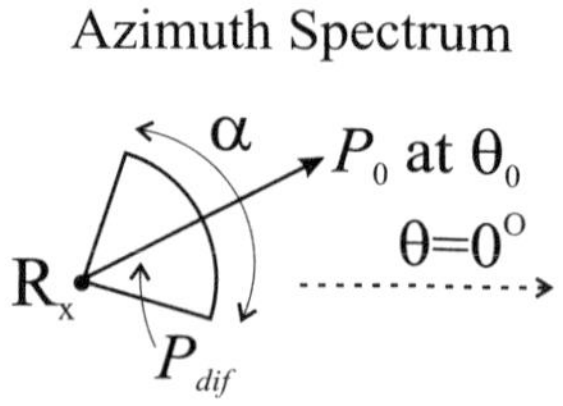

9. For a specified location, local area propagation is described by an azimuth spectrum that resembles the sector propagation model with a large specular component arriving *opposite* the center of one sector. Based on this configuration, answer the following questions.

a. What is the K-factor of the Rician PDF describing this fading?

b. What are the multipath shape factors as a function of P_{dif}, P_1, θ_0, and α?

c. You are told that a channel has the following parameters: $[\Lambda = 0.65,\ \gamma = 0.8,\ \theta_{\max} = -80°\ ]$. Construct an equivalent spatial channel model based on the azimuth spectrum of this problem.

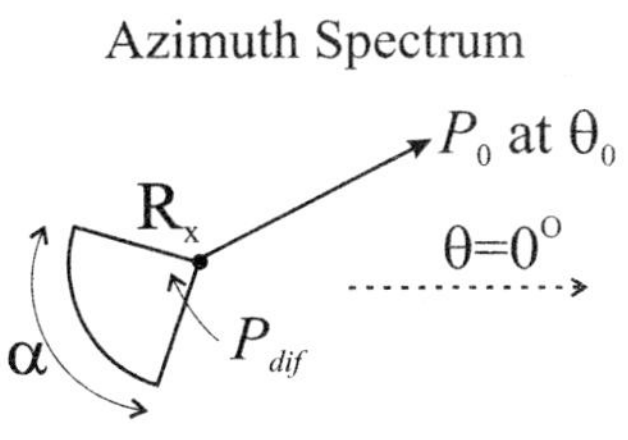

10. Some definitions of angle spread rely on the second centered moment of the azimuth spectrum $p(\theta)$. This definition - analogous to the RMS spectrum spreads in previous chapters - is defined below:

$$\sigma_\theta = \sqrt{\frac{\int_{-\pi}^{\pi} (\theta - \bar{\theta})^2 p(\theta)\, d\theta}{\int_{-\pi}^{\pi} p(\theta)\, d\theta}}$$

where $\bar{\theta}$ is the centroid of the azimuth spectrum (the direction of peak multipath arrival in a symmetric spectrum). Use this definition for angle spread to answer the following questions:

a. Calculate the angle spread for the sector model of Section 6.4.2 based on the above equation.

b. What are the benefits and drawbacks of this definition of angle spread compared to the shape factor Λ?

c. Derive an equation for converting σ_θ to Λ for the sector model.

d. Discuss the merits of using the resulting equation in (c) as a general purpose formula for converting angle spread values in the literature to shape factor values.

11. Prove the following shape factor theorems:

a. Rotational independence of angle spread and constriction: If $p_B(\theta) = p_A(\theta + \theta_0)$, then $\Lambda_B = \Lambda_A$, $\gamma_B = \gamma_A$, and $\theta^B_{\max} = \theta^A_{\max} + \theta_0$.

b. Scalar independence of shape factors: If $p_B(\theta) = \alpha p_A(\theta)$, then $\Lambda_B = \Lambda_A$, $\gamma_B = \gamma_A$, and $\theta^B_{\max} = \theta^A_{\max}$.

c. Reflection independence of shape factors: If $p_B(\theta) = p_A(-\theta)$, then $\Lambda_B = \Lambda_A$, $\gamma_B = \gamma_A$, and $\theta^B_{\max} = -\theta^A_{\max}$.

12. An azimuth spectrum is said to have N-fold radial symmetry, where N is the largest integer that satisfies the following relationship:

$$p(\theta) = p(\theta + \Delta\theta) = p(\theta + 2\Delta\theta) = \cdots p(\theta + N\Delta\theta) \quad \text{for } \Delta\theta = \frac{2\pi}{N}$$

Use this definition of N-fold radial symmetry to prove the following:

a. $\Lambda = 1$ for all $p(\theta)$ with $N > 1$

b. $\gamma = 0$ for all $p(\theta)$ with $N > 2$

13. A classical problem in UHF radio propagation is the single ground bounce scenario between a transmitter and a receiver operating with a line of sight link. This scenario is sketched below:

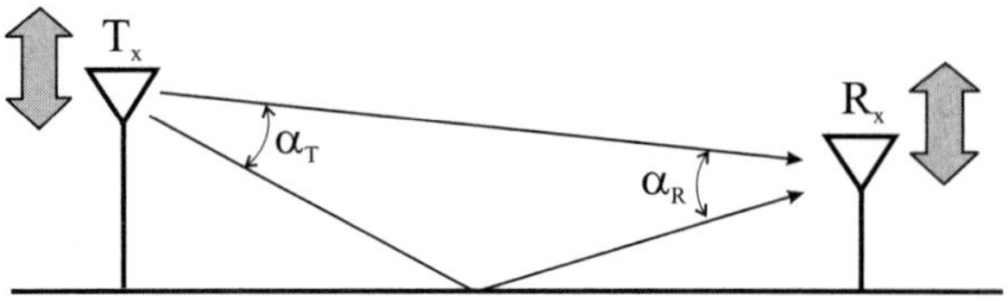

The separation angle between the line-of-sight and ground bounce waves is given by α_T at the transmitter (AOD) and α_R at the receiver (AOA). Since the ground reflection occurs at grazing incidence, its power is equal to the line-of-sight component. Answer the following questions based on this multipath scenario:

a. Shape factors have been used to describe propagation that lies only in the azimuth plane. How can shape factors be used to describe the above scenario containing waves with different elevation angles?

b. Assuming SLAC conditions are met, write an equation for σ_k for vertical movement of either the transmitter or receiver antenna in terms of $\alpha_{R,T}$ and wavelength of radiation, λ. (You may approximate $\theta_{\max}$ to be pointing in the vertical direction).

c. How small must $\alpha_{R,T}$ become in order for the wavenumber spread to be $(.1)\frac{2\pi}{\lambda}$? $(.05)\frac{2\pi}{\lambda}$? $(.01)\frac{2\pi}{\lambda}$?

d. **Challenge:** How does transmitter and receiver antenna height in the above scenario affect the validity of shape factor application?

14. Here is an application from optics based on an experiment you can do at your desk. Shine a laser pointer device onto a semismooth surface (like a piece of paper or a desk top). Although the surface may be smooth to the touch, the wavelength of laser light is extremely small. Most surfaces appear "electromagnetically" rough to a laser beam, so that diffuse scattering results at the interface of the smooth surface. This diffuse scattering produces the familiar speckle pattern seen on the illuminated area of the semismooth surface.

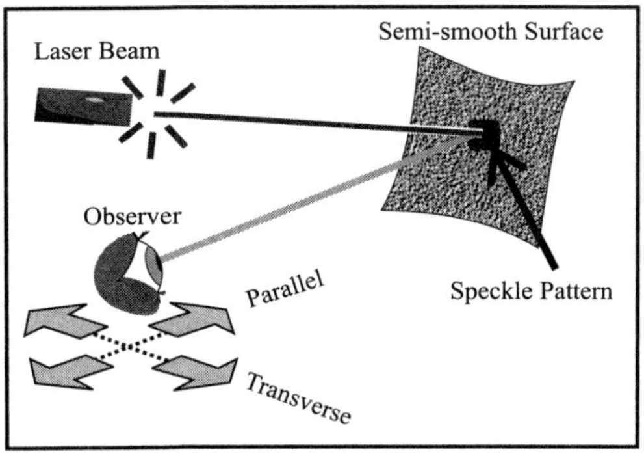

The human retina, with its dense arrangement of light-sensing cells, is like an array of receiver antennas in space. Normally, the wavelength of light (in the nanometers) is much too small for the human eye to make out the fading of a speckle pattern. However, the illuminated area on the semismooth surface has an extremely small *angular spread*, so the fading pattern is wide enough in space for a human eye to discern.

Use this scenario to answer the following questions:

a. First, develop a small-angle approximation for angular spread and angular constriction, based on the expressions in Equation (6.4.4). Use the following approximation for a sinc function with small argument: $\text{sn}(x) \approx 1 - \frac{(\pi x)^2}{6}$.

b. The head of an observer moves back and forth slowly, first in the transverse direction and then in the parallel direction. Using the shape factors in part (a), explain why the observer sees a change in the speckle pattern occur much more rapidly in the transverse direction as opposed to the parallel direction.

c. Many cellular phone base stations experience propagation from user terminals that is spread in azimuth over a very narrow sector. Using this optical example for an analogy, explain the orientation dependence of spatial fading at the top of a base station tower.

Chapter 7

SECOND-ORDER CHANNEL STATISTICS

Until now, we have discussed only second-order statistics such as fading rate variances and spectrum spreads. These metrics are the cornerstone of any fading analysis, but they are also somewhat nebulous. For link design, there are important statistics we would like to derive that affect performance: How often does a fading channel drop below a specific signal threshold? How long does the fading channel stay beneath this threshold? What are rigorous definitions for channel coherence?

This chapter presents statistical techniques for characterizing small-scale envelope fading of *Rayleigh* channels as a function of time, space, and frequency. Statistics such as level crossing rates, average fade duration, and correlation properties are discussed. The analysis of this chapter shows that all of these statistics may be characterized in terms of the *spectrum spreads* of the wireless multipath channel and, in the case of spatial selectivity, the *shape factors*. The contents of this chapter include

- Section 7.1: Analysis for the Rayleigh level crossing problem.
- Section 7.2: Unit autocovariance as a functions of time, space, and frequency.
- Section 7.3: Classical spatial channel models using shape factors.
- Section 7.4: Statistics and principles for wideband channels.
- Section 7.5: Summary of important concepts.

The techniques of this chapter may be extended to non-Rayleigh fading situations, although the mathematics quickly becomes difficult outside the Rayleigh case. For the sake of simplicity, illustrations in this chapter will involve the diffusely propagating multipath waves of Rayleigh fading.

7.1 The Level-Crossing Problem

A PDF is useful for describing the overall duration a received signal envelope falls beneath a certain threshold - perhaps a signal-to-noise or signal-to-interference ratio threshold that is acceptable to a communications link. The PDF, however, does not answer questions about *how frequently* the random envelope process crosses a threshold or *what duration* it spends below this threshold, once crossed. Such metrics are vital to the performance of error-correction coding, spatial diversity, frequency hopping, and all other communication techniques affected by second-order fading statistics.

7.1.1 Level-Crossing Rate

For a stochastic process as a function of time, the *level crossing rate* is defined as the average number of crossings per second that the process drops beneath a specified threshold level. The level crossing rate, N_t, for a general time-varying process may be calculated from the joint PDF of envelope, R, and its first time derivative, $\frac{dR}{dt}$ [Pap91]:

$$N_t = \int_0^\infty \dot{\rho} f_{R\dot{R}}(R, \dot{\rho}) d\dot{\rho} \tag{7.1.1}$$

The value R is the envelope threshold of interest and $f_{R\dot{R}}(\rho, \dot{\rho})$ is the joint envelope, $R(t)$, and time derivative of envelope, $\dot{R}(t)$, PDF.

One of the convenient aspects of Rayleigh-fading processes is the relatively simple formula for the joint PDF in Equation (7.1.1). For a Rayleigh process, the time derivative of envelope follows a Gaussian distribution that is independent of the Rayleigh-distributed envelope PDF. The joint PDF for this case may be written as

$$f_{R\dot{R}}(\rho, \dot{\rho}) = \underbrace{\frac{2\rho}{P_{\text{dif}}} \exp\left(-\frac{\rho^2}{P_{\text{dif}}}\right)}_{f_R(\rho)} \underbrace{\frac{1}{\sigma_t\sqrt{\pi}} \exp\left(-\frac{\dot{\rho}^2}{\sigma_t^2}\right)}_{f_{\dot{R}}(\dot{\rho})} \tag{7.1.2}$$

where P_{dif} is the average power of the signal and σ_t^2 is the temporal fading rate variance, as discussed in Section 3.4.5. Recall also that we may express the temporal fading rate variance using the fundamental spectrum spread theorem of Section 3.4.6:

$$\sigma_t^2 = P_{\text{dif}}\sigma_\omega^2 \tag{7.1.3}$$

where σ_ω is the RMS Doppler spread. Using this result, the joint PDF in Equation (7.1.2), and the original level crossing rate equation of Equation (7.1.1), we may write the final Rayleigh fading level crossing rate expression as

$$N_t = \frac{\sigma_\omega}{\sqrt{\pi}} \rho_{\text{rms}} \exp\left(-\rho_{\text{rms}}^2\right) \tag{7.1.4}$$

where ρ_{rms} is the envelope threshold normalized against the RMS envelope value ($\rho_{\text{rms}}^2 = R^2/P_{\text{dif}}$). Figure 7.1 contains a sample fading channel that illustrates the level crossings, fade durations, and the envelope threshold. Example 7.1 provides a sample calculation of temporal level crossing rate as applied to a back-of-the-envelope analysis of a packet data link.

> *Note: Normalized Channels*
>
> Equation (7.1.4) uses the normalized channel convention (NCC), since the threshold specification ρ_{rms} referenced against the RMS value. Thus, ρ_{rms} is unitless. From here onward, we often normalize the channel threshold against its mean-squared power level to avoid the messy unit conventions.

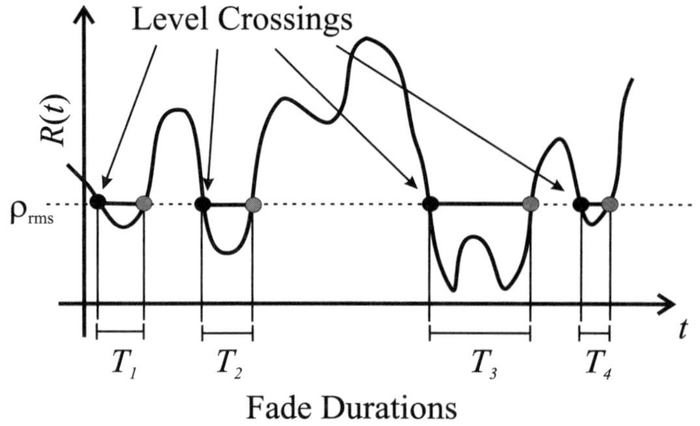

Figure 7.1 Envelope process with level crossings, fade durations, and threshold level ρ_{rms}.

Example 7.1: Level Crossings in Time

Problem: Transmitted data bits in a digital communications link are often grouped into data frames or packets that have self-contained error-correction block codes to correct data received in error. One such narrowband wireless system transmits a data packet every 20 ms through a temporal Rayleigh-fading channel with Doppler spread of 5 Hz. If the received signal strength drops below 10 dB of the average received power level during a packet transmission, the error-correction code will be overwhelmed and the entire packet will be lost. The system cannot tolerate more than one of these "catastrophic fades" for every five data packets transmitted. Does this system meet that requirement?

Solution: A Doppler spread of 5 Hz implies $\sigma_\omega = 2\pi \times (5\,\text{Hz}) = 31.4$ rad/s. The threshold of 10 dB below average received power implies $\rho_{\text{rms}} = \sqrt{0.1}$.

We can substitute these values into Equation (7.1.4):

$$\begin{aligned} N_t &= \frac{31.4}{\sqrt{\pi}}\sqrt{0.1}\exp\left(-0.1\right) \\ &= 5.06 \text{ crossings/s} \end{aligned}$$

If there are 5.06 threshold crossings every second, then there will be an average of one crossing for every 200 ms of transmission time. The system under consideration sends a data packet once every 20 ms; over a duration of 200 ms, 10 data packets will be sent. Under the assumption that a fade wipes out only one data packet (as opposed to consecutive data packets), then wiping out 1 in 10 packets is much better than the design specification of 1 in 5.

7.1.2 Average Fade Duration

The calculation of the *average fade duration* statistic is similar to the level crossing rate. The average fade duration is the amount of time that a received envelope spends underneath a threshold level once the envelope has crossed that level. This concept is graphically illustrated in Figure 7.1.

For a given threshold level R the average fade duration, $\bar{t}$, is calculated by the following:

$$\bar{t} = \frac{1}{N_t}\int_0^R f_R(\rho)d\rho \tag{7.1.5}$$

For the temporal Rayleigh-fading case, the average fade duration is given by

$$\bar{t} = \frac{\sqrt{\pi}}{\sigma_\omega \rho_{\text{rms}}}\left[\exp\left(\rho^2_{\text{rms}}\right) - 1\right] \tag{7.1.6}$$

The average fade duration is useful for describing the "burstiness" of bit errors in a temporal fading channel. See Example 7.2 for a sample calculation of average fade duration.

> *Note: Averages Versus Distributions*
>
> Notice that we are focusing on the *average* level crossing rate and fade duration. According to Rice's pioneering work, the averages for these quantities have closed-form solutions [Ric44]. Closed-form solutions for the exact *distribution* of fade durations, however, were only possible by making oversimplified assumptions. To date, effective analytical treatment of the distribution of Rayleigh fades is an unsolved problem in communication analysis.

Example 7.2: Average Temporal Fade Duration

Problem: In Example 7.1, we assumed that once a threshold of 10 dB below average power was crossed, the fade duration would be less than the duration of one data packet, which was given as 20 ms. Based on the scenario of Example 7.1, is this assumption valid?

Solution: For Example 7.1, the following parameters defined the temporal channel fading: $\sigma_\omega = 31.4$ rad/s and $\rho_{\text{rms}} = \sqrt{0.1}$. We can substitute these values into Equation (7.1.6):

$$\begin{aligned} \bar{t} &= \frac{\sqrt{\pi}}{31.4\sqrt{0.1}} \left[\exp(0.1) - 1\right] \\ &= 18.8 \text{ ms} \end{aligned}$$

The average fade duration of 18.8 ms is less than the 20 ms duration between data packets. A deep fade will typically wipe out one, possibly two, data packets for this communications link. The original assumption holds.

7.1.3 Level Crossing in Frequency

Traditionally, most level crossing analysis is performed on time-varying fading channels. However, for a fixed static channel, we may also define level crossing rates in terms of crossings per Hz. Applying duality ($\sigma_\omega \longrightarrow 2\pi\sigma_\tau$), we may write the level crossing rate for frequency selective channels as

$$N_f = 2\sqrt{\pi}\sigma_\tau \rho_{\text{rms}} \exp\left(-\rho_{\text{rms}}^2\right) \tag{7.1.7}$$

where σ_τ is the delay spread. Equation (7.1.7) is an excellent measure of the frequency selectivity of a dispersive wireless channel.

Likewise, an *average fade bandwidth*, $\bar{f}$, may also be calculated in terms of the delay spread. Applying duality produces the following expression:

$$\bar{f} = \frac{\exp\left(\rho_{\text{rms}}^2\right) - 1}{2\sqrt{\pi}\sigma_\tau \rho_{\text{rms}}} \tag{7.1.8}$$

The average fade bandwidth is a useful parameter for wireless systems that employ frequency-hopping. In such systems, successive frequency hops should, on average, change the transmitted carrier frequency in excess of the average fade bandwidth to maintain an acceptable signal-to-noise ratio.

7.1.4 Level Crossing in Space

Following in the footsteps of the frequency analysis, it is also possible to do a level crossing analysis on a static narrowband received envelope as a function of space.

Once again, applying duality ($\sigma_\omega \longrightarrow \sigma_k$) produces an expression for the level crossing rate in units of *crossings per unit distance*,

$$N_r = \frac{\sigma_k}{\sqrt{\pi}} \rho_{\rm rms} \exp\left(-\rho_{\rm rms}^2\right) \tag{7.1.9}$$

and also for the *average fade length*,

$$\bar{r} = \frac{\sqrt{\pi}}{\sigma_k \rho_{\rm rms}} \left[\exp\left(\rho_{\rm rms}^2\right) - 1\right] \tag{7.1.10}$$

Both the spatial level crossing rate and the average fade distance depend on the wavenumber spread, σ_k.

The expressions in Equation (7.1.9) and Equation (7.1.10) may also be written in terms of shape factors. By recognizing the relationship between wavenumber spread and the multipath shape factors of Equation (6.3.5), we can write new equations for spatial level crossing rate and average fade length:

$$N_r = \frac{\sqrt{2\pi}\Lambda\rho_{\rm rms}}{\lambda} \sqrt{1 + \gamma\cos\left[2(\theta_R - \theta_{\max})\right]} \exp\left(-\rho_{\rm rms}^2\right) \tag{7.1.11}$$

$$\bar{r} = \frac{\lambda\left[\exp\left(\rho_{\rm rms}^2\right) - 1\right]}{\sqrt{2\pi}\rho_{\rm rms}\Lambda\sqrt{1 + \gamma\cos\left[2(\theta_R - \theta_{\max})\right]}} \tag{7.1.12}$$

where λ is the wavelength of propagation and θ_R is the azimuthal orientation.

7.2 Envelope Unit Autocovariance

Recall that the unit autocovariance of a received envelope as a function of time, frequency, or space determines the coherence of a wireless channel. The unit autocovariance functions developed in this section may be used to develop analytical expressions for coherence time, bandwidth, and distance.

7.2.1 Temporal Unit Autocovariance

The unit autocovariance of a time-varying stochastic process may be computed by subtracting the mean value from the autocorrelation and normalizing the result. The autocorrelation in turn is calculated from its Doppler spectrum by performing an inverse Fourier transform. For the complex time-varying channel, $\tilde{h}(t)$, this natural progression is illustrated below:

$$S_{\tilde{h}}(\omega) \xrightarrow{\text{Fourier Transform}} C_{\tilde{h}}(\Delta t) \xrightarrow{\text{Normalize}} \varrho_{\tilde{h}}(\Delta t) \tag{7.2.1}$$

Thus, if the Doppler spectrum of $\tilde{h}(t)$ is known, it is possible to calculate the unit autocovariance, $\varrho_{\tilde{h}}(\Delta t)$.

The Doppler spectrum for received *envelope*, $S_R(\omega)$, however, is extremely difficult to calculate - even if we know the *channel* Doppler spectrum, $S_{\tilde{h}}(\omega)$. We would like to use a similar method,

$$S_R(\omega) \xrightarrow{\text{Fourier Transform}} C_R(\Delta t) \xrightarrow{\text{Normalize}} \varrho_R(\Delta t) \tag{7.2.2}$$

but this is extremely difficult, since the act of taking an envelope of a random process is a highly nonlinear operation. Thus, derivation of $S_R(\omega)$ from $S_{\tilde{h}}(\omega)$ is difficult, even though envelope and complex baseband channel satisfy the simple (but non-linear) relationship $R(t) = |\tilde{h}(t)|$.

Rather than develop an expression for envelope unit autocovariance directly (which itself requires approximations), we will use the approximate analytical technique developed in Appendix 7.A. This technique produces an expression for envelope unit autocovariance based *solely* on the RMS spectral spread of the process. For the temporal envelope unit autocovariance, $\varrho_R(\Delta t)$, the technique produces the following expression:

$$\varrho_R(\Delta t) \approx \exp\left(-\frac{23}{2\pi^2}\sigma_\omega^2\, \Delta t^2\right) \tag{7.2.3}$$

where σ_ω is the RMS Doppler spread. Adapted from the original technique in [Dur99a], Equation (7.2.3) produces near-exact values for unit autocovariance over the most useful range of the function - when Δt is small. Despite the approximations used to construct Equation (7.2.3), it only becomes inaccurate after a duration, Δt, in which the time-varying envelopes have become uncorrelated. Therefore, Equation (7.2.3) is extremely useful for engineering wireless modem designs that are affected by temporal correlation.

We may also note that with an expression for envelope unit autocovariance, we may now provide a formal expression for *temporal coherence.* The value *coherence time*, T_c, is the duration over which the envelope of a temporal fading channel appears to be a constant, as discussed in Section 2.2. Another way of stating this definition is that coherence time, T_c, is the value of separation time, Δt, for which time-varying envelopes become *uncorrelated.*

Defined in this manner, the exact coherence time for a time-varying channel will depend on how strictly the engineer chooses to define "uncorrelated." A strict definition of coherence time may arise if we require the signal envelope to be 90% correlated ($\varrho_R(T_c) = 0.90$). A more relaxed decorrelation standard will lead to a larger T_c. Unless specified otherwise, this work will use the quantity $\exp(-1)$ to describe envelopes that have sufficiently decorrelated. In fact, the time that satisfies this definition is often called the *correlation time* in statistics [Bro97]. Therefore, our generic formal definition for coherence time becomes

$$T_c = \frac{2\pi}{\sqrt{46}\sigma_\omega} = \frac{1.08}{\sigma_\omega} \tag{7.2.4}$$

Note that this definition of coherence time depends solely and inversely on the RMS Doppler spread, σ_ω. The other fine details and structure of the Doppler spectrum, $S_{\tilde{h}}(\omega)$, do not greatly effect temporal channel coherence. Note that this independence is noted and empirically shown by Fleury in [Fle96].

> *Note: Temporal Rayleigh Fading*
>
> For fixed, narrowband receivers, temporal fading in a wireless channel is due to the motion of scatterers. Rayleigh fading as a function of time requires temporally diffuse propagation, which physically means that the propagation environment consists of numerous scatterers, *all of which are in motion.* One such scenario occurs for mm-wave propagation through a rain storm, where drops of water scatter radio waves to the receiver [Xu99]. However, most cases of temporal Rayleigh fading in terrestrial wireless propagation are due to receivers *moving* through small-scale spatial Rayleigh fading. Recognize, therefore, that *temporal* Rayleigh fading in the absence of receiver (or transmitter) motion is much more rare than *frequency* or *spatial* Rayleigh fading.

7.2.2 Frequency Unit Autocovariance

Once again, applying duality to the temporal analysis produces similar expressions for frequency unit autocovariance. Realizing that $\sigma_\omega \longrightarrow 2\pi\sigma_\tau$ and $\Delta t \longrightarrow \Delta f$, the envelope unit autocovariance as a function of frequency separation becomes

$$\varrho_R(\Delta f) \approx \exp\left(-46\sigma_\tau^2\,\Delta f^2\right) \tag{7.2.5}$$

where σ_τ is the RMS delay spread. Again, note the independence of the approximate unit autocovariance in Equation (7.2.5) on the fine structure of the delay spectrum, $S_{\tilde{h}}(\tau)$; only the RMS delay spread is needed to characterize the small-Δf correlation behavior.

Our formal definition for *coherence bandwidth*, B_c, will be based on the frequency separation Δf that drives the envelope unit autocovariance in Equation (7.2.5) to a level of $\exp(-1)$. Following this definition,

$$B_c = \frac{1}{\sqrt{46}\sigma_\tau} = \frac{1}{6.78\sigma_\tau} \tag{7.2.6}$$

Note how similar this definition of coherence bandwidth is to the arbitrary rule of thumb,

$$B_c \approx \frac{1}{5\sigma_\tau} \tag{7.2.7}$$

which was discussed in Section 3.4.4.

7.2.3 Spatial Unit Autocovariance

Of course, the correlation analysis must include the spatial unit autocovariance in order to be complete. Applying the duality results $\Delta t \longrightarrow \Delta r$ and $\sigma_\omega \longrightarrow \sigma_k$, we obtain

$$\varrho_R(\Delta r) \approx \exp\left(-\frac{23}{2\pi^2}\sigma_k^2\,\Delta r^2\right) \tag{7.2.8}$$

where σ_k is the RMS wavenumber spread. *Coherence distance*, D_c, is the separation distance in space over which a fading channel appears to be unchanged. Using our general correlation criterion for channel coherence, $\varrho_R(D_c) = \exp(-1)$, the expression for coherence distance becomes

$$D_c = \frac{2\pi}{\sqrt{46}\sigma_k} = \frac{1.08}{\sigma_k} \tag{7.2.9}$$

Note the insensitivity of this expression to the fine structure of the wavenumber spectrum $S_{\tilde{h}}(k)$. Only the RMS wavenumber spread is important.

Multipath shape factor theory may be applied to Equation (7.2.8) to generate equivalent envelope unit autocovariance functions that depend on the geometrical properties of multipath angle-of-arrival. Furthermore, the shape factor theory allows us to study correlation properties as a function of azimuthal orientation, θ_R, in space instead of one-dimensional position:

$$\varrho_R(\Delta r, \theta_R) \approx \exp\left(-23\Lambda^2\left[1+\gamma\cos(2[\theta_R - \theta_{\max}])\right]\left[\frac{\Delta r}{\lambda}\right]^2\right) \tag{7.2.10}$$

Equation (7.2.10) allows us to estimate the envelope correlation between two points in space separated by a distance Δr along an azimuthal direction θ_R, as illustrated in the geometry of Figure 7.2. The orientation-dependent coherence distance is

$$D_c(\theta_R) = \frac{\lambda}{\Lambda\sqrt{23(1+\gamma\cos[2(\theta_R - \theta_{\max})])}} \tag{7.2.11}$$

There are few rules of thumb for channel coherence in the literature, other than the case of omnidirectional diffuse propagation ($\Lambda = 1$, $\gamma = 0$). A convenient definition for the coherence distance, D_c, is the value that satisfies the equation $\rho_{\text{rms}}(D_c) = 0.5$ [Ste94]. The classical value for coherence distance in an omnidirectional Rayleigh channel is given by

$$D_c \approx \frac{9\lambda}{16\pi} \tag{7.2.12}$$

where λ is the wavelength of radiation. Using the generalized autocovariance function of Equation (7.2.10) with this criterion for correlation leads to the definition

$$D_c(\theta_R) \approx \frac{\lambda\sqrt{\ln 2}}{\Lambda\sqrt{23(1+\gamma\cos[2(\theta_R - \theta_{\max})])}} \tag{7.2.13}$$

For omnidirectional propagation, Equation (7.2.13) differs from Equation (7.2.12) by only -3.0% [Dur00b]. Space correlation is discussed in greater detail in Chapter 10.

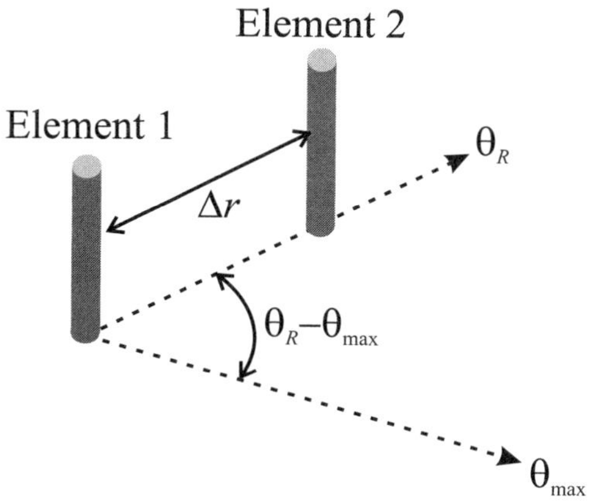

Figure 7.2 The geometry of two identical, co-polar antenna elements separated in space by Δr.

7.2.4 Joint Unit Autocovariance

With the increased focus on wideband mobile or multi-antenna signaling techniques, wireless researchers may find it helpful to characterize a joint unit autocovariance for received envelope as a function of spatial *and* frequency separation. For modest bandwidth systems, it is sufficient to treat spatial and frequency decorrelation as independent phenomena. Therefore, the joint unit autocovariance, $\varrho_R(\Delta f, \Delta r)$, may be written as the product of the individual unit autocovariance functions [Fle00]:

$$\varrho_R(\Delta f, \Delta r) \approx \varrho_R(\Delta f)\varrho_R(\Delta r) \tag{7.2.14}$$

Equation (7.2.14) describes how the received envelope changes as a function of frequency separation, Δf, and spatial separation, Δr. Plugging Equation (7.2.5) and Equation (7.2.8) into Equation (7.2.14) produces

$$\varrho_R(\Delta f, \Delta r) \approx \exp\left(-\frac{23}{2\pi^2}\left[4\pi^2\sigma_\tau^2\,\Delta f^2 + \sigma_k^2\,\Delta r^2\right]\right) \tag{7.2.15}$$

Of course, shape factors may be applied to better understand the spatial correlation properties as a function of displacement and orientation. The final unit autocovariance for space–frequency varying envelopes becomes

$$\varrho_R(\Delta f, \Delta r, \theta_R) \approx \exp\left(-46\sigma_\tau^2\,\Delta f^2 - 23\Lambda^2\left[1 + \gamma\cos(2[\theta_R - \theta_{\max}])\right]\left[\frac{\Delta r}{\lambda}\right]^2\right) \tag{7.2.16}$$

Equation (7.2.16) may prove invaluable for describing correlation properties (and hence performance) of wireless systems that have large bandwidths and multiple transmitter-receiver antennas. The powerful approximations help characterize the complicated phenomenon of joint space–frequency fading.

7.2.5 Second-Order Statistic Summary

Clearly there are numerous level crossing and correlation statistics to present for channels that vary with respect to time, frequency, or space. Despite the symmetries involved in each calculation, it is difficult to keep track of the exact mathematical formulas. To solve this problem, Table 7.1 presents an organized summary of the second-order Rayleigh fading statistics discussed in this chapter.

7.3 Classical Spatial Channel Models

This section shows several examples of how level crossing rate, average fade duration, and unit autocovariance relationships derived using shape factors compare to three well-known cases of propagation that have analytical solutions [Jak74].

7.3.1 Classical Models

The first case of multipath propagation that we will study corresponds to a narrowband receiver operating in a local area with multipath arriving from all directions such that the angular distribution of power, $p(\theta)$, is a constant. The receiver antenna is assumed to be an omnidirectional whip, oriented perpendicular to the ground. Due to the vertical electric-field polarization of the whip antenna, this propagation scenario is referred to as the E_z-case [Cla68].

The second two cases correspond to the same narrowband receiver in the same omnidirectional multipath channel, but with a small loop antenna mounted atop the receiver such that the plane of the loop is perpendicular to the ground. The antenna pattern of the small loop antenna attenuates the arriving multipath such that the angular distribution of power becomes

$$p(\theta) = A \sin^2 \theta \tag{7.3.1}$$

where A is some arbitrary gain constant. Unlike the omnidirectional E_z-case, the statistics of this propagation scenario will depend on the direction of travel by the receiver. The H_x-case will refer to a receiver traveling in a direction perpendicular to the main lobes of the loop antenna pattern ($\theta = 0$). The H_y-case will refer to a receiver traveling in a direction parallel to the main lobes ($\theta = \frac{\pi}{2}$). Figure 7.3 illustrates the E_z, H_x, and H_y cases for the modeled receiver antennas.

7.3.2 Channel Model Solutions

The first step is to calculate the three spatial parameters from the angular distribution of power, $p(\theta)$, using Equation (6.3.2), Equation (6.3.3), and Equation (6.3.4). The spatial parameters for the E_z-case are $\Lambda = 1$, $\gamma = 0$, and $\theta_{\max} = 0$. Since this case is omnidirectional, the angular spread is at a maximum ($\Lambda = 1$) and the angular constriction is at a minimum ($\gamma = 0$). For the H_x- and H_y-cases, the spatial parameters are $\Lambda = 1$, $\gamma = \frac{1}{2}$, and $\theta_{\max} = \frac{\pi}{2}$. Since the impinging multipath have no clear bias in *one* direction, the angular spread is at a maximum, just as

Table 7.1 Second-Order Statistics for Rayleigh Envelopes in Time, Frequency, and Space

TEMPORAL LCR, N_t	AF DURATION, $\bar{t}$	UNIT AUTOCOVARIANCE, $\varrho_R(\Delta t)$
$\frac{\sigma_\omega}{\sqrt{\pi}}\rho_{\rm rms}\exp\left(-\rho_{\rm rms}^2\right)$	$\frac{\sqrt{\pi}}{\sigma_\omega \rho_{\rm rms}}\left[\exp\left(\rho_{\rm rms}^2\right)-1\right]$	$\exp\left(-\frac{23}{2\pi^2}\sigma_\omega^2\,\Delta t^2\right)$
FREQUENCY LCR, N_f	AF BANDWIDTH, $\bar{f}$	UNIT AUTOCOVARIANCE, $\varrho_R(\Delta f)$
$2\sqrt{\pi}\sigma_\tau\rho_{\rm rms}\exp\left(-\rho_{\rm rms}^2\right)$	$\frac{\exp\left(\rho_{\rm rms}^2\right)-1}{2\sqrt{\pi}\sigma_\tau\rho_{\rm rms}}$	$\exp\left(-46\sigma_\tau^2\,\Delta f^2\right)$
SPATIAL LCR, N_r	AF LENGTH, $\bar{r}$	UNIT AUTOCOVARIANCE, $\varrho_R(\Delta r)$
$\frac{\sigma_k}{\sqrt{\pi}}\rho_{\rm rms}\exp\left(-\rho_{\rm rms}^2\right)$ or $\frac{\sqrt{2\pi}\Lambda\rho_{\rm rms}\exp\left(-\rho_{\rm rms}^2\right)}{\lambda}\sqrt{1+\gamma\cos\left[2(\theta_R-\theta_{\rm max})\right]}$	$\frac{\sqrt{\pi}}{\sigma_k \rho_{\rm rms}}\left[\exp\left(\rho_{\rm rms}^2\right)-1\right]$ or $\frac{\lambda\left[\exp\left(\rho_{\rm rms}^2\right)-1\right]}{\sqrt{2\pi}\rho_{\rm rms}\Lambda\sqrt{1+\gamma\cos[2(\theta_R-\theta_{\rm max})]}}$	$\exp\left(-\frac{23}{2\pi^2}\sigma_k^2\,\Delta r^2\right)$ or $\exp\left(-23\Lambda^2\left[1+\gamma\cos(2[\theta_R-\theta_{\rm max}])\right]\left[\frac{\Delta r}{\lambda}\right]^2\right)$

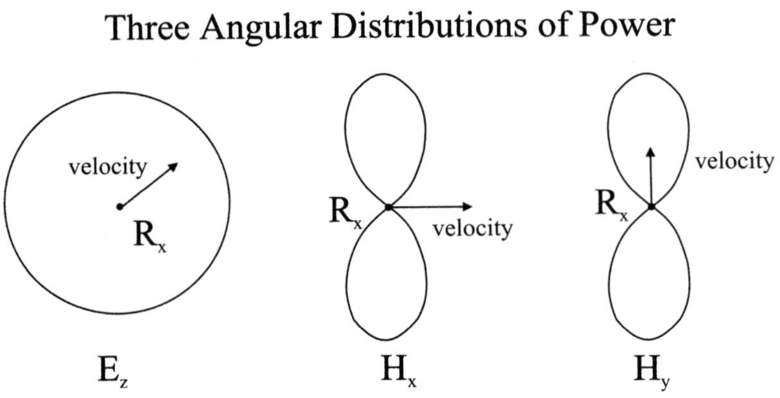

Figure 7.3 Top view of azimuth propagation for three different multipath-induced mobile-fading scenarios.

in the E_z-case. However, there is clearly a bias in *two* directions, resulting in a middle-valued angular constriction of $\gamma = \frac{1}{2}$.

After substitution of these parameters into Equation (7.1.11) along with the appropriate direction of mobile travel, the level crossing rates for the three cases become

$$E_z: \quad N_t = \frac{\sqrt{2\pi}v\rho_{\text{rms}}}{\lambda}\exp\left(-\rho_{\text{rms}}^2\right) \tag{7.3.2}$$

$$H_x: \quad N_t = \frac{\sqrt{\pi}v\rho_{\text{rms}}}{\lambda}\exp\left(-\rho_{\text{rms}}^2\right) \tag{7.3.3}$$

$$H_y: \quad N_t = \frac{\sqrt{3\pi}v\rho_{\text{rms}}}{\lambda}\exp\left(-\rho_{\text{rms}}^2\right) \tag{7.3.4}$$

Note that although we are studying spatial selectivity, we use the classical literature convention of studying a *temporal* level crossing rate, N_t, by making the substitution $r = vt$, where v is the speed of the mobile receiver. The corresponding average fade durations are

$$E_z: \quad \bar{t} = \frac{\lambda}{\sqrt{2\pi}v\rho_{\text{rms}}}\left[\exp\left(\rho_{\text{rms}}^2\right) - 1\right] \tag{7.3.5}$$

$$H_x: \quad \bar{t} = \frac{\lambda}{\sqrt{\pi}v\rho_{\text{rms}}}\left[\exp\left(\rho_{\text{rms}}^2\right) - 1\right] \tag{7.3.6}$$

$$H_y: \quad \bar{t} = \frac{\lambda}{\sqrt{3\pi}v\rho_{\text{rms}}}\left[\exp\left(\rho_{\text{rms}}^2\right) - 1\right] \tag{7.3.7}$$

These expressions exactly match the original solutions presented by Clarke in [Cla68].

Now substitute the channel shape factors into the approximate spatial autocovariance functions in Equation (7.2.10). The results for the three cases are

$$E_z: \quad \varrho_R(\Delta r) = \exp\left(-23\left[\frac{\Delta r}{\lambda}\right]^2\right) \tag{7.3.8}$$

$$H_x: \quad \varrho_R(\Delta r) = \exp\left(-11.5\left[\frac{\Delta r}{\lambda}\right]^2\right) \tag{7.3.9}$$

$$H_y: \quad \varrho_R(\Delta r) = \exp\left(-34.5\left[\frac{\Delta r}{\lambda}\right]^2\right) \tag{7.3.10}$$

These three functions are compared to their more rigorous analytical solutions in Figure 7.4, Figure 7.5, and Figure 7.6. (See Appendix 7.B for a derivation of Clarke's unit autocovariance functions for envelope.) Note that all three model the spatial autocovariance function consistent with the approximation made in the derivation of Equation (7.2.10). The behavior is nearly exact for values of Δr equal to or less than a correlation distance.

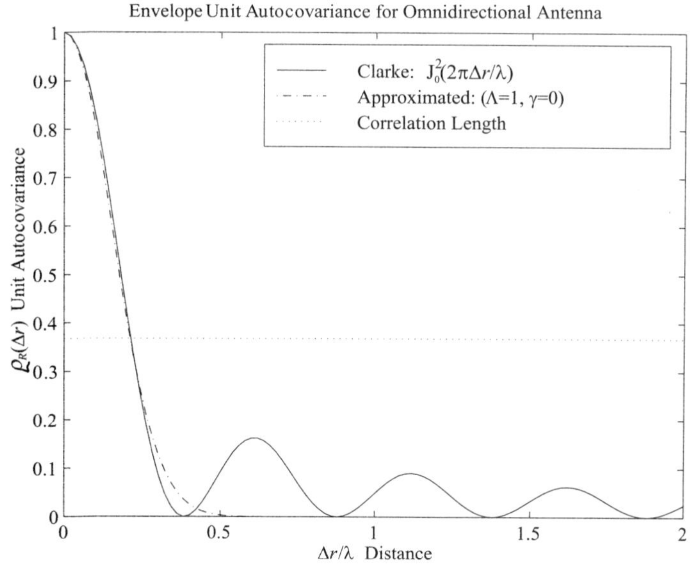

Figure 7.4 Comparison of Clarke theoretical and approximate envelope autocovariance functions for E_z-case.

7.3.3 Additional Comments

The shape factor technique for finding fading statistics is an intuitive way to relate the physical channel characteristics to the fading behavior. In the previous

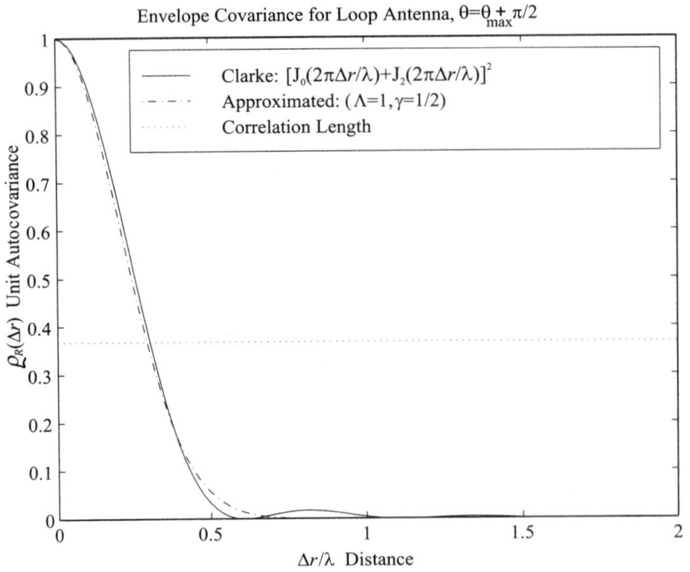

Figure 7.5 Comparison of Clarke theoretical and approximate envelope autocovariance functions for H_x-case.

examples, the spatial parameters may be calculated analytically or even estimated intuitively by simply looking at the distributions of multipath power in Figure 7.3. The use of spatial parameters to find level crossing rate, average fade duration, and spatial autocovariance is quite simple when compared to the full analytical solutions of the E_z-, H_x-, and H_y-cases presented in [Jak74]. The proposed solution is also more comprehensive. For example, once the shape factors have been found, Equation (7.1.11), Equation (7.1.12), and Equation (7.2.10) provide statistics for *all* directions of travel for the H_x- and H_y-cases, and not just specific directions such as $\theta_R = 0$ or $\theta_R = \frac{\pi}{2}$. Thus, specific fading behaviors for various directions of receiver motion are modeled easily.

The solution form of Equation (7.1.11), Equation (7.1.12), and Equation (7.2.10) reveals an interesting property about statistics in Rayleigh-fading channels. Since the three shape factors depend on only low-order Fourier coefficients, many of the second-order statistics of Rayleigh-fading channels are insensitive to the higher order multipath structure. The general biases of angular spread and angular constriction truly dominate the space and time evolution of these fading processes.

7.4 Properties of Wideband Channels

Radio channels are inherently band-limited, implying that they may be sampled at the *Nyquist* rate. This result simplifies the simulation and analysis of complicated wideband channels. This section discusses resulting properties of this useful characteristic.

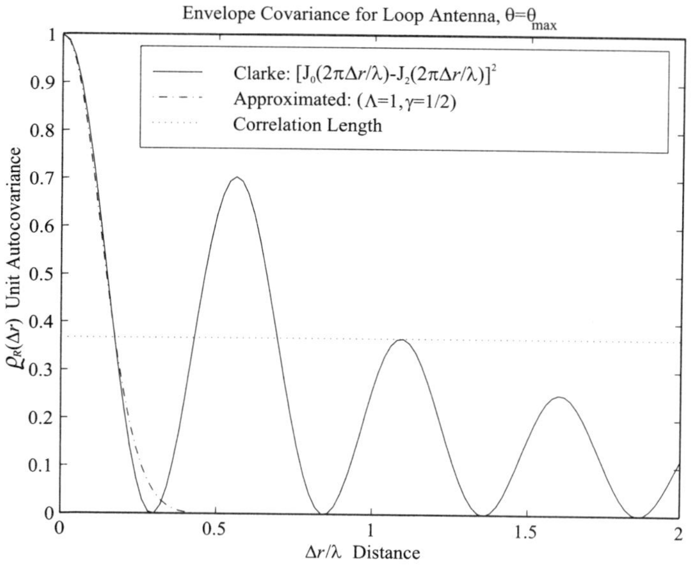

Figure 7.6 Comparison of Clarke theoretical and approximate envelope autocovariance functions for H_y-case.

7.4.1 Discrete Wideband Channels

Since radio systems operate in a band-limited channel, the channel itself may be sampled just like the received signal. This sampling allows the representation of channel effects using the discrete tap-delay line of Figure 7.7. As is the convention in digital signal processing, the boxes of z^{-1} represent the time delay of one symbol period (according to the popular z-transform). The tap-delay line filter takes samples of the transmitted signal (digital symbols) as inputs and produces samples of the received signal as outputs. The characteristics of the time dispersion are controlled by the complex coefficients, $\{\tilde{H}_i\}$.

The general model for the impulse response of the channel, $\tilde{H}(\tau)$, is based on the following equation:

$$\tilde{H}(\tau) = \sum_{i=1}^{M} \tilde{H}_i \delta(\tau - [i-1]\Delta\tau) \tag{7.4.1}$$

where τ is the unit of propagation delay and $\Delta\tau$ is the sampling period of the channel; in a system simulation, the engineer chooses $\Delta\tau = \frac{1}{B}$, where B is the Nyquist bandwidth of the received signal. Each $\Delta\tau$-separated sample is called a *delay bin* [Rap02a].

In a discrete digital transmission, a set of input symbols, $\{\tilde{x}_k\}$, must be convolved with the channel $\tilde{H}(\tau)$ to produce a set of output symbols, $\{\tilde{y}_k\}$. The operation of

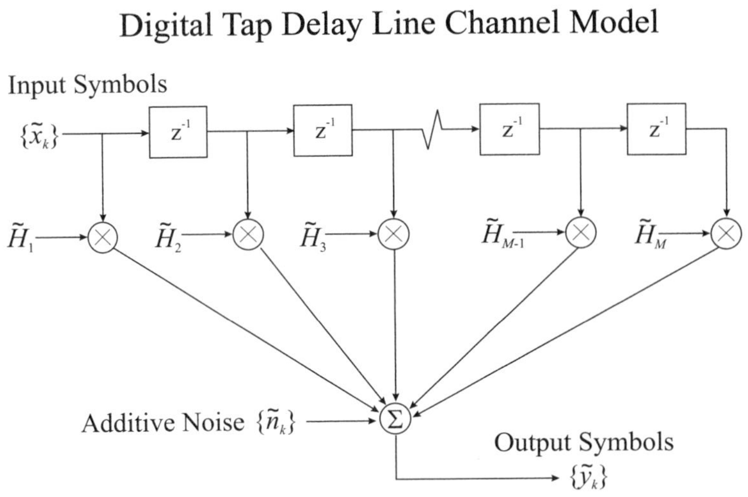

Figure 7.7 A discrete, tap-delay line channel model with additive white Gaussian noise.

continuous convolution is represented by the symbol ⊗ in the following equation:

$$\tilde{y}(\tau) = \frac{1}{2}\tilde{x}(\tau) \otimes \tilde{H}(\tau) \tag{7.4.2}$$

The discrete case is calculated for each output sample, $\tilde{y}_k$, in the following manner:

$$\tilde{y}_k = \frac{1}{2}\sum_{i=1}^{M} \tilde{H}_i \tilde{x}_{k-i+1} \tag{7.4.3}$$

The operation of Equation (7.4.3) mathematically represents the operation of the tap-delay filter of Figure 7.7. Note that Equation (7.4.2) and Equation (7.4.3) only apply if the channel is time-invariant (i.e., the coefficients $\{H_i\}$ do *not* change with respect to time).

> *Note: Multipath Components*
>
> Wireless engineers often speak of multipath *components* in wideband channel description. They are usually referring to the group of multipath waves that arrive within a single, specified delay bin at the receiver.

7.4.2 Time-Varying Wideband Channels

If the channel is *time-varying*, then the channel representation becomes more complicated. The additional dependency of time, t, must be added so that we study $\tilde{H}(\tau; t)$ instead of just $\tilde{H}(\tau)$. Instead of a set of channel samples, $\{\tilde{H}_i\}$, as in Figure 7.7, we would require a *matrix* of channel samples, $\tilde{\mathbf{H}}$. The rows of the channel matrix represent samples with respect to delay - the delay bins - each separated by $\Delta\tau$ units. The columns of the channel matrix represent samples with respect to time, separated by Δt units. This structure is illustrated in Figure 7.8.

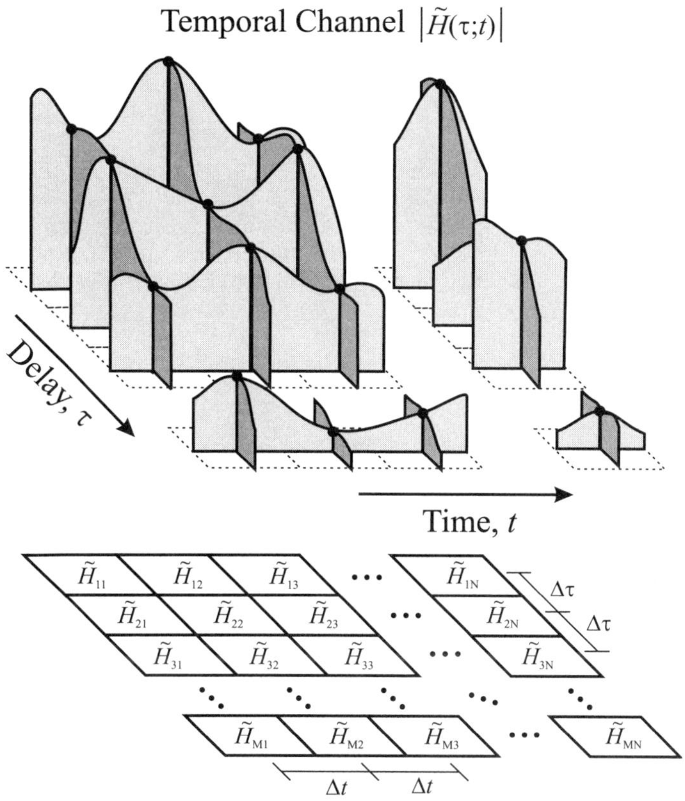

Figure 7.8 An example of how the complex, baseband radio channel is sampled in time and delay to form a matrix.

Figure 7.8 demonstrates a principle of wideband channel modeling: discrete separability. Looking at Figure 7.8 from left to right, it appears to be a layered collection of narrowband, time-varying processes. In fact, much of wideband channel modeling and computer simulation takes advantage of this property [Has93].

Each delay bin within the wideband channel may be treated as an independent *narrowband* random channel process with unique fading properties. Thus, the fading is described not by a single PDF, but by a collection of independent PDFs:

$$\tilde{H}(\tau;t) \Longrightarrow \{\tilde{h}_i(t)\} \longrightarrow \{f_{Ri}(\rho)\}$$

where the subscript i denotes the corresponding delay bin. This separability also applies to the power spectrum. The wideband, time-varying channel is described by a collection of independent Doppler spectra, each with its own distinct Doppler spread:

$$S_{\tilde{h}}(\tau, \omega) \Longrightarrow \{S_{\tilde{h}i}(\omega)\} \longrightarrow \{\sigma_{\omega i}\}$$

So everything we have learned about characterizing channels with a single time–Doppler dependency - level crossing rates, average fade durations, correlation behavior, and so on - still applies to the wideband channel. Analysis only needs iteration through each delay bin.

Note: Independence Between Delay Bins

The statistical independence of multipath components in different bins is a common assumption in wideband channel modeling. In our terminology, an I-SLAC model clearly fits this assumption, but it is not the *only* type of SLAC model with independent channel bins in delay. The condition of independence between delay bins in a SLAC model is weaker than the requirement of independent phases for *all* constituent waves.

Of course, what is true for time-varying wideband channels is also true for space-varying wideband channels. Each space-varying wideband channel is separated into a discrete set of independent space-varying channels. The first-order fading is described by a collection of independent PDFs:

$$\tilde{H}(\tau; r) \Longrightarrow \{\tilde{h}_i(r)\} \longrightarrow \{f_{Ri}(\rho)\}$$

Likewise, the delay–wavenumber spectrum may be broken into a set of unique wavenumber spectra,

$$S_{\tilde{h}}(\tau, k) \Longrightarrow \{S_{\tilde{h}i}(k)\} \longrightarrow \{\sigma_{ki}\}$$

or even a set of azimuth spectra with corresponding shape factors,

$$S_{\tilde{h}}(\tau, k) \Longrightarrow \{p_i(\theta)\} \longrightarrow \{\Lambda_i, \gamma_i, \theta_{\max i}\}$$

Once again, this divide-and-conquer approach simplifies channels with multiple dependencies so that single-dependency analysis can be used.

7.4.3 Discrete Transmission

Almost all cases of broadband communications involve *slow-fading* wireless channels, where the rate at which the channel changes with respect to time is much slower than the time it takes to transmit a digital symbol. In more formal terms, the coherence time, T_C, of the channel is much larger than the symbol period of transmitted digital symbols. In fact, it is also likely that an entire *frame* or *packet* of data may be transmitted well within the coherence time of a time-varying wideband channel. This is a useful assumption because many digital systems operate on a frame-by-frame or packet-based transmission scheme.

The matrix representation of Figure 7.8 lends itself to a frame-by-frame simulation in a time-varying wideband channel. If there are K transmitted symbols in a frame of data, then the total time to transmit that frame is $K\Delta\tau$. When this total frame time is less than the coherence time, we may employ a frame-by-frame convolution of simulation inputs with the channel matrix. Since the channel remains relatively constant over the frame period, the first frame of input symbols is convolved with the first column of channel samples in the channel matrix. The next frame of input symbols is convolved with the second column, and so on, until the simulation is finished.

To describe this with more formal mathematics, assume we have an M-by-N channel matrix, $\tilde{\mathbf{H}}$. The time sampling interval, Δt, is chosen to be the time between frame transmissions. The output symbol $\tilde{y}_{kj}$ - the kth output symbol of the jth frame - may be calculated using the following equation:

$$\tilde{y}_{kj} = \frac{1}{2}\sum_{i=1}^{M} \tilde{H}_{ij}\tilde{x}_{[k-i+1]j} \tag{7.4.4}$$

In this notation, $\tilde{x}_{kj}$ is the kth input signal of the jth frame. Note that Equation (7.4.4) and the frame-by-frame method works best for data frames or packets that transmit in regular or semiregular intervals, Δt. On the other hand, the method is trivial to adapt for data frames of varying length.

7.4.4 Notes on Temporal Modeling

Stationary radio channels are susceptible to small temporal variations due to motion within the propagation environment. For outdoor environments, the moving scatterers are most likely tree leaves rustling in the wind, automobile traffic, and pedestrian groups. For indoor environments, moving scatterers include elevators, doors, and moving people. As discussed in Chapter 2, the motion of these scatterers introduces a small, fluctuating component, $\delta\tilde{H}(\tau;t)$, to a larger nonfluctuating component, $\tilde{H}(\tau)$:

$$\tilde{H}(\tau;t) = \tilde{H}(\tau) + \delta\tilde{H}(\tau;t)$$

The nonfluctuating component, $\tilde{H}(\tau)$, is the static channel, and the fluctuating component, $\delta\tilde{H}(\tau;t)$, is the transient channel. For temporal channel modeling, we are most interested in modeling the transient channel in a way that represents the physics of moving channel objects.

The wideband Doppler spectrum of the transient channel is written as $S_{\delta\tilde{h}}(\tau,\omega)$. If the channel is time-invariant, then the Doppler spectrum vanishes; if the channel varies rapidly with respect to time, then there is a large amount of power and frequency content added into the Doppler spectrum. Thus, the amount of fluctuation depends on the power and frequency of the Doppler spectrum; in turn, the power and the frequency of the Doppler spectrum depend on the size and speed of the moving scatterer. Larger objects introduce components with larger powers into the Doppler spectrum. Faster objects introduce components with higher Doppler frequencies into the Doppler spectrum.

Most scatterers that move in a radio channel generate very small frequency shifts in the Doppler spectrum of the transient channel. For example, Figure 7.9 illustrates the typical Doppler frequencies of scatterers in an outdoor environment at a radio carrier frequency of 2 GHz. Large scatterers, such as terrain and buildings, produce no movement and only contribute to the static channel. Tree leaves blowing in the wind produce the largest amounts of transient scatter, but are slow in speed and produce only 2 Hz to 4 Hz of Doppler frequency. Contrast that with highway traffic, which creates larger Doppler shifts of up to 100 Hz to 200 Hz in the 2 GHz band; the total power of this contribution, however, is low.

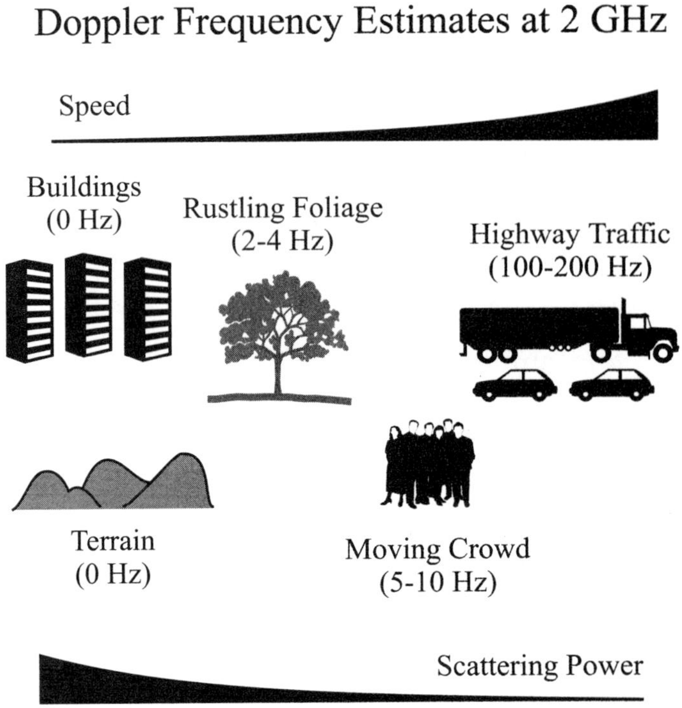

Figure 7.9 Scatterers that contribute the most power are likely to have the smallest Doppler frequencies.

In fact, there is a general trend among scatterers in temporal channel modeling. We refer to this trend as the big-slow rule, which is summarized below:

> **Big-Slow Rule:** In most radio channels, bigger objects scatter more power but are less likely to move.

This simple rule is true for any wireless link, indoor or outdoor. Despite its simplicity, it provides powerful insight for organizing a temporal channel model.

For example, it provides some hints about what a realistic transient Doppler spectrum might look like. Transient Doppler spectra should be symmetric and monotonically decreasing about $\omega = 0$. The symmetry arises because positive and negative Doppler frequencies (corresponding to forward and backward motion) should occur in similar degrees. The monotonicity arises from the big-slow rule: As the magnitude of Doppler frequency increases, the power contribution in the transient channel diminishes. (A Gaussian Doppler spectrum is a great starting point because it not only has symmetric and monotonic properties, but it is also very simple analytically to generate and characterize.) Also, we can expect a time-varying radio link to follow a Rician distribution, as discussed in the next section.

7.4.5 Rician Fading in Time-Varying Channels

The bulk of transient power arriving at a receiver is a conglomerate of many waves scattered by small moving objects. Thus, we can assume with impunity that the transient power is *diffuse*. Recall, however, that at each delay time, τ, this diffuse, Rayleigh power will be added to a static channel, $\tilde{H}(\tau)$. The envelope of a *complete* channel, $\tilde{H}(\tau; t)$, at any given point in delay time will follow a Rician distribution. A Rician distribution arises whenever a Rayleigh-fading complex voltage is added to a specular (i.e., static) voltage value [Dur99c].

Recall that the shape of a Rician envelope distribution is characterized by a K-factor, the ratio of specular power to diffuse power. Our wideband K-factor as a function of delay is best defined by the following formula:

$$K(\tau) = \int_{\tau-\frac{1}{2}\Delta\tau}^{\tau+\frac{1}{2}\Delta\tau} S_{\tilde{h}}(\tau)\, d\tau \Bigg/ \int_{\tau-\frac{1}{2}\Delta\tau}^{\tau+\frac{1}{2}\Delta\tau} S_{\delta\tilde{h}}(\tau)\, d\tau \qquad (7.4.5)$$

where $\Delta\tau$ is the resolution or averaging window, often the symbol period of the communication link, as in Equation (7.4.1). A K-factor of 0 indicates that the static portion of the channel has vanished at τ and that only transient power remains. Since transient power is Rayleigh-distributed, the envelope distribution of the complete channel is also Rayleigh for this case. An extremely large K-factor indicates that power in the static portion is large compared to the transient power; such a channel does not experience deep temporal fades.

A computer-generated example of a wideband, temporal channel model is shown in Figure 7.10. The RMS delay spread is 30 ns, and the Doppler spread of each delay bin is 5 Hz. The ratio of the static channel power to the diffuse transient channel power (the total Rician K-factor) for Figure 7.10 is 10 dB. Figure 7.10 is a good template for visualizing how wideband temporal channels may be broken into a set of simpler time-varying channels.

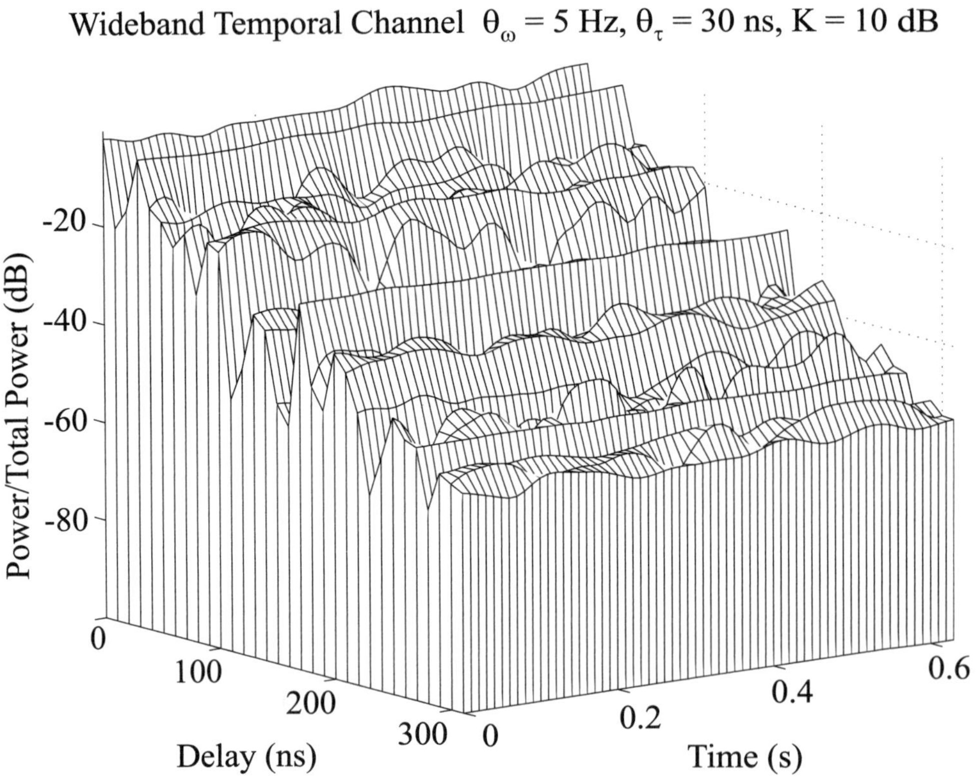

Figure 7.10 Computer-generated example of a wideband temporal channel $|\tilde{H}(\tau; t)|$ with Rician $K = 10$ dB.

> *Note: Racking up the Ks*
>
> For the case of *temporal* fading with fixed receiver/transmitter position, each location within a local area may have a fading envelope PDF with a different Rician K-factor; the diffuse transient power is, on average, the same, but the specular component follows the I-SLAC PDFs describing the spatial and frequency fading - which may *also* be Rician, but with a different K-factor!

7.5 Chapter Summary

This chapter presented a number of second-order fading statistics involving the envelopes of Rayleigh-fading channels. Below is a summary of the key points of the chapter:

- Level crossing statistics for Rayleigh-fading channels are solely dependent on RMS spectrum spreads.

- ▷ The *level crossing rate* describes how often an envelope threshold is crossed.
- ▷ The *average fade duration* describes how often a signal spends below a threshold, once crossed.
- ▷ Level crossing statistical expressions found for one channel dependency lead to expressions for all dependencies if duality is applied.

■ Correlation statistics for Rayleigh-fading channels may be calculated from RMS spectrum spreads as well.

- ▷ The resulting unit autocovariance functions are approximate, but they characterize the correlation behavior over the most important regions of the function.
- ▷ Formal definitions for coherence time, bandwidth, and distance may be based on these unit autocovariance functions.
- ▷ Using our unified approach, *joint* space–frequency autocovariance functions are also easy to approximate.

■ The use of *shape factors* in our statistics allows characterization of level crossing rate, average fade length, and spatial unit autocovariance that depends solely on several geometrical properties of multipath angle-of-arrival. Shape factor expressions match well with several analytical baseline cases of multipath propagation.

■ Wideband channels may be represented by a discrete set of samples, corresponding to regularly spaced *delay bins.*

- ▷ By extension, time- or space-varying wideband channels may be simplified and viewed as a collection of narrowband channels.
- ▷ These principles were illustrated by developing wideband, time-varying channel models.

Many of the statistics discussed in this chapter have useful applications in the design and performance prediction of wireless modems. Designs involving frequency hopping, diversity, space–time coding, adaptive equalization, interleaving, or error-correction coding all require knowledge of these received envelope statistics.

PROBLEMS

1. Suppose that a set of $\{t_i\}$ denotes all the points in time, t, in which a normalized envelope process is equal to an RMS threshold level, $R(t) = \rho_{\text{rms}}$. Let us define a *crossing-counter* function that registers a unit-amplitude delta function whenever the envelope process crosses this threshold:

$$\varphi(t) = \sum_i \delta(t - t_i)$$

a. Prove the following equality:

$$\varphi(t) = \left|\frac{dR(t)}{dt}\right| \delta(R(t) - \rho_{\text{rms}})$$

b. Derive an expression based on $\varphi(t)$ for counting the ρ_{rms} threshold crossings of $R(t)$ on the interval $[b, a]$.

c. **Challenge:** Use your expression in (b) to derive Equation (7.1.1).

2. Write expressions for temporal level crossing rate and average fade duration for a Rayleigh-fading envelope in terms of *power* thresholds instead of envelope thresholds.

3. A digital communication system transmits packets of data that are, on average, 600 μs in duration. This link will operate in Rayleigh fading and can tolerate signal fading up to -20 dB below average received power before packets are dropped. The Doppler spread of the time-varying wireless channel is 30 Hz ($\sigma_\omega = 60\pi$ rad/s).

a. What is the coherence time, T_c, of this channel (based on $\varrho_R < 0.5$)?

b. How often does the channel fade beneath an acceptable signal strength threshold?

c. If packets are sent continuously (one right after another), estimate what percentage of packets contain unacceptable signal fades.

d. What is the average fade duration for unacceptable signal strengths?

e. If a signal fade occurs within a single packet, estimate the percentage of time, on average, that the packet spends with unacceptable signal strength.

4. A mobile handset receiver is traveling through space at a rate of 5 m/s. The channel is Rayleigh fading with an omnidirectional azimuth spectrum and the carrier frequency is 850 MHz.

a. What is the coherence distance for this channel? The coherence time (based on $\varrho_R < \exp(-1)$)?

b. What is the level crossing rate for a threshold of 8 dB beneath the RMS signal envelope?

c. What is the average fade duration for a threshold of 8 dB beneath the RMS signal envelope?

d. Repeat parts (b) and (c) for signal thresholds of 12 dB and 18 dB beneath the RMS signal envelope.

5. Repeat the previous problem with the following changes:

a. The mobile receiver is moving with a speed of 10 m/s at a carrier frequency of 1500 MHz.

b. The mobile receiver is moving with a speed of 10 m/s at a carrier frequency of 850 MHz and is now using a directional antenna with a 30° sectored antenna pattern. The motion of the receiver is in the direction of the beam center.

6. For the omnidirectional multipath azimuth spectrum, Jakes derives the following approximation for the wavenumber spectrum of narrowband, spatial envelope fading [Jak74]:

$$S_R(k) = \frac{1}{k_0}\mathrm{K}\left(\sqrt{1-\left(\frac{k}{2k_0}\right)^2}\right)\ \mathrm{u}(2k_0 - |k|)$$

where $\mathrm{K}(\cdot)$ is the complete elliptical integral of the first kind (see Appendix A.5). Answer the following questions in reference to this PSD:

a. What is the wavenumber spectrum of this envelope process using the approximate unit autocovariance in Equation (7.2.8)?

b. Plot the approximations from Jakes and part (a) on the same graph.

c. When plotted as unit autocovariance functions (in Figure 7.4), these functions are in close agreement over their most useful range. Other than their RMS widths, the functions you plot in (b) should have markedly different shapes. How do you explain this disagreement in the PSDs?

7. Assume that a space-varying envelope process, $R(r)$, and its space-derivative, $R'(r)$, are independent processes. Now consider the following form for spatial unit autocovariance in terms of shape factors:

$$\varrho_R(\Delta r) \approx \exp\left(-\frac{\pi^2\Lambda^2\left[1+\gamma\cos\left(2[\theta_R-\theta_{\max}]\right)\right]}{2\left[1-\mu^2\right]}\left[\frac{\Delta r}{\lambda}\right]^2\right) \qquad \text{where } \mu = \frac{\mathrm{E}\{R\}}{\sqrt{\mathrm{E}\{R^2\}}}$$

Answer the following questions:

a. Prove that the above approximation is valid (follow Appendix 7.A for help).

b. Solve for χ with Rayleigh statistics and show that the above autocovariance function converges to Equation (7.2.10).

c. Derive a similar frequency autocovariance for multipath with delay spread σ_τ.

d. Derive a similar temporal autocovariance for multipath with Doppler spread σ_ω.

e. Derive a spatial unit autocovariance for the Rician model of Figure 6.13 (Chapter 6). To approximate the mean of a Rician process, use Equation (7.C.1) in Appendix 7.C.

f. Show that your expression in (e) approaches classical behavior when $K = 0$.

g. Simulate the correlation properties of the Rician spatial model and compare to the approximation derived in (e). How well does this approximate autocovariance compare to simulated results?

h. Explain why this is a bad approximation for TWDP fading.

8. You are to design a wireless receiver with both frequency hopping and a linear equalizer to combat frequency selectivity. The bandwidth of the signal is B, and the average frequency hop length is f_{hop}.

a. A linear equalizer can be very sensitive to frequency coherence. The frequency correlation over the signal bandwidth must be higher than 0.9 for the linear equalizer to function properly. What relationship between B and RMS delay spread, σ_τ, does this imply?

b. For a frequency-hopping system to work properly, the frequency correlation over the average hop length must be less than 0.5. What relationship between f_{hop} and RMS delay spread does this imply?

c. What is the minimum average hop length, f_{hop}, in terms of the signal bandwidth, B?

d. Your wireless receiver is to operate in an indoor environment where the RMS delay spread never exceeds 100 ns. What is the most frequency-compact (smallest f_{hop} and B) that fits this design specification?

9. Use Equation (7.2.11) - the coherence distance in terms of shape factors - to answer the following questions:

a. In which azimuthal direction do we observe the shortest correlation distance, D_{min}? Write an expression for D_{min}.

b. In which azimuthal direction do we observe the longest correlation distance, D_{max}? Write an expression for D_{max}.

c. Based on your answers in (a) and (b), derive an expression for D_{av}, the average correlation distance based on the Pythagorian means of the shortest and longest correlation distances ($D_{\text{av}} = \sqrt{D_{\text{min}}^2 + D_{\text{max}}^2}$). How many shape factors does D_{av} depend on? Why is this so?

10. In Chapter 5, we studied the various envelope distributions for space–frequency channel models. In Section 7.4, we studied the Rician nature of temporal fading. Now consider the following channel model of space, time, and frequency:

$$\tilde{h}(f,\vec{r},t) = \underbrace{\tilde{h}(f,\vec{r})}_{\text{static}} + \underbrace{\delta\tilde{h}(f,\vec{r},t)}_{\text{transient}}$$

Answer the following questions, assuming that the total channel power has been normalized to 1:

a. The static channel follows a two-wave SLAC distribution with $\Delta = 0.5$, and the static and transient channels have equal average power. Write a PDF that describes the fading of the entire channel envelope, $R(f,\vec{r},t)$.

b. The static channel follows a Rician distribution with $K = 4$ dB. The average power in the static channel is twice the average power in the transient channel. Write a PDF that describes the fading of the entire channel envelope, $R(f,\vec{r},t)$.

c. The total envelope fading of the space–time–frequency channel follows a Rayleigh PDF. If measured at specific points in space, the fading appears to be Rician with an RMS K-factor value of 8 dB. Write a PDF that describes the fading of the transient channel envelope, $\delta R(f,\vec{r},t)$.

d. If measured at specific points in space, the envelope fading of the channel follows a Rayleigh PDF. Write a PDF that describes the fading of the entire channel envelope, $R(f,\vec{r},t)$.

7.A Approximate Spatial Autocovariance

Unit autocovariance of a WSS time-varying signal envelope process, $R(t)$, is defined as follows [Tur95]:

$$\varrho_R(\Delta t) = \frac{\overbrace{\mathrm{E}\left\{R(t)R(t+\Delta t)\right\}}^{C_R(\Delta t)} - \left(\mathrm{E}\left\{R\right\}\right)^2}{\mathrm{E}\left\{R^2\right\} - \left(\mathrm{E}\left\{R\right\}\right)^2} \tag{7.A.1}$$

We may express $\varrho_R(\Delta t)$ as the following Maclaurin series:

$$\varrho_R(\Delta t) = \sum_{n=0}^{\infty} \frac{\Delta t^{2n}}{(2n)!} \left.\frac{d^{2n}\varrho_R(t)}{dt^{2n}}\right|_{t=0} \tag{7.A.2}$$

Equation (7.A.2) contains only even powers of Δt, since any real autocovariance function is an even function. The differentiation of an autocorrelation function satisfies the following relationship [Pap91]:

$$\left.\frac{d^{2n}C_R(\Delta t)}{d\Delta t^{2n}}\right|_{\Delta t=0} = (-1)^n \mathrm{E}\left\{\left(\frac{d^n R(t)}{dt^n}\right)^2\right\} \tag{7.A.3}$$

and is useful for re-expressing the Maclaurin series:

$$\begin{aligned} \varrho_R(\Delta t) &= 1 + \frac{\sum\limits_{n=1}^{\infty} \frac{(-1)^n \Delta t^{2n}}{(2n)!} \mathrm{E}\left\{\left(\frac{d^n R(t)}{dt^n}\right)^2\right\}}{\mathrm{E}\left\{R^2\right\} - \left(\mathrm{E}\left\{R\right\}\right)^2} \\ &= 1 - \frac{\mathrm{E}\left\{\left(\frac{dR(t)}{dt}\right)^2\right\}}{2\left[\mathrm{E}\left\{R^2\right\} - \left(\mathrm{E}\left\{R\right\}\right)^2\right]} \Delta t^2 + \cdots \end{aligned} \tag{7.A.4}$$

For an envelope that follows a Rayleigh PDF, the denominator of Equation (7.A.4) evaluates to

$$\mathrm{E}\left\{R^2\right\} - \left(\mathrm{E}\left\{R\right\}\right)^2 = \left(1 - \frac{\pi}{4}\right) P_{\text{dif}} \tag{7.A.5}$$

Rayleigh fading processes also have the following unique property [Dur00b]:

$$\mathrm{E}\left\{\left(\frac{dR(t)}{dt}\right)^2\right\} = \frac{1}{2}\mathrm{E}\left\{\left|\frac{d\tilde{h}(t)}{dt}\right|^2\right\} = \frac{1}{2}\sigma_t^2 = \frac{1}{2}\sigma_\omega^2 P_{\text{dif}} \tag{7.A.6}$$

Now consider $\varrho_R(\Delta t)$ approximated by an arbitrary Gaussian function and its Maclaurin expansion:

$$\varrho_R(\Delta t) \approx \exp\left[-a\Delta t^2\right] = \sum_{n=0}^{\infty} \frac{(-1)^n a^n}{n!} \Delta t^{2n} = 1 - a\Delta t^2 + \cdots \tag{7.A.7}$$

A Gaussian function is chosen as a generic approximation to the true unit autocovariance, since it is a convenient and well-behaved correlation function. The appropriate constant a is chosen by setting equal the second terms of Equation (7.A.4) and Equation (7.A.7), ensuring that the behavior of both autocovariance functions is identical for small Δt:

$$a = \frac{1}{2\pi^2} \underbrace{\left[\frac{2\pi^2}{4-\pi}\right]}_{\approx 23.00} \sigma_\omega^2 \tag{7.A.8}$$

Therefore, the approximate unit autocovariance depends only on the RMS Doppler spread. The final result is presented in the main text as Equation (7.2.3).

7.B Classical Envelope Autocovariance

If the complete PSD of a complex baseband random process is known, there is a useful approximation for calculating the unit autocovariance of its *envelope*, R. As an example, let us say we have a random, time-varying channel, $\tilde{h}(t)$, and its temporal autocorrelation, $C_{\tilde{h}}(\Delta t)$, is known. The operation of taking the envelope, $R(t) = |\tilde{h}(t)|$, produces a random process with autocorrelation function, $C_R(\Delta t)$. Although there is no simple, exact relationship between complex channel and envelope autocorrelation functions, Jakes showed in [Jak74] that for a Rayleigh-fading process they are related by the following infinite series:

$$C_R(\Delta t) = \frac{\pi C_{\tilde{h}}(0)}{4}\left[1 + \frac{1}{4}\left|\frac{C_{\tilde{h}}(\Delta t)}{C_{\tilde{h}}(0)}\right|^2 + \frac{1}{64}\left|\frac{C_{\tilde{h}}(\Delta t)}{C_{\tilde{h}}(0)}\right|^4 + \cdots\right] \tag{7.B.1}$$

Since each term of the series is progressively smaller, the envelope autocorrelation function is relatively accurate if all but the first two terms are thrown away. Since the first term is a constant offset, a convenient approximation for the unit autocovariance of a Rayleigh envelope process is given by

$$\varrho_R(\Delta t) \approx \left|\frac{C_{\tilde{h}}(\Delta t)}{C_{\tilde{h}}(0)}\right|^2 \tag{7.B.2}$$

This approximation is equally valid for time-, space-, and frequency-varying channels. In contrast to the approximation developed in Appendix 7.A, Equation (7.B.2) requires complete knowledge of the baseband channel autocorrelation.

To illustrate the use of Equation (7.B.2), Table 7.2 presents the unit autocovariance approximations of three space-varying envelopes - the E_z, H_x, and H_y cases discussed in Section 7.3. From left to right, Table 7.2 shows how Clarke derived his original unit autocovariance expressions from the azimuth spectrum of incoming multipath.

Table 7.2 Overview of Classical Spatial Characterizations ($k_0 = \frac{2\pi}{\lambda}$) [Cla68]

$p(\theta) =$	$S_{\tilde{h}}(k) =$	$C_{\tilde{h}}(\Delta r) =$	$\varrho_R(\Delta r) \approx$
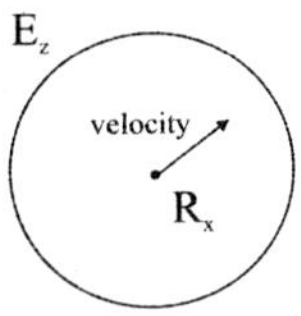	$\frac{\mathrm{u}(k_0 - \lvert k \rvert)}{\sqrt{k_0^2 - k^2}}$	$\frac{1}{2}\mathrm{J}_0\left(k_0 \Delta r\right)$	$\mathrm{J}_0^2\left(k_0 \Delta r\right)$
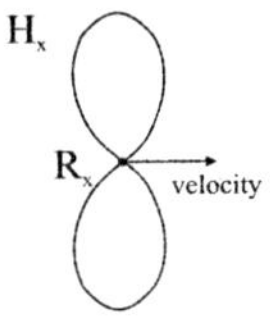	$\sqrt{k_0^2 - k^2}\,\mathrm{u}(k_0 - \lvert k \rvert)$	$\frac{k_0}{2\Delta r}\mathrm{J}_1(k_0 \Delta r)$	$\left[\mathrm{J}_0(k_0 \Delta r) + \mathrm{J}_2(k_0 \Delta r)\right]^2$
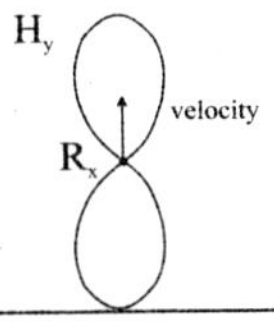	$\frac{k^2\,\mathrm{u}(k_0 - \lvert k \rvert)}{\sqrt{k_0^2 - k^2}}$	$\frac{k_0^2}{4}\left[\mathrm{J}_0(k_0 \Delta r) - \mathrm{J}_2(k_0 \Delta r)\right]$	$\left[\mathrm{J}_0(k_0 \Delta r) - \mathrm{J}_2(k_0 \Delta r)\right]^2$

7.C Rician Mean Approximation

The following algebraic expression is a useful way to characterize the mean of a Rician distribution as a function of K-factor:

$$\mu = \frac{\mathrm{E}\{R\}}{\sqrt{\mathrm{E}\{R^2\}}} \approx \frac{\sqrt{4K^2 + 6K + \pi}}{2(K+1)} \tag{7.C.1}$$

Equation (7.C.1) is exact for the pure Rayleigh case ($K = 0$) and for the limiting case of $K \longrightarrow \infty$. Elsewhere, the error in the approximation is always less than $\pm\%0.5$. Figure 7.11 graphs the approximation and the exact value, which was computed using small-step numerical integration. It is doubtful that an approximation more accurate than Equation (7.C.1) exists without greatly complicating the mathematics.

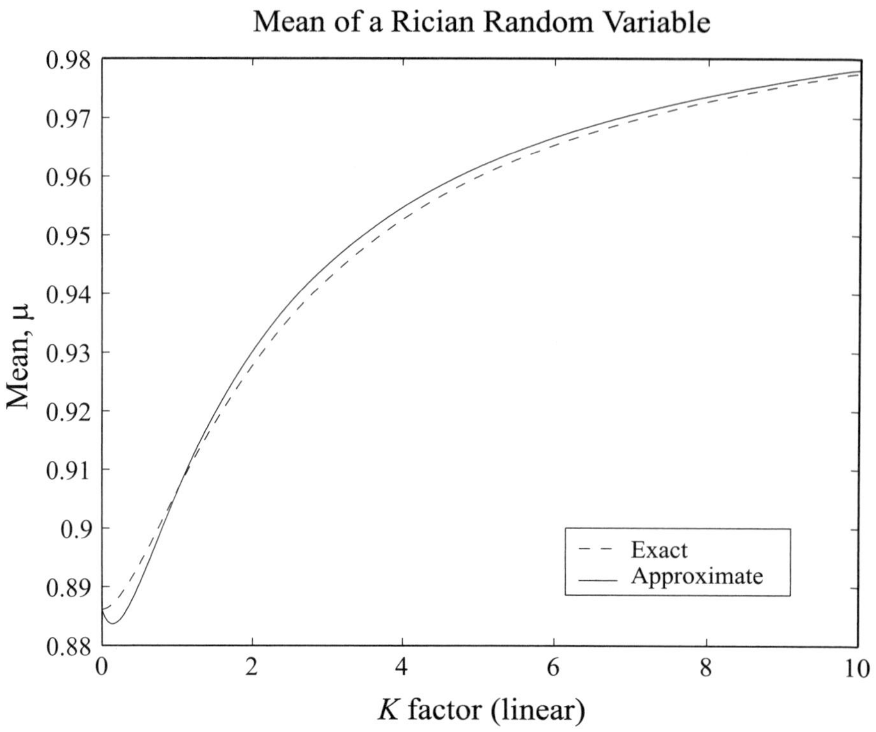

Figure 7.11 Comparison of the exact Rician mean and the algebraic approximation.

Chapter 8

DIVERSITY

Diversity is the best technique for overcoming multipath fading in a wireless channel without adding inordinate complexity to the receiver unit [Jak74]. Rather than rely on a single fading wireless channel, diversity receivers exploit multiple copies of the same signal received from *different* channels. Each channel in such a receiver is called a diversity *branch*. The more branches available to a diversity receiver, the more resistant to noise and bit errors the overall communications link. The reason: One branch may fail, but it is very unlikely that all branches will fail simultaneously.

This chapter is a general overview of diversity concepts. The different types of diversity as well as the range of combining algorithms are discussed. A summary of the contents of this chapter follows:

- Section 8.1: Overview of diversity types.
- Section 8.2: Summary of different branch combining techniques.
- Section 8.3: BER and channel capacity calculation.
- Section 8.4: Summary of important concepts.

Two of the most important things to remember in this chapter are the failure points of diversity: branch correlation and unequal branch power. These two types of pathological channel conditions will return to haunt us in the analysis of MIMO systems in Chapter 9 and antenna designs for multipath channels in Chapter 10.

8.1 Diversity Concept

There are many different types of diversity, although a few general principles govern the performance of any diversity design. This section develops the basic concepts and unifying characteristics behind multiple branch diversity schemes.

8.1.1 The Role of Diversity

The success or failure of wireless communications comes down to the *capacity* of the link, best measured in terms of bit rate for digital systems. It is the wireless channel that places the ultimate capacity limitation on the network. For a single *additive white Gaussian noise* (AWGN) channel, Shannon derived the upper-limit capacity in bits-per-second to be dependent on the signal-to-interference+noise ratio (SINR) and transmission bandwidth, B [Cov91]:

$$C = B \log_2 (1 + \text{SINR}) \ \text{bits/s} \tag{8.1.1}$$

where SINR is either ensemble- or time-averaged power levels of desired signal, $\tilde{y}(t)$, noise, $\tilde{n}(t)$, and interference, $\tilde{w}(t)$:

$$\text{SINR} = \frac{\mathrm{E}\left\{|\tilde{y}(t)|^2\right\}}{\mathrm{E}\left\{|\tilde{n}(t)|^2\right\} + \mathrm{E}\left\{|\tilde{w}(t)|^2\right\}} \tag{8.1.2}$$

This formulation assumes that interference, $\tilde{w}(t)$, is approximately Gaussian in its statistics - like the noise $\tilde{n}(t)$. Although the wireless channel is more complicated than the AWGN channel, there is basic truth here: For a given bandwidth, the SINR determines the absolute channel capacity. If a receiver does not receive enough signal power, no modulation or equalization technique will produce an acceptable data rate.

The channel capacity expression of Equation (8.1.1) demonstrates why fading is so crippling in a wireless link. Signal fades will lower the SINR to a point where reliable, fixed-rate digital communications cannot occur. The goal of diversity - whatever the type - is to use multiple signal branches to keep the overall SINR above this minimum acceptable level.

> *Note: The Elusive Shannon Limit*
>
> The Shannon limit on channel capacity of a wireless link is almost trivial to calculate. *Achieving* that limit requires a momentous engineering effort. Communications engineers are more fortunate than most, however, because at the very least they *know* when they have found the best design.

8.1.2 Antenna Diversity

In this book, we divide the different types of diversity into two categories: *antenna* diversity and *temporal* diversity. The goal of antenna diversity is to overcome *spatial selectivity* in the wireless channel caused by small-scale fading. Below is a summary of the different types of antenna diversity:

Space Diversity: A receiver that uses space diversity has multiple, similar antennas (co-polar monopole antennas, for example), each providing a separate channel for wireless communications. In a pure space diversity design, the pattern and polarization of each antenna element is identical.

Polarization Diversity: This diversity scheme uses two or more differently polarized antennas to provide independent channels at the receiver.

Pattern Diversity: This type of diversity uses antennas with different patterns. These varied antenna patterns recombine multipath in a way that produces a different channel for each pattern, even if the antennas are colocated.

Transmit Diversity: Transmit diversity is not so much a unique type of antenna diversity as a philosophy for using *other* types of antenna diversity. Rather than use multiple antennas at the receiver for space, polarization, and/or pattern diversity, it is possible to use multiple antennas at the *transmitter* to control fading experienced by the receiver end. Transmit diversity often requires some closed-loop control between the receiver and transmitter to obtain the channel state, although we will discuss methods for circumventing this limitation in Chapter 9.

It should be noted that an antenna diversity design can incorporate *any combination* of these diversity types to overcome fading in a spatially-selective multipath channel.

Indeed, a wireless link could incorporate all four types, using transmit diversity that has cross-polar, spatially-separated antenna elements with different patterns. Clearly there is a great deal of freedom in designing antenna diversity for a wireless system.

> *Note: Is It Pattern or Is It Space?*
>
> Array theory shows us that coherent combination of identical antenna elements separated in space is equivalent to changing the overall pattern of a single arrayed antenna. For this reason, many engineers refer to this form of space diversity as either pattern diversity or *switched-beam* diversity. For the purposes of this book, we will refer to this strictly as *space* diversity. We reserve the term *pattern* diversity for schemes whose individual antenna elements have different spatial patterns. Not only does this distinction make physical sense, but the end-to-end performance of pattern diversity systems is quite different from arrayed antenna elements.

8.1.3 Temporal Diversity

There are other types of multipath fading that involve time-varying or frequency-varying wireless channels. Diversity schemes that seek to overcome this type of fading fall under the category of *temporal* diversity. Temporal diversity includes any scheme that obtains additional channels of information using signal processing. There are three common types of temporal diversity:

Frequency Diversity: This is the type of diversity used in either *orthogonal frequency division multiplexing* (OFDM) or *frequency-hopped spread spectrum* systems. The goal is to exploit a large open or shared frequency band and send multiple signals at different carrier frequencies [Wu95].

Code Diversity: Like frequency diversity, code diversity is a method for obtaining multiple signal branches to fight the frequency-selective nature of the channel. Used in *direct-sequence spread spectrum* systems, code diversity takes advantage of the unique multipath-filtering properties of a *Rake* equalizer. Copies of the transmitted signal that arrive with largely different time delays at the receiver can be separated by the "fingers" of a Rake equalizer into independent signals, resulting in diversity branches [Lib99].

Time Diversity: Time diversity is a method for overcoming fades in a time-varying channel (a mobile wireless channel, for example). In practice, a receiver that uses *error-correction coding* and *interleaving* is actually using a form of time diversity, although most communications engineers do not think of these techniques in terms of diversity.

Like antenna diversity, combinations of temporal diversity techniques are possible.

In general, temporal diversity techniques are understood very well by the communications industry because they are suited for *baseband* signal processing. Antenna diversity techniques must be coupled with the *radio frequency* hardware of a receiver, and have not been exploited by the wireless industry to a large extent. Antenna diversity also requires a more intimate understanding of multipath propagation.

It is important to study antenna and temporal diversity together, however. First, regardless from where the signal branches come, there is a uniform approach to diversity combining and performance estimation. Second, the presence of one diversity technique affects the performance of another. For example, the code diversity gain attained by a spread spectrum receiver may also mitigate the small-scale spatial fading, thereby eliminating the need for antenna diversity.

8.1.4 Diversity Failure

There are two potential failure points for diversity, and in general practice they are mutually exclusive. These problems depend on the signal powers of the individual diversity branches with respect to one another.

Types of diversity that have signal branches with equal average power are best. However, such types of diversity that guarantee equal power branches are vulnerable to *branch correlation*. Remember that the whole reason for having diversity is to use other strong signal branches to compensate for fades in a single branch. If correlation exists between diversity branches, then there is a much larger probability of *concurrent fading* of all available signals, providing no true diversity gain. Space, frequency, and time diversity are most susceptible to branch correlation.

Some types of diversity, such as code, polarization, and pattern, are *not* susceptible to branch correlation. It is a cruel twist of physics, however, that these diversity types are subject to an entirely different problem: unequal average powers. A wireless channel is unpredictable, so the strength of each diversity branch will never be known before operation. For a given overall received power level, diversity works best if that power is, on average, split evenly between the branches. If one diversity branch contains the bulk of the average received power, then there is little that the other diversity branch or branches can do to overcome a deep fade in the dominant branch. And overcoming deep fades is the whole point of diversity.

So, signal correlation and dominant branches are the two key failure points for a diversity design - and every type of diversity will be vulnerable to one of these problems. There are other factors that an engineer may have to consider in choosing a diversity type for a wireless system. A summary of diversity types and their drawbacks (roughly) are listed in Table 8.1.

Table 8.1 Different Types of Diversity Used by a Receiver

Diversity Type	Unequal Power Branches	Correlation Possible	Closed-Loop Control	Multiple Antennas Needed	Large Bandwidth Needed
Time		X			*
Frequency		X			X
Code	X				X
Space		X		X	
Transmit		X	X	†	
Polarization	X			X	
Pattern	X	X		X	

* One could argue that redundant bit transmission requires increased bandwidth to maintain a constant *real* data rate.

† Multiple antennas are used at the *transmitter*.

8.2 Combining Techniques

The previous section discussed *where* a receiver can get different diversity branches. This is only half of the story. If N diversity branches are available to a receiver, then the receiver must combine or select these branches in a way that increases the performance of the communications link. This section discusses the various combining techniques that are used in wireless communications. Use Figure 8.1 to classify the different combining techniques.

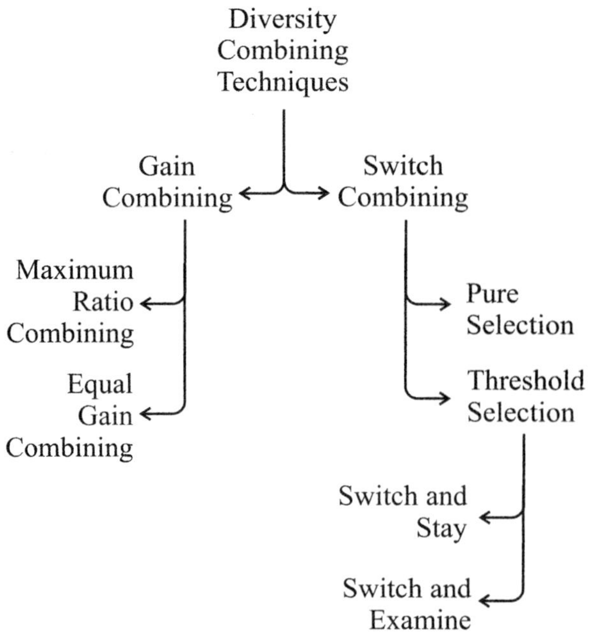

Figure 8.1 A graphical overview of diversity-combining algorithms.

8.2.1 Gain Combining

Gain-combining techniques result in outstanding link performance at the price of a complicated, expensive receiver structure. In gain combining, the final signal is the weighted sum of *all* diversity branches available to the receiver. How the signals are weighted determines the type of gain combining. The two most common schemes are listed below:

Maximum Ratio Combining (MRC): The goal of MRC is to weight individual diversity branches in a way that maximizes the output signal's SINR. For reasons discussed in Section 8.1.1, maximizing SINR is the critical step for maximizing channel capacity. Thus, MRC is the best diversity-combining technique in terms of link performance. MRC is complicated, though, because it requires an SINR estimation algorithm to compute branch weights.

Equal Gain Combining (EGC): This technique, while not optimal, is much simpler than MRC. EGC simply sums the signal envelopes from each branch with equal weights to produce a single output.

In practice, MRC and EGC are expensive techniques to incorporate into mass-produced receivers. Gain-combining techniques require simultaneous demodulation of each diversity branch. A receiver that has radio frequency hardware and signal processing for N antenna diversity branches becomes extremely expensive.

Mathematically, gain-combining algorithms are represented as the sum of N signals using a set of complex weights $\{\tilde{\alpha}_i\}$. The output of a combining algorithm, $\tilde{z}(t)$, for a simple narrowband N-diversity scheme is given by

$$\tilde{z}(t) \;=\; \sum_{i=1}^{N} \tilde{\alpha}_i\, \tilde{y}_i(t) \;=\; \sum_{i=1}^{N} \tilde{\alpha}_i \left[\tfrac{1}{2}\tilde{h}_i\, \tilde{x}(t) + \tilde{n}_i(t)\right] \qquad (8.2.1)$$

where $\{\tilde{y}_i(t)\}$ are the received complex signal branches, $\{\tilde{n}_i(t)\}$ are the noise (or interference) signals on each branch, $\tilde{x}(t)$ is the original transmitted signal, and $\{\tilde{h}_i\}$ are the channel weights of each branch (this case shows static channel coefficients to simplify our exposition). While the combining weights, $\{\tilde{\alpha}_i\}$, may be complex-valued, their phases are always chosen to remove the relative phase shifts introduced by the channel weights, $\{\tilde{h}_i\}$. Thus, the combining weights always have the following form, regardless of combining algorithm:

$$\tilde{\alpha}_i = \alpha_i \frac{\tilde{h}_i^*}{R_i} \qquad \left[\text{Recall: } \frac{\tilde{h}_i}{R_i} = \exp\left(j \arg\{\tilde{h}_i\}\right)\right] \qquad (8.2.2)$$

where R_i is the envelope, $|\tilde{h}_i|$, of the ith channel. The choice of real-valued weight magnitudes, $\{\alpha_i\}$, depends on the combining method.

In comparative studies where the incoming branch signals contain the same average noise power, the set of combination weights, $\{\alpha_i\}$, are normalized to have the following property:

$$\sum_{i=1}^{N} \alpha_i^2 = 1 \qquad (8.2.3)$$

This normalization assures that the total noise power in the diversity branches is identical to the noise power in the diversity output signal, $\tilde{z}(t)$. Thus, any gain from combining is due strictly to signal-processing and not to raw amplification of noise *and* desired signals.

For the EGC case, simply set all values of $\alpha_i = \frac{1}{\sqrt{N}}$ to combine N signal branches. The resulting EGC output, $\tilde{z}(t)$, may be written as

$$\tilde{z}(t) = \frac{1}{\sqrt{N}} \sum_{i=1}^{N} \left[\frac{1}{2} R_i \tilde{x}(t) + \tilde{n}_i(t)\right] \qquad (8.2.4)$$

For the MRC case, each branch weight takes on a value proportional to its own envelope (assuming similar noise levels in each signal branch). In terms of Equation (8.2.1) and Equation (8.2.3), the MRC weights are given by the expression

$$\alpha_i = \frac{R_i}{\sqrt{\sum_{i=1}^{N} R_i^2}} \qquad (8.2.5)$$

With these MRC weights, we can reduce Equation (8.2.1) to

$$\tilde{z}(t) = \frac{1}{2}\tilde{x}(t)\underbrace{\sqrt{\sum_{i=1}^{N} R_i^2}}_{R} + \frac{\sum_{i=1}^{N} R_i\,\tilde{n}_i(t)}{\sqrt{\sum_{i=1}^{N} R_i^2}} \tag{8.2.6}$$

In this case, the value R is the effective channel envelope of the input signal, $\tilde{z}(t)$. Contrary to gain combining, the switch combining diversity techniques in the following section do not require weights.

> *Note: MMSE Combining*
>
> Some texts discuss a diversity algorithm called *minimum mean-squared error* (MMSE) combining. This is essentially the same as MRC. The term MMSE specifies a type of algorithm for calculating the MRC weights by sending a known training sequence through the channel to sound out the channel weights $\{\tilde{h}_i\}$. The MMSE algorithm selects the weights that maximize the SINR of $\tilde{z}(t)$, as required by MRC.

8.2.2 Signal Envelope for Gain Combining

Since MRC is the optimum combining scheme for any general N-branch diversity scheme, it is a useful baseline for understanding the operation and gains of diversity. Studying Equation (8.2.6), we can see that the transmitted signal, $\tilde{x}(t)$, is multiplied by a real-valued envelope, R, which is the RMS combination of the fading branch envelopes, $\{R_i\}$. In this way, the total signal envelope, R, smoothes over the deep fades that may exist in the individual branch envelopes.

As a canonical case, consider signal branches that have uncorrelated Rayleigh-fading envelopes with equal average power, P_{dif}. For this scenario, the power of the MRC output, $P = R^2$, follows a *Chi-squared* distribution of order $2N$ [Rap02a]:

$$f_P(p) = \frac{p^{N-1}}{P_{\mathrm{dif}}^N (N-1)!}\exp\left(-\frac{p}{P_{\mathrm{dif}}}\right)\mathrm{u}(p) \tag{8.2.7}$$

Following the mapping in Equation (5.2.2), the MRC envelope PDF is given by

$$f_R(\rho) = \frac{2\rho^{2N-1}}{P_{\mathrm{dif}}^N (N-1)!}\exp\left(-\frac{\rho^2}{P_{\mathrm{dif}}}\right)\mathrm{u}(\rho) \tag{8.2.8}$$

For the case of $N = 1$ (no diversity) we can see that Equation (8.2.8) collapses to a simple Rayleigh PDF.

Figure 8.2 plots the CDFs for MRC diversity output using branches of 1, 2, 3, and 4. The horizontal axis of the graph shows SINR gain with respect to SINR_0, which is the signal-to-interference+noise ratio of an unfaded signal with power equal to the average signal strength of one branch. Note the significant performance gains for MRC diversity using 2, 3, and 4 branches. Not only do the additional branches provide a significant gain in average power, but the depth of fading reduces dramatically compared to the non-diversity case of $N = 1$.

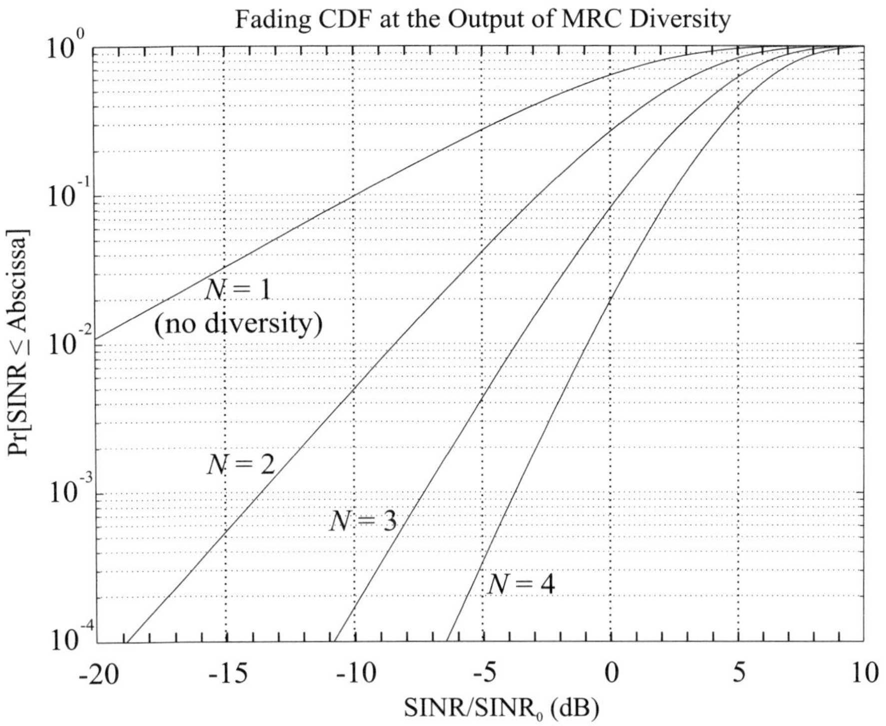

Figure 8.2 CDF of the fading SINR (Chi-squared distribution) at the output of MRC diversity for N = 1, 2, 3, and 4 uncorrelated Rayleigh-fading branches.

8.2.3 Switch Combining

It is much cheaper to build a single-channel receiver only capable of demodulating one diversity branch at a time. The performance of such a receiver is worse, on average, than a gain-combining receiver, but the cost savings are substantial. The goal of selection combining is to choose one of the N-diversity branches that maximizes the SINR. Due to cost and complexity, not every selection-combining technique achieves this goal.

Pure Selection: Pure selection combining is the best way to implement selection diversity: Simply pick the branch with the highest SINR. There must be a way to quickly discern SINR on all of the branches, so there is some complexity and cost associated with this technique.

Threshold Selection - Switch and Examine: This type of combining is one of the only true "single channel receiver" techniques, making it inexpensive to implement. The receiver monitors the current diversity branch, and as soon as the signal strength drops below a precalculated threshold, the receiver switches to another branch. The key drawback: There is no guarantee that the next branch will have a signal strength above the threshold. In fact, if none of the branches have a signal strength above the threshold at a particular instant, there will be a large number of pointless and noisy switches between the diversity branches.

Threshold Selection - Switch and Stay: The answer to the pointless switching problem of the previous combining technique is the *switch and stay* implementation of threshold selection. For this technique, the receiver switches to a new branch and will not switch again until the current SINR crosses the threshold from high to low. The drawback here is that a receiver may "park" on a faded diversity branch, having to wait for the received signal to increase and then fade again before the next switch is made.

The trade-offs of both gain-combining and selection-combining schemes are summarized in Table 8.2.

8.2.4 Two-Branch Example

To illustrate the differences between the diversity-combining algorithms, consider a two-branch diversity system that uses the spatial fading signals of Figure 8.3. The signals shown in Figure 8.3 are independently generated Rayleigh-fading signals with the same time-averaged SINR_0. The second-order statistics of the two signals derive from an omnidirectional multipath angle spectrum.

Figure 8.4 shows the corresponding diversity output signals using EGC, MRC, and pure selection algorithms. Note that MRC gain combining always provides the best output SINR. EGC consistently (but not always) outperforms pure selection combining. Recognize also that the pure selection signal represents the upper limit of what is achievable using threshold selection schemes.

8.3 BER and Capacity

If we know where to find diversity branches and how to combine them, then we are nearly home in our understanding of how to combat the effects of multipath fading. We must shift our analysis, however, to the end-to-end performance of a wireless link. This section discusses how to calculate bit-error rates (BER) for fading channels with and without diversity.

8.3.1 BER for Nonfading Channels

Before we can discuss the BER for fading channels, we need a recipe book for calculating the BER for nonfading channels. Table 8.3 is an example of such a recipe book, recording the BER for typical modulation schemes. These are all standard results from communication theory, all of which are derived in any good digital communications textbook [Pro89]. For any given modulation scheme, the BER

Table 8.2 Overview of Diversity-Combining Techniques

TECHNIQUE	HOW IT WORKS	ADVANTAGES	DISADVANTAGES
Maximum Ratio Combining (MRC)	N-diversity branches combined to maximize total SINR	Leads to optimal performance of an N-branch communication system	Requires simultaneous demodulation of N channels Requires SINR estimation algorithms
Equal Gain Combining (EGC)	N-diversity branches are summed equally	Near-optimal performance of an N-branch communication system Does not require SINR estimate	Requires simultaneous demodulation of N channels
Pure Selection	Pick the best of N available channels	Best performance for a switch-combining receiver	Requires monitoring signals on N branches
Threshold Selection, Switch and Examine	Pick another branch if current branch drops below a threshold	Uses simple, single-channel receiver	Simultaneous branch fades result in noisy bursts of rapid switching Most practical for $N = 2$
Threshold Selection, Switch and Stay	Pick another branch if current branch drops below threshold and stay, once switched	Uses simple, single-channel receiver	Newly switched branch could be worse than previous branch Most practical for $N = 2$

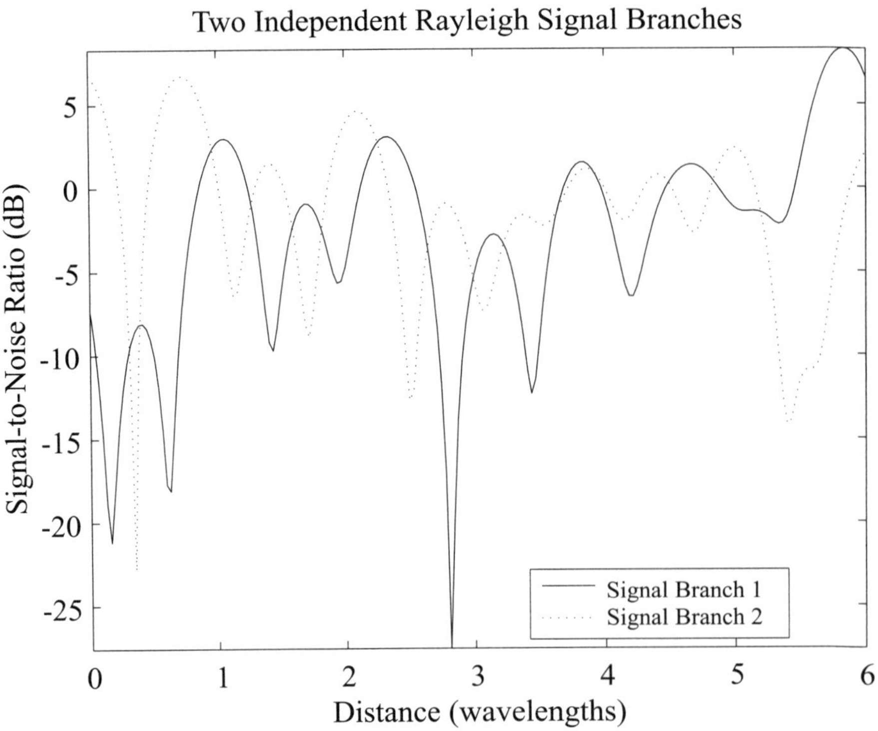

Figure 8.3 A comparison of outputs from EGC, MRC, and pure selection diversity-combining techniques.

expression depends solely on the received signal strength relative to the undesired signal power, whether it be noise or uncancelled interference.

> *Note: What Are the Units of BER?*
>
> The expressions for Table 8.3 are unitless. What wireless engineers call the BER is actually the probability of bit error (the term *rate* is a little misleading). So if a system has a BER of 0.002, that means that an average of 1 in 500 bits is received in error.

8.3.2 BER for Fading Channels

Before we construct a BER expression for a realistic wireless link, we must select the type of fading on the link. The type of fading is determined by the PDF of the received signal envelope, as discussed in Chapter 5.

We now have enough tools to calculate the BER for a specified modulation scheme and channel fading. Here is a basic procedure to follow:

Table 8.3 BER Expressions for Various Modulation Types*

MODULATION TYPE	BER
Coherent Amplitude-Shift Keying (ASK)	$Q\left(\sqrt{\mathrm{SINR}/2}\right)$
Coherent Frequency-Shift Keying (FSK)	$Q\left(\sqrt{\mathrm{SINR}}\right)$
Noncoherent FSK	$\frac{1}{2}\exp\left(-\mathrm{SINR}/2\right)$
Coh. Binary Phase-Shift Keying (BPSK)	$Q\left(\sqrt{2\,\mathrm{SINR}}\right)$
Differential Phase-Shift Keying (DPSK)	$\frac{1}{2}\exp\left(-\mathrm{SINR}\right)$
Quadrature Phase-Shift Keying (QPSK)	$Q\left(\sqrt{2\,\mathrm{SINR}}\right)$
Minimum Shift Keying (MSK)	$Q\left(\sqrt{1.7\,\mathrm{SINR}}\right)$
Gaussian Minimum-Shift Keying (GMSK)	$\approx Q\left(\sqrt{1.36\,\mathrm{SINR}}\right)$
Coherent M-ary PSK	$\approx 2Q\left(\sqrt{4\,\mathrm{SINR}}\sin\frac{\pi}{2M}\right)$
M-ary Quad. Amplitude Mod. (M-QAM)	$\approx 4\left(1-\frac{1}{\sqrt{M}}\right)Q\left(\sqrt{3\,\mathrm{SINR}/(M-1)}\right)$
Spread Spectrum with N Interferers (SS)	$\approx Q\left(\left[\sqrt{\frac{K-1}{3N}+0.5\,\mathrm{SINR}^{-1}}\right]^{-1}\right)\dagger$

* Expressions assume interference statistically resembles Gaussian noise.

† K is the processing gain of the spread spectrum modulation.

$Q(z)$ Q-function; see Appendix A.

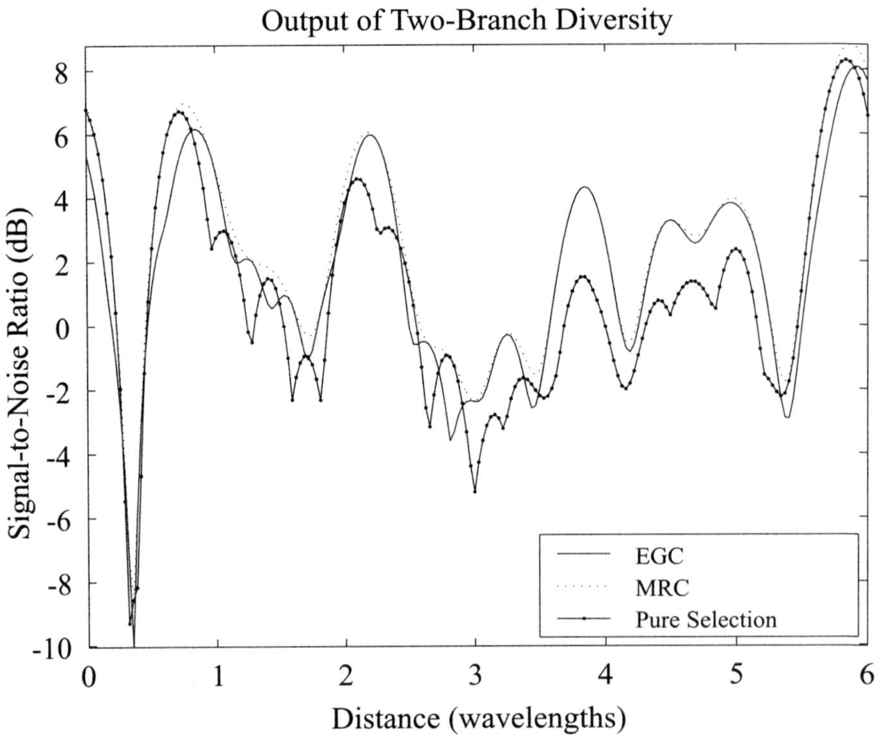

Figure 8.4 A comparison of outputs from EGC, MRC, and pure selection diversity-combining techniques.

1. From Table 8.3, choose the appropriate probability of bit error as a function of SINR. We will designate this probability of bit error as BER(SINR), a function of received signal strength.

2. Choose the fading distribution for the link. The easiest choice is the Rayleigh distribution, but any valid envelope PDF is possible.

3. Choose the amount of noise and/or in-band interference power, P_N, in the system. The instantaneous SINR for the receiver is then

$$\text{SINR} = \frac{R^2}{P_N}$$

which assumes a constant P_N and a randomly fluctuating signal envelope, R, that varies according to a fading PDF $f_R(\rho)$.

4. Now condition the probability of bit error, BER(SINR), on the expression for SINR in the previous step. Take the ensemble average to produce the

following expression:

$$\text{BER} = \text{E}\left\{\text{BER}\left(\frac{R^2}{P_N}\right)\right\} = \int_0^\infty \text{BER}\left(\frac{\rho^2}{P_N}\right) f_R(\rho)\, d\rho \qquad (8.3.1)$$

Solution of the integral in Equation (8.3.1) produces a single, unitless number that is the final average BER for the fading channel. For some simple expressions, Equation (8.3.1) has analytical solutions that can be found in good digital communications textbooks (see [Rap02a, p.340–343]). The general case, though, requires the expression to be evaluated numerically.

Example 8.1: BER for a GSM Handset

Problem: A GSM (global system for mobile communications) receiver is operating in Rayleigh fading with an average SINR of 15 dB. Set up the expression to compute the BER of this system.

Solution: A GSM receiver uses GMSK to modulate bits, so the BER as a function of SINR is

$$\text{BER(SINR)} \approx \text{Q}\left(\sqrt{1.36\,\text{SINR}}\right)$$

The average SINR is 15 dB, which is 31.6 in the linear scale (always use linear scale values in these calculations). If P_T is the average power received by a handset, then the noise power is $P_N = P_T/31.6$, making the SINR

$$\text{SINR} = \frac{31.6\,R^2}{P_T}$$

Thus, we can write the average BER as

$$\text{BER} = \int_0^\infty \text{Q}\left(\sqrt{(1.36)\left(\frac{31.6\rho^2}{P_T}\right)}\right) \frac{2\rho}{P_T} \exp\left(-\frac{\rho^2}{P_T}\right) d\rho$$

Make a change of integration, $u = \frac{31.6\rho^2}{P_T}$:

$$\text{BER} = \frac{1}{31.6}\int_0^\infty \text{Q}\left(\sqrt{1.36u}\right) \exp\left(-\frac{u}{31.6}\right) du = 0.0112$$

The BER of an *unfaded* channel with a 15 dB SINR is given by

$$\text{BER(SINR)} \approx \text{Q}\left(\sqrt{1.36\,[31.6]}\right) = 1.6 \times 10^{-11}$$

There are considerably fewer bit errors in the unfaded channel with the same average SINR.

8.3.3 Fading Channel Capacity

The BER analysis gives a good estimate of receiver performance for a given fading channel. However, there are many other aspects of receiver processing that affect performance in addition to modulation scheme. These processing techniques include interleaving, error-correction coding, equalization, data framing, multiple access schemes, and even networking protocols. In the end, it is data *throughput* that determines the value of a communications link. Lower BER and higher bandwidth leads to higher throughput, but there is no simple way to translate BER into a final throughput without exhaustive simulation of every system detail.

Therefore, the channel capacity limit of Equation (8.1.1) may be a useful alternative measure of wireless link quality. Recall that this fundamental relationship places the absolute mathematical limit on bits-per-second throughput, independent of all the gory system details. In fact, we can condition the channel capacity equation on a fading distribution in the same way we did for BER expressions. This produces the following equation:

$$C = \mathrm{E}\left\{C(\mathrm{SINR})\right\} = B \int_0^\infty \log_2\left(1 + \frac{\rho^2}{P_N}\right) f_R(\rho)\, d\rho \quad \text{bits/s} \tag{8.3.2}$$

Equation (8.3.2) is another useful way to characterize a communications link. It is a top-down approach compared to the nuts-and-bolts workings of BER Equation (8.3.1).

8.3.4 Empirical BER and Capacity

Another difficulty with calculating the analytical expressions of Equation (8.3.1) and Equation (8.3.2) is the necessity of characterizing the fading PDF. In analysis, realism often takes a back seat to convenience, as the majority of people rely exclusively on the Rayleigh PDF to characterize multipath fading. However, there are numerous other fading PDFs and there is always debate among engineers as to which are most common in real life.

For this reason, it would be better to use either simulated or measured fading distributions in the calculations to achieve greater understanding and realism. If, instead of a PDF, we have a collection of M measured or simulated samples of received power, $\{P_i\}$, then the BER and channel capacity expressions could be averaged over these results:

$$\mathrm{BER} \approx \frac{1}{M} \sum_{i=1}^{M} \mathrm{BER}\left(\frac{P_i}{P_N}\right) \tag{8.3.3}$$

$$C \approx \frac{B}{M} \sum_{i=1}^{M} \log_2\left(1 + \frac{P_i}{P_N}\right) \quad \text{bits/s} \tag{8.3.4}$$

These equations use a set of powers, $\{P_i\}$, measured in the field or generated by simulation. Even more powerful is the ability to take multiple simulated or measured fading profiles and run the data through a software implementation of diversity-combining algorithms. Indeed, an engineer using this technique can discern link behavior of sophisticated diversity schemes in realistic environments without resorting to exhaustive simulation or prototyping entire systems in hardware.

8.3.5 Diversity Gain for Multiple Branches

The performance gain for a system that uses diversity can be quite dramatic. Consider the plots of average BER versus SINR for different numbers of MRC diversity branches in Figure 8.5. The curves are based on QPSK modulation through uncorrelated Rayleigh fading branches with equal average power. From Figure 8.5 we see how adding diversity branches decreases the BER for a given level of SINR.

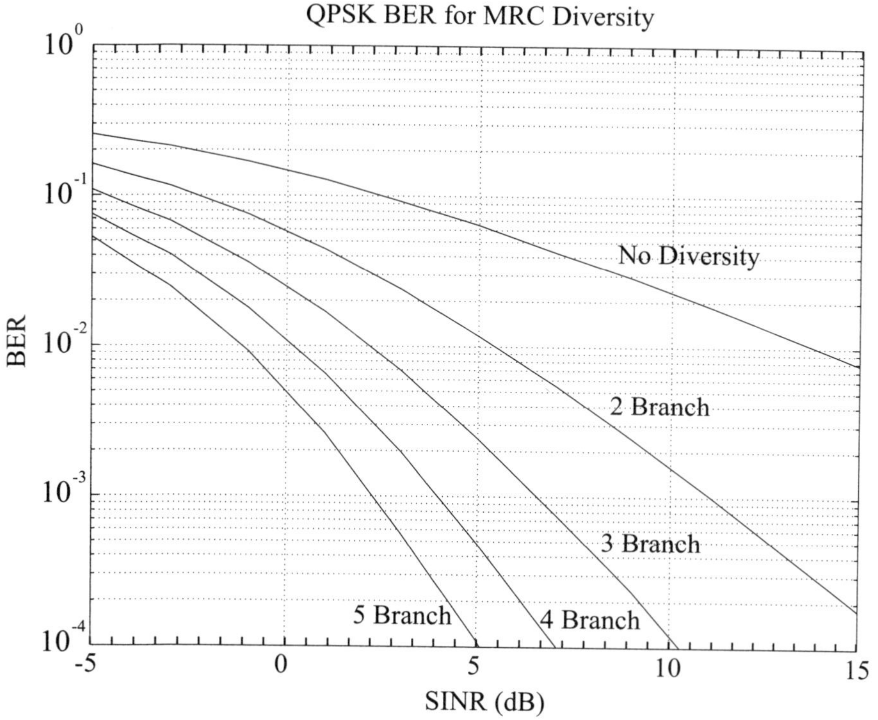

Figure 8.5 QPSK BER for MRC diversity using multiple branches of uncorrelated Rayleigh-fading channels with equal average power.

At this point, it is useful to define what wireless engineers mean by *diversity gain* - a value that quantifies the performance improvement afforded by a diversity scheme in a fading channel. For a given modulation scheme, there is usually a maximum BER threshold (and hence a minimum SINR) beyond which the wireless link fails. When diversity is added, a lower SINR achieves the same maximum BER as the case without diversity. As Figure 8.5 shows, the difference between the SINR for achieving the target BER threshold with and without diversity is the diversity gain, usually measured in dB. Example 8.2 shows how to calculate diversity gain.

Example 8.2: Diversity Gain for QPSK

Problem: A QPSK link must maintain a BER of 0.01 in a Rayleigh-fading channel. What is the diversity gain possible for this link using MRC on two, three, four, and five uncorrelated signal branches?

Solution: As measured from Figure 8.5, the SINR value corresponding to a BER of 0.01 for QPSK without diversity is 13.9 dB. We may also measure these SINR values for multibranch MRC diversity and the diversity gain that results:

Branches	SINR (dB)	Gain (dB)
2	5.5	8.4
3	2.2	11.7
4	0.1	13.8
5	−1.3	15.2

Notice that even two-branch diversity adds significant diversity gain. While adding more branches also increases diversity gain, there are diminishing returns.

8.3.6 Illustration of Branch Correlation on Diversity

We now have enough principles to illustrate the effects of signal branch correlation on the quality of fading wireless links. Figure 8.6, Figure 8.7, and Figure 8.8 show the effects of correlation on Shannon channel capacity for different levels of envelope correlation. Since Shannon channel capacity, C, in a fading channel is a random variable, these figures plot the CDF, $F_C(c)$. Each curve is normalized against the channel capacity, C_0, for a single, unfaded channel. Channels with average $\text{SINR}_0 = 10$ dB are used in the analysis, so $C_0 = 3.46$ bits/s/Hz.

Figure 8.6, Figure 8.7, and Figure 8.8 were constructed using simulated Rayleigh-fading channels with identical average signal strengths. The channels were generated as a function of space using the omnidirectional multipath angle spectrum - but since the graphs show only first-order capacity statistics, they are general for any type of Rayleigh fading for any set of diversity branches taken from time, frequency, or space. The different levels of envelope correlation, ϱ_R, for each graph are based on unit autocovariance, as described in Section 7.2.

First note the performance of MRC in correlated Rayleigh-fading channels in Figure 8.6. For the ideal, uncorrelated case ($\varrho_R = 0$), almost 75% of the fading channel experiences channel capacity greater than an unfaded single branch. As expected, this capacity degenerates as ϱ_R approaches 1, particularly in the low-capacity portion of Figure 8.6. For the highest case of branch correlation ($\varrho_R = 1$), a capacity drop below $0.1C_0$ occurs for about 1.5% of all cases.

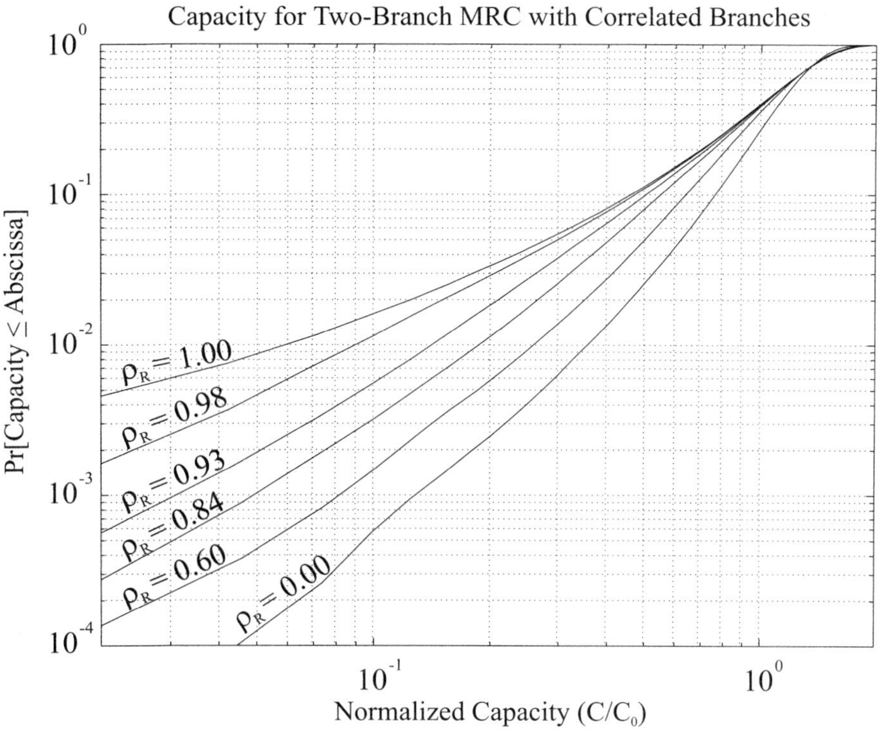

Figure 8.6 CDF of Shannon capacity for MRC diversity using two correlated Rayleigh branches, each with average SINR = 10 dB.

Figure 8.6 is a good reminder of where diversity schemes are most sensitive to branch correlation. There are substantial differences in the capacity distributions for signal correlations of 1.0, 0.98, 0.93, and 0.84; yet there is only a modest change between correlations of 0 and 0.6! The principle illustrated here is that performance metrics of a wireless link are sensitive only to high levels of correlation. This leads to many textbooks and researchers claiming that signal branches are virtually "uncorrelated" for $|\rho_R| < 0.4$.

> *Note: No More Hand-Waving*
>
> Recall that the Gaussian approximations made to unit autocorrelation in Section 7.2 required throwing away higher order behavior, which led to inaccuracies for low values of correlation ($|\varrho_R| \leq 0.3$). We made the hand-waving assertion that this region of inaccuracy had little effect on wireless link performance; now we may finally justify this claim with Figure 8.6, Figure 8.7, and Figure 8.8.

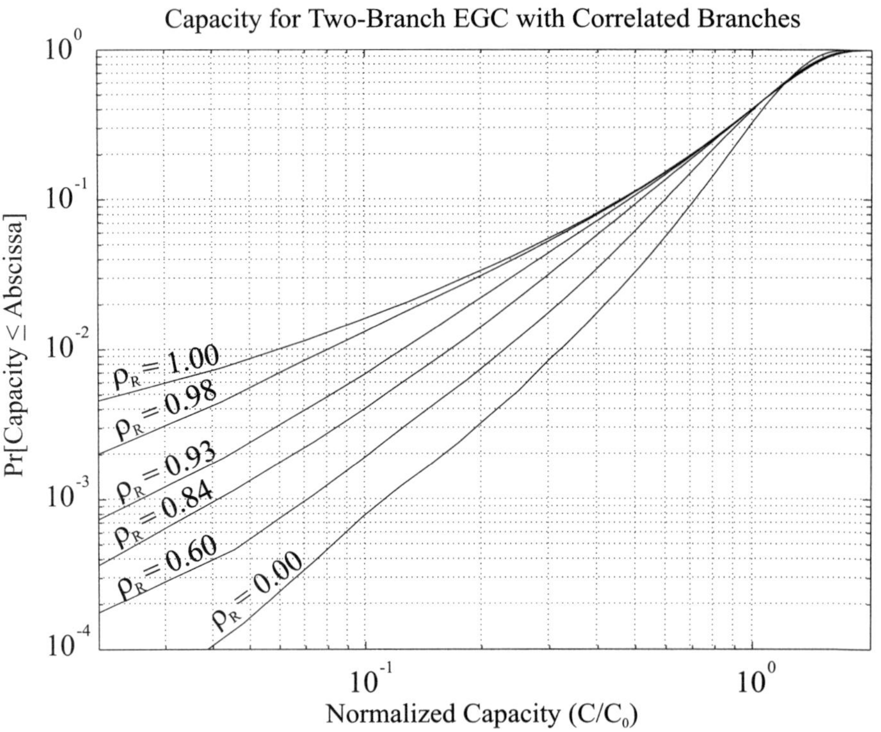

Figure 8.7 CDF of Shannon capacity for EGC diversity using two correlated Rayleigh branches, each with average SINR = 10 dB.

The EGC cases of Figure 8.7 and the pure selection cases of Figure 8.8 behave in similar manners. Both were generated from signal branches with the same average SINR (10 dB) as the MRC graph of Figure 8.6. Note the following behavior:

- For the case of perfect correlation ($\varrho = 1$), there is no difference between EGC and MRC.
- Pure selection is especially sensitive to correlated branches, experiencing capacity drops below $0.1C_0$ twice as often as MRC and EGC.
- For all correlation values, both EGC and pure selection diversity are sub-optimal when compared to the case of MRC diversity.

These graphs show the importance for obtaining uncorrelated signal branches in fading channels, regardless of diversity type or combining technique. For antenna diversity, there must be sufficient spacings between co-polar antenna elements (as discussed later in Chapter 10). For frequency diversity, there must be sufficient spacing between the carrier frequencies of different signal sources. For temporal diversity, there must be sufficient spacing in time (usually in the form of *interleaving*) between consecutively coded bits.

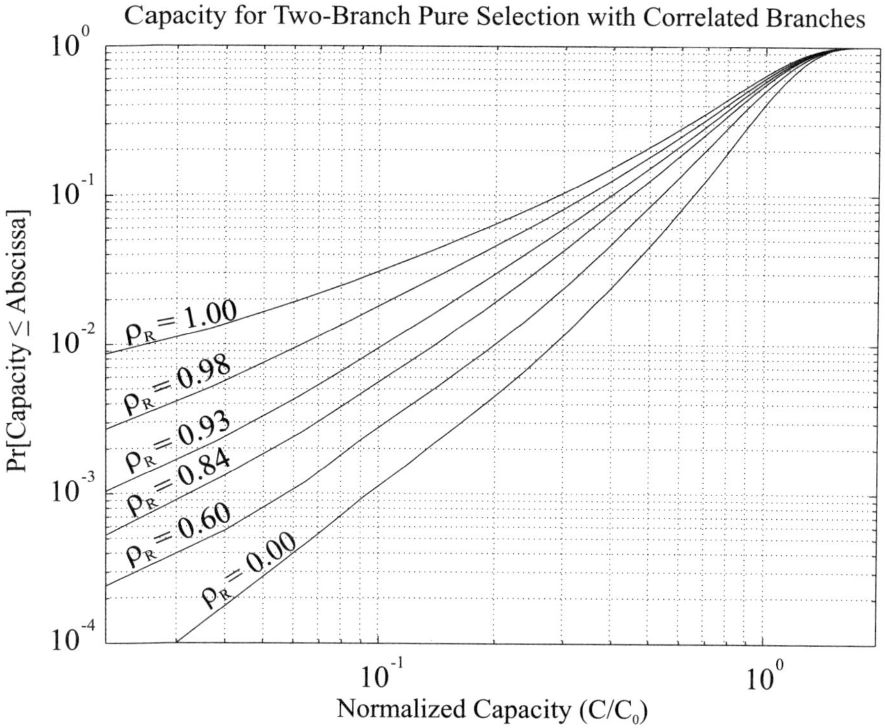

Figure 8.8 CDF of Shannon capacity for pure selection diversity using two correlated Rayleigh branches, each with average SINR = 10 dB.

8.3.7 Illustration of Unequal Branch Power on Diversity

Let us also consider the quantitative effects of unequal power on a two-branch diversity scheme. For this analysis, we generate two independent Rayleigh-fading signals, $\tilde{h}_1(t)$ and $\tilde{h}_2(t)$. Although there is no signal correlation between elements, the signals are scaled so that they have an unequal average envelope, as described by the familiar Δ-parameter (from Chapter 5):

$$\Delta = \frac{2R_1R_2}{R_1^2 + R_2^2} \quad \text{for } R_1 = \mathrm{E}\left\{|\tilde{h}_1(t)|\right\}, \quad R_2 = \mathrm{E}\left\{|\tilde{h}_2(t)|\right\}$$

A value of $\Delta = 1$ implies that both signals have equal average signal strength. A value of $\Delta = 0$ implies one of the signals has vanished completely.

Like the correlation analysis, we plot first-order statistics of Shannon channel capacity, normalized against the capacity of an unfaded channel with SINR = 10 dB ($C_0 = 3.46$ bits/s/Hz). For example, the case of $\Delta = 0.5$ corresponds to diversity combining between a branch with SINR = 10 dB and a branch with SINR = −1.44 dB. The SINR of the weaker signal branch, for this example, is related to Δ by the following equation:

$$\mathrm{SINR}_2 = 10\ \mathrm{dB} \quad - \quad 20\log_{10}\left(\frac{1}{\Delta} - \sqrt{\frac{1}{\Delta^2} - 1}\right) \tag{8.3.5}$$

The graphs are general to any two-branch diversity scheme in a Rayleigh-fading channel, regardless of whether the branches are obtained from time, frequency, or space.

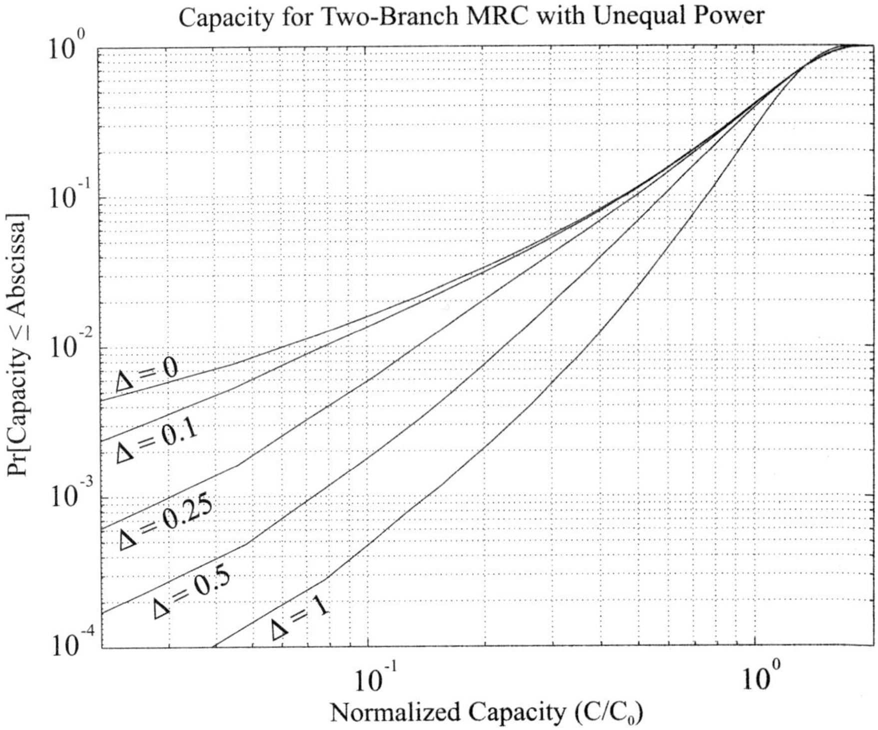

Figure 8.9 CDF of Shannon capacity for MRC diversity using two Rayleigh-fading branches with unequal average signal strength.

Figure 8.9 shows the CDF of Shannon channel capacity in a Rayleigh-fading environment using two-branch MRC diversity. The case where signal power is distributed in even amounts corresponds to $\Delta = 1$; clearly this outperforms all other cases. Capacity drops below $0.1C_0$ have less than a 0.05% occurrence. Contrast this to the case of $\Delta = 0$, which represents the absence of another diversity branch; for this case, capacity drops below $0.1C_0$ have more than a 1.5% occurrence - a 30-fold increase! Note that the case of $\Delta = 1$ is identical to the case of $\varrho_R = 0$ in the previous section; both cases represent combining of uncorrelated, equal-powered Rayleigh envelopes.

Although not as pronounced as the case of correlation sensitivity, the performance of MRC is more sensitive for lower values of Δ. Over the interval $\Delta = [0, 0.5]$, there is a much more dramatic change in capacity distribution than over the interval $\Delta = [0.5, 1.0]$.

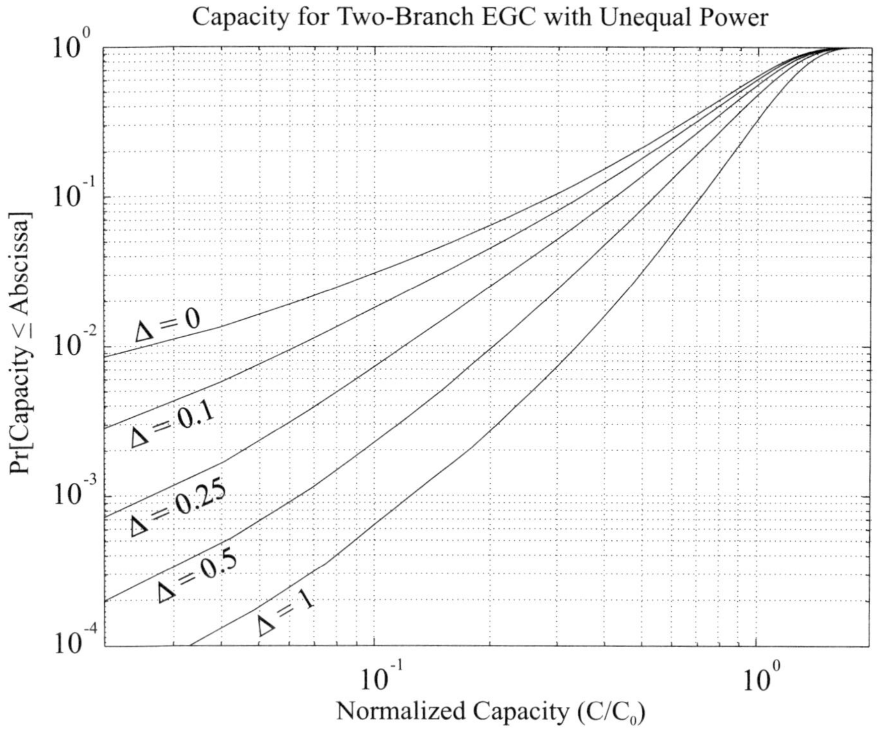

Figure 8.10 CDF of Shannon capacity for EGC diversity using two Rayleigh-fading branches with unequal average signal strength.

Figure 8.10 shows similar plots for EGC diversity, and Figure 8.11 shows similar plots for diversity with pure selection. Each graph follows trends similar to the MRC diversity exhibited by Figure 8.9. Again note that the case of $\Delta = 1$ is identical to the case of $\varrho_R = 0$ in the previous section.

While EGC generally outperforms pure selection diversity, this is *not* true for small Δ. This may be counterintuitive, unless the details of EGC and pure selection algorithms are considered. Recall that EGC is an indiscriminate addition of two signals. For small Δ, the weaker signal is mostly noise; EGC degrades the only reliable signal branch by adding a worthless, noisy signal branch. The algorithm of pure selection - which essentially *ignores* the weaker signal - is better suited for branches with low Δ. Due to this property as well as its increased complexity, EGC diversity is inferior to pure selection diversity unless the receiver is guaranteed that

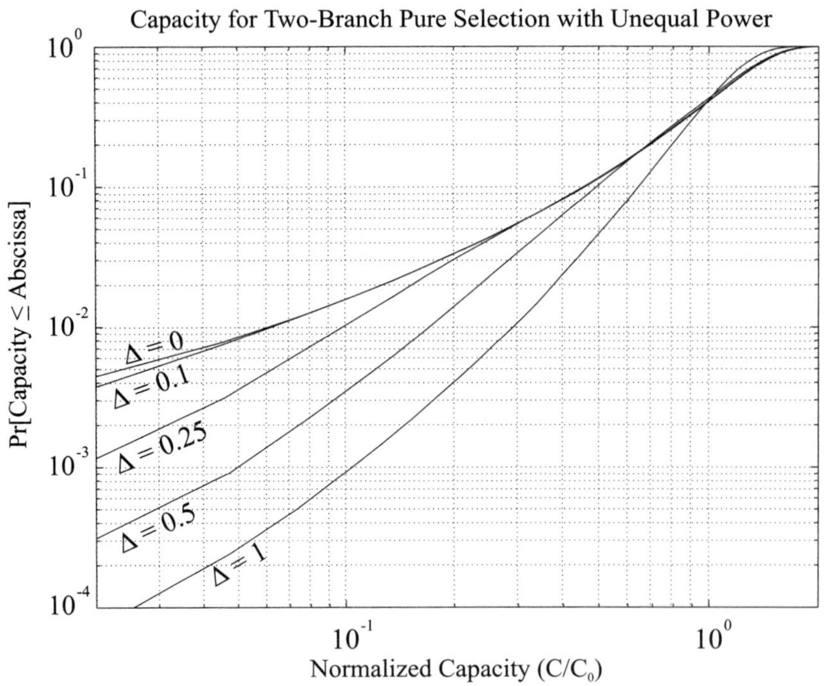

Figure 8.11 CDF of Shannon capacity for pure selection diversity using two Rayleigh-fading branches with unequal average signal strength.

the diversity branches have similar average power levels.

8.4 Chapter Summary

- Signal branches for diversity may be taken from a variety of sources that fall within two basic categories:
  - ▷ Antenna diversity schemes extract multiple signal branches from antennas at either the transmitter or receiver. These schemes include space, pattern, polarization, and transmit diversity.
  - ▷ Temporal diversity schemes extract multiple signal branches from time and frequency channel dependencies. These schemes include time diversity, frequency diversity, and code diversity.
- If multiple signal branches are available in a radio link, they may be combined in a variety of ways:
  - ▷ Gain combining produces a fade-resistant signal by simultaneously adding signal branches with different weights. The weights depend on whether

Maximum Ratio Combining (MRC) or Equal Gain Combining (EGC) is used.

- ▷ Switch combining produces a composite fade-resistant signal by switching between the multiple signal branches, avoiding the branches with deep fades. Pure selection and threshold selection are two examples of switching algorithms.

- Performance metrics for diversity improvement may be measured by the following methods:
  - ▷ diversity gain (measured in units of dB)
  - ▷ Shannon channel capacity
  - ▷ bit-error rate (BER)

- There are two crucial failure points for any diversity scheme:
  - ▷ correlation between diversity branches
  - ▷ unequal average power between diversity branches

Despite its complexity, the study of *multiple-input, multiple-output* (MIMO) wireless systems requires only a small advancement - once the concepts of diversity are understood.

PROBLEMS

1. Using the U-SLAC model for representing the wireless channel (Equation (4.4.1) with independent random phases), prove the following claims about different types of diversity:

a. Diversity branches taken from local area channels at different points in space, $\tilde{h}(\vec{r}_1)$ and $\tilde{h}(\vec{r}_2)$, have equal average power but are vulnerable to correlation.

b. Diversity branches taken from local area channels at different frequencies, $\tilde{h}(f_1)$ and $\tilde{h}(f_2)$, have equal average power but are vulnerable to correlation.

c. Diversity branches taken from local area channels at different wavevectors, $\tilde{H}(\vec{k}_1)$ and $\tilde{H}(\vec{k}_2)$, are uncorrelated but are vulnerable to unequal average power. (Diversity branches from different wavevectors is one way of viewing switched-beam antenna diversity using an array.)

d. Diversity branches taken from local area channels at different delays, $\tilde{H}(\tau_1)$ and $\tilde{H}(\tau_2)$, are uncorrelated but are vulnerable to unequal average power.

2. Derive the envelope distribution for the output of N-branch MRC diversity by showing that the root-mean sum of independent Rayleigh-fading envelopes with average power P_{dif} follows Equation (8.2.8). Hint: Use characteristic functions and Equation (5.2.1).

3. Use the CDF graph of ideal MRC diversity output in Figure 8.2 to answer the following questions:

a. How often does three-branch MRC output fall −10 dB below SINR_0?

b. How often does two-branch MRC output fall −15 dB below SINR_0?

c. Note that for the deep fading region, the curves in Figure 8.2 become straight lines. Using extrapolation, estimate the SINR differences for 0.01% fading levels in all four cases.

4. An idealized delay spectrum may take the shape of an exponentially decaying function:

Exponential Delay Spectrum

$$S_{\tilde{h}}(\tau) = S_0 \exp\left(-\frac{\tau}{\sigma_\tau}\right) \mathrm{u}(\tau)$$

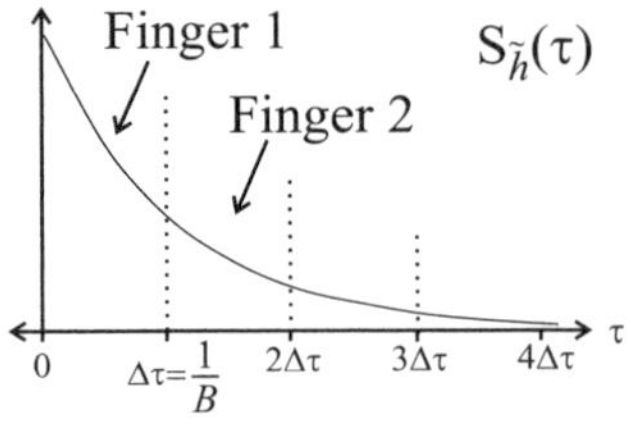

where σ_τ is the RMS delay spread. If spread spectrum modulation with bandwidth, B is used, a *Rake receiver* may be used to obtain code diversity. This is possible because a Rake receiver uses signal-processing *fingers* to isolate channel power within a delay bin of width $\tau = \frac{1}{B}$. If the Rake receiver has two or more fingers, then it has multiple signal branches for diversity. In all likelihood, a two-fingered Rake receiver will isolate the most power from the first two delay bins in the delay spectrum (as our exponential sketch shows). Signals from these two bins will be uncorrelated but will not have the same average power. Using the exponential delay spectrum, answer the following questions:

a. Calculate the RMS envelope in the diversity branches available to the Rake receiver as a function of bandwidth, B, and delay spread, σ_τ. Use the following formulas:

$$R_1 = \sqrt{\int_0^{\Delta\tau} S_{\tilde{h}}(\tau)\, d\tau} \qquad R_2 = \sqrt{\int_{\Delta\tau}^{2\Delta\tau} S_{\tilde{h}}(\tau)\, d\tau}$$

where R_1 is the RMS envelope for finger number 1 and R_2 is the RMS envelope for finger number 2.

b. Calculate the Δ difference parameter that describes how the two fingers differ from one another:

$$\Delta = \frac{2R_1R_2}{R_1^2 + R_2^2}$$

The answer should be a function of bandwidth, B, and delay spread, σ_τ.

c. To provide minimal effective diversity, we require Δ to have a minimum value of 0.25. Under this condition, establish a relationship between bandwidth, B, and delay spread, σ_τ.

d. An indoor wireless channel experiences a typical RMS delay spread of 30 ns. What is the minimum bandwidth, B, that provides sufficient code diversity, based on the criterion developed in (c)?

e. An outdoor wireless channel (from a high-tier base station tower to a mobile user) experiences a typical RMS delay spread of 200 ns. What is the minimum bandwidth, B, that provides sufficient code diversity, based on the criterion developed in (c)?

f. The IS-95 cellular standard uses a spread spectrum signal with 1.25 MHz bandwidth to modulate a digital voice signal. Based on your analysis of (d) and (e), is code diversity an effective form of diversity for this wireless standard?

g. The third-generation wideband CDMA standard uses a spread spectrum signal with 3.84 MHz bandwidth to modulate a digital voice and data signal. Based on your analysis of (d) and (e), is code diversity an effective form of diversity for this wireless standard?

5. Write a computer program to make plots of multiple diversity branches with uncorrelated Rayleigh fading with equal average power. Plot cases of no diversity, and two, three, and four branches on the same graphs for the following cases:

a. BER of QPSK modulation versus SINR for EGC combining.

b. BER of QPSK modulation versus SINR for pure selection combining.

c. Shannon capacity versus SINR for MRC combining.

d. Shannon capacity versus SINR for EGC combining.

e. Shannon capacity versus SINR for pure selection combining.

f. Select one of the previous cases (a)–(e) and repeat the analysis using uncorrelated *Rician* fading channels with $K = 5$ dB.

g. Select one of the previous cases (a)–(e) and repeat the analysis using uncorrelated *TWDP* fading channels with $K = 10$ dB and $\Delta = 1$.

6. Assuming uncorrelated, equal-powered Rayleigh branches, calculate the diversity gain possible for the following scenarios:

a. QPSK modulation for BER = 0.001 using two-branch MRC.

b. ASK modulation for BER = 0.0001 using two-branch pure selection.

c. DPSK modulation for BER = 0.001 using three-branch EGC.

d. Non-coherent FSK modulation for BER = 0.0001 using three-branch MRC.

e. GMSK modulation for BER = 0.01 using three-branch pure selection.

f. Select one of the previous cases (a)–(e) and repeat the analysis using uncorrelated *Rician* fading channels with $K = 5$ dB.

g. Select one of the previous cases (a)–(e) and repeat the analysis using uncorrelated *TWDP* fading channels with $K = 10$ dB and $\Delta = 1$.

7. Write a computer program to make CDF plots for the following cases of unequal average power with $\Delta = [0, 0.1, 0.25, 0.5, 1]$:

a. Shannon capacity CDFs for EGC diversity for SINR = 15 dB.

b. Shannon capacity CDFs for pure selection diversity for SINR = 5 dB.

c. BER CDFs for QPSK modulation using MRC diversity for SINR = 5 dB.

d. BER CDFs for QPSK modulation using pure selection diversity for SINR = 10 dB.

e. BER CDFs for noncoherent FSK modulation using MRC diversity for SINR = 5 dB.

f. BER CDFs for ASK modulation using EGC diversity for SINR = 10 dB.

Chapter 9

MIMO CHANNELS

Concepts of *multiple-input, multiple-output* (MIMO) signaling go beyond mere antenna diversity. In a MIMO wireless link, a data stream is broken into separate signals and sent through separable multipath channels in space. Although practical considerations, such as multiple antenna integration and signal processing, make MIMO signaling a serious technical challenge, many engineers regard such techniques as the future of band-limited wireless communications. Indeed, incredible amounts of data may be squeezed into a tiny amount of signal spectrum through clever use of space.

This chapter presents a basic introduction to MIMO concepts and signal processing, emphasizing the potential gains in Shannon channel capacity through increasing the spatial ports on a transmitter and a receiver. The major topics discussed in this section are listed below:

- Section 9.1: Channel capacity for conventional antenna arrays.
- Section 9.2: Overview of the MIMO concept.
- Section 9.3: Practical example of MIMO architecture.
- Section 9.4: Basics of space–time block coding.
- Section 9.5: Summary of important concepts.

Despite the complexity, MIMO channels are similar to diversity schemes, discussed in Chapter 8, since they are impaired by the same types of channels. In fact, familiarity with both diversity and MIMO signaling will provide a unique appreciation for Chapter 10, which directly addresses the issues of antenna design for multipath environments.

9.1 Conventional Multiple Antenna Systems

This section discusses the Shannon-limit performance of multiple antenna systems that attempt to maximize the signal power through the radio channel. These basic scenarios must be understood before making the leap to space–time block codes and MIMO architecture.

9.1.1 Single-Input, Single-Output (SISO)

Our baseline for comparison will always be the simple SISO channel of Figure 9.1. For a given channel and a given transmitter input power of P_T, there will be a signal at the receiver that has an average signal-to-noise+interference ratio of SINR_0. From this we can at least estimate the Shannon limit on channel capacity to be

$$C = B \log_2 (1 + \mathrm{SINR}_0) \tag{9.1.1}$$

where B is the available bandwidth. Clearly SINR_0 is the critical parameter affecting channel capacity. The next sections explore how increasing the number of transmitter and receiver antennas affects channel capacity.

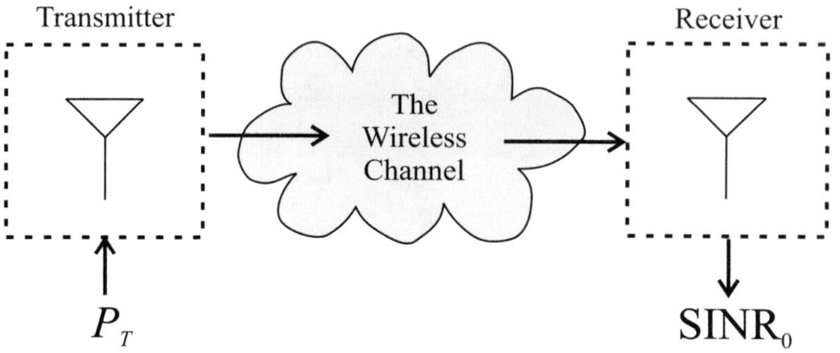

Figure 9.1 A SISO antenna configuration.

> *Note: Symbols Versus Bits*
>
> Recall from digital communication theory that available signal bandwidth limits the *symbol* rate, but not necessarily the bit rate. Binary symbols are the most familiar to engineers, but it is possible to transmit more than one bit on a single complex baseband symbol. While symbols with more than two states are more vulnerable to noise and interference, high SINR makes their reliable transmission possible. Thus, the use of nonbinary symbols (along with the redundancy of error-correction codes) helps communication systems approach Shannon-limit capacity in band-limited systems.

9.1.2 Single-Input, Multiple-Output (SIMO)

For the SIMO system of Figure 9.2, N antennas are used at the receiver, producing N various copies of the faded signal at the receiver. If these signals are, on average, the same amplitude, then they may be added coherently to produce an N^2 increase in signal power. Of course, there are N sets of noise/interference that will add together as well. Fortunately, noise and interference add incoherently to create only an N-fold increase in noise power. Thus, there is still a net overall increase in SINR:

$$\text{SINR} \approx \frac{N^2[\text{Signal Power}]}{N[\text{Noise+Interference}]} = N\text{SINR}_0 \tag{9.1.2}$$

Following this logic, the channel capacity for this system is approximately

$$C \approx B \log_2 (1 + N\text{SINR}_0) \tag{9.1.3}$$

which is somewhat higher than the SISO case.

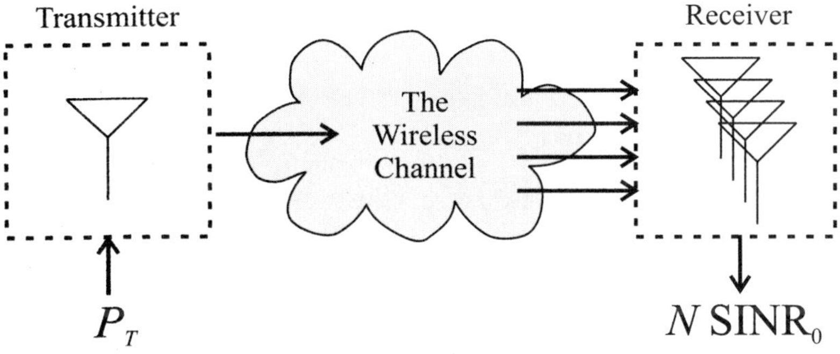

Figure 9.2 A SIMO antenna configuration.

Before we continue, it is useful to discuss the physical interpretation of this transmitter-receiver configuration. Exactly what does it mean, physically, to add channels coherently? This operation depends on the type of wireless channel. While the overall capacity gains of a SIMO receiver are similar, the physical interpretation of a receiver operating in an open-space channel without multipath is markedly different than a receiver operating in a multipath channel that has severe small-scale fading. Consider these two scenarios:

Open-Space: Coherent signal addition for a SIMO system in an open-space channel follows a beam-steering analogy. The receiver antenna elements are phased together to form an array with maximum gain in the direction of signal arrival. This peak gain causes the subsequent increase in SINR and channel capacity.

Multipath: In a multipath channel, direction of signal arrival may be dispersed throughout the azimuth, which makes the analogy of beam-steering useless.

Rather, the implementation of coherent signal addition uses diversity, combining the branches from each antenna using a maximum ratio combining (MRC) algorithm.

The approximate gain in capacity predicted by Equation (9.1.3) is appropriate for each propagation scenario, although the physical interpretation is different for each case.

> *Note: Open Space Versus Line of Sight*
>
> The term *line of sight* describes a transmitter and a receiver with antennas that have a clear, unobstructed path of radio wave propagation between them (i.e., they can "see" each other). In this section we use the term *open space* to emphasize the absence of multipath power. After all, a line of sight link often has many multipath components accompanying the strong direct path.

9.1.3 Multiple-Input, Single-Output (MISO)

For the MISO system in Figure 9.3, M antennas are used at the transmitter. Thus, the total transmitted power is divided into the M transmitter branches. Although the power per transmitter antenna drops, the transmitted signals may be phased at the transmitter antennas in such a way that they add coherently *at the receiver*; the actual received signal would have (approximately) a net M-fold increase in SINR over the SISO case. Because there is just one receiver channel, the received noise/interference power should be the same. The overall increase in SINR becomes

$$\text{SINR} \approx \frac{M^2[\text{Signal Power}/M]}{\text{Noise+Interference}} = M\text{SINR}_0 \tag{9.1.4}$$

Following this logic, the channel capacity for this system is

$$C \approx B \log_2 (1 + M\text{SINR}_0) \tag{9.1.5}$$

which is, as for SIMO, higher than the SISO case. The physical interpretation for each propagation scenario in this MIMO system is similar to the SIMO system as well:

Open-Space: For this case, the beam-steering occurs at the transmitter instead of the receiver. The M transmitter antennas are phased as an array so that the peak gain is towards the direction of the receiver. It is this array directivity that increases SINR and raises channel capacity.

Multipath: Once again, a rich multipath environment makes the analogy of beam-steering difficult. Instead, we think of coherent signal combining as *transmit diversity*. With closed-loop feedback, the transmitter can change the phases and signal weighting on the M transmitter antennas to maximize receiver SINR and channel capacity.

The physical interpretations show how each scenario achieves the SINR gains.

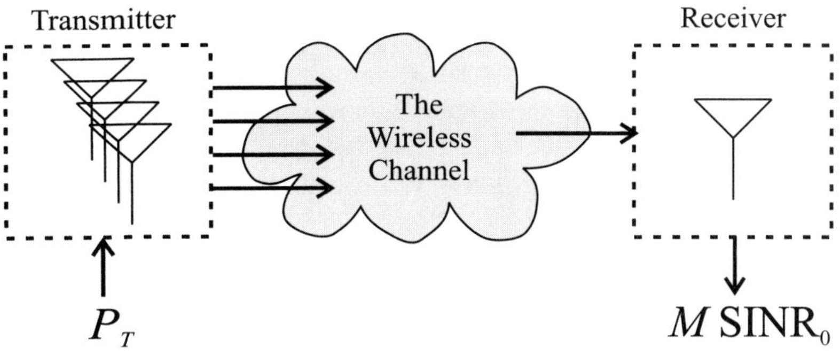

Figure 9.3 A MISO antenna configuration.

9.1.4 Multiple-Input, Multiple-Output (MIMO)

For MIMO systems like Figure 9.4, there are M transmitter antennas *and* N receiver antennas. We may view this as a combination of the SIMO case and the MISO case: Signals are transmitted and received with the phasings at each antenna that maximizes total signal power through the wireless channel. The result is an (approximate) MN-fold increase in the SINR of the received signal.

$$C \approx B \log_2 (1 + MN\mathrm{SINR}_0) \qquad (9.1.6)$$

which is much larger than the channel capacity for SIMO or MISO cases alone.

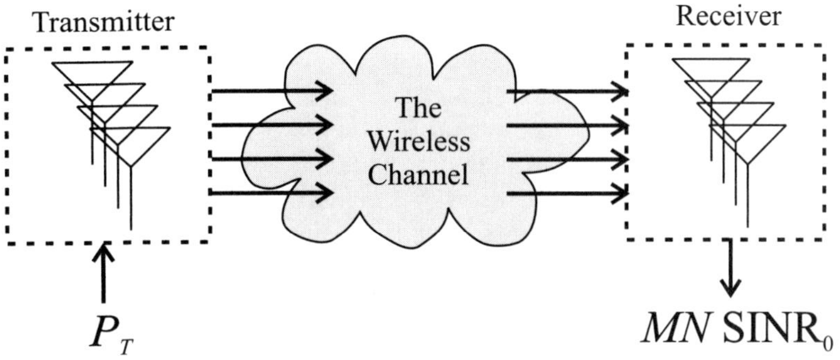

Figure 9.4 A MIMO antenna configuration.

The physical interpretation for different multipath scenarios in a conventional multi-antenna system is a combination of the SIMO and MISO cases. Consider the following channels:

Open-Space: Both transmitter and receiver use their multiple antennas in a phased-array configuration. The transmitter steers its beam in the direction of the receiver and vice versa. By taking advantage of the peak gain for both transmitter and receiver arrays, the SINR rises substantially.

Multipath: To achieve the maximum channel capacity, transmit diversity and space diversity with MRC at the receiver are used to achieve the highest possible SINR in the overall link.

Up until this point, the physical interpretation of the MIMO channel is conventional. Like the other multi-antenna techniques, the premise of Equation (9.1.6) is maximizing channel capacity by maximizing the SINR for a single transmitted signal. Unlike the previous cases, though, there is another physical interpretation of a MIMO channel that leads to even *higher* capacity than Equation (9.1.6).

9.2 Separating Channels in Multipath

A shift in perspective led to the development of *truly* multiple-input, multiple-output systems that have channel capacity higher than the SINR-maximizing techniques of Section 9.1. Our discussion of Equation (9.1.6) was based on a system that maximizes the SINR of a *single* transmitted signal. Dramatic capacity increases are possible if we consider different signals sent through each transmitter antenna.

9.2.1 MIMO Channel Matrix

Before proceeding, we need to develop a matrix channel model for a MIMO system. The baseband representation of Figure 9.5 is the general MIMO channel representation. The structure of Figure 9.5 models the individual transfer functions, $\tilde{H}_{ij}(\tau; t)$, between the set of signals, $\{\tilde{x}_i(t)\}$, sent into each of the M transmitter antennas and the set of signals, $\{\tilde{y}_j(t)\}$, received at the N receiver antennas. Furthermore, these channels may be functions of time, t, to model a time-varying channel, and functions of delay, τ, to model the dispersion incurred by wideband transmission.

It is a useful book-keeping practice to use a vector/matrix notation to keep track of all the transmitted and received signals in this complicated system. A vector of received signals, $\tilde{\mathrm{y}}(t)$ (as seen at the inputs of the N receiver antennas), may be calculated from the vector of transmitted signals, $\tilde{\mathrm{x}}(t)$. The output vector is related to the input vector by the channel transfer matrix, $\tilde{\mathbf{H}}(\tau; t)$, according to the following equation:

$$\tilde{\mathrm{y}}(t) = \frac{1}{2} \int_{-\infty}^{+\infty} \tilde{\mathbf{H}}(\tau; t) \tilde{\mathrm{x}}(\tau) \, d\tau \tag{9.2.1}$$

where

$$\tilde{\mathrm{y}}(t) = \begin{bmatrix} \tilde{y}_1(t) \\ \tilde{y}_2(t) \\ \vdots \\ \tilde{y}_N(t) \end{bmatrix}, \qquad \tilde{\mathrm{x}}(t) = \begin{bmatrix} \tilde{x}_1(t) \\ \tilde{x}_2(t) \\ \vdots \\ \tilde{x}_M(t) \end{bmatrix},$$

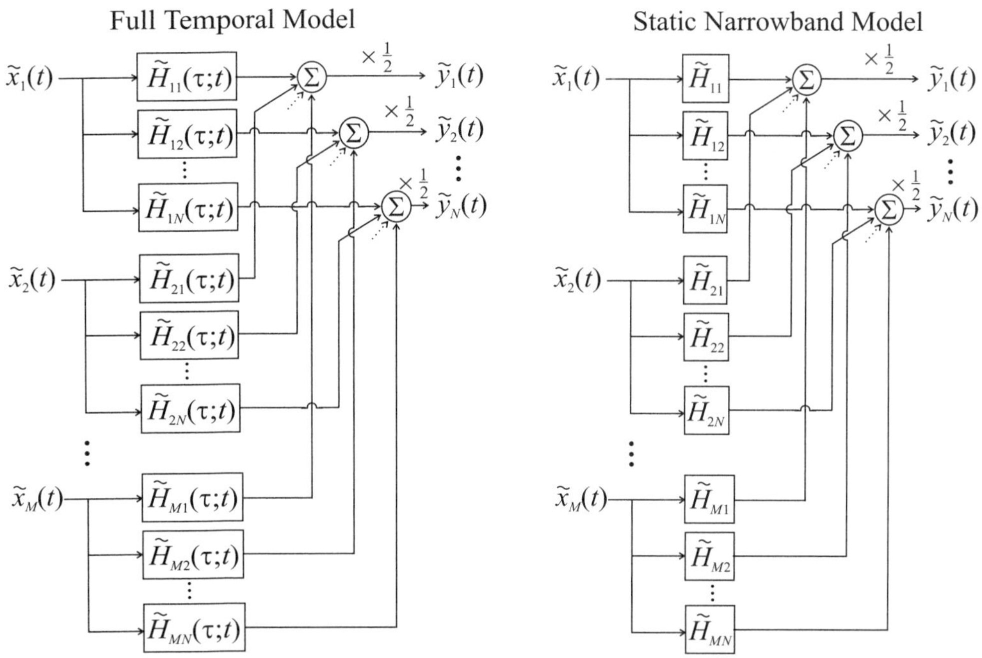

Figure 9.5 Block diagram of the different channels that exist in a MIMO radio link.

$$\tilde{\mathbf{H}}(\tau;t) = \begin{bmatrix} \tilde{H}_{11}(\tau;t) & \tilde{H}_{21}(\tau;t) & \cdots & \tilde{H}_{M1}(\tau;t) \\ \tilde{H}_{12}(\tau;t) & \tilde{H}_{22}(\tau;t) & \cdots & \tilde{H}_{M2}(\tau;t) \\ \vdots & \vdots & \ddots & \vdots \\ \tilde{H}_{1N}(\tau;t) & \tilde{H}_{2N}(\tau;t) & \cdots & \tilde{H}_{MN}(\tau;t) \end{bmatrix} \tag{9.2.2}$$

In this representation, $\tilde{H}_{ij}(\tau;t)$ is the CIR from the ith transmitter antenna to the jth receiver antenna. Without much loss in generality, it is easier to understand MIMO signal processing by using a *narrowband*, time-invariant channel model. In this case, there are no frequency or time dependencies in the channel, resulting in a constant $\tilde{\mathbf{H}}$ matrix:

$$\tilde{\mathbf{H}} = \begin{bmatrix} \tilde{H}_{11} & \tilde{H}_{21} & \cdots & \tilde{H}_{M1} \\ \tilde{H}_{12} & \tilde{H}_{22} & \cdots & \tilde{H}_{M2} \\ \vdots & \vdots & \ddots & \vdots \\ \tilde{H}_{1N} & \tilde{H}_{2N} & \cdots & \tilde{H}_{MN} \end{bmatrix} \tag{9.2.3}$$

This simplifies the calculation of the received signal set to

$$\tilde{\mathrm{y}}(t) = \frac{1}{2}\tilde{\mathbf{H}}\,\tilde{\mathrm{x}}(t) \tag{9.2.4}$$

This equation will be the basis of the MIMO capacity analysis.

Note: The Many Names of MIMO

In the research literature, you will find various MIMO discussions under some of these alternative names:

- **MTMR (Multiple-Transmitter antenna, Multiple-Receiver antenna):** Emphasizes MIMO for wireless communications, since MIMO is also possible for other types of communications.
- **MEA (Multi-Element Antenna):** Similar in emphasis to MTMR.
- **Spatial Multiplexing:** Different spatial channels used to send different bit streams.
- **BLAST (Bell Labs Layered Space–Time architecture):** One of the earliest MIMO prototype systems.

9.2.2 Processing the MIMO Signal

A MIMO system is capable of signal processing at the transmitter and receiver to produce the set of received signals with highest overall capacity. Linear recombination - represented as a matrix multiplication - is performed on transmitted and received signal vectors. The operations discussed in the following text correspond to the block diagram in Figure 9.6.

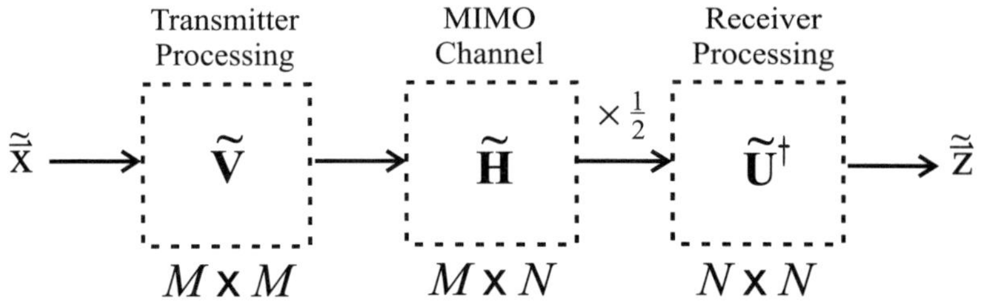

Figure 9.6 The matrix representation of signal processing operations in a MIMO system.

At the Transmitter

The magnitude of a signal vector represents total signal power. Let us use a linear signal processing operation, $\tilde{\mathbf{V}}$, on the transmitted signal vector, $\tilde{\vec{x}}(t)$, to produce a new set of signals to be fed into the array of transmitter antennas, $\tilde{\mathbf{V}}\tilde{\vec{x}}(t)$. Furthermore, let us assert that $\tilde{\mathbf{V}}$ is an M-by-M *unitary matrix* with the property $\tilde{\mathbf{V}}\tilde{\mathbf{V}}^\dagger = \mathbf{I}$. Unitary matrices do not change the geometrical length of vectors, so we are not adding any power to the total transmitted signal. (This is important because we want to show capacity gain due solely to MIMO signal processing and not due to any transmitter amplification.)

> *Note: Hermitian Operator* †
>
> The symbol † stands for the conjugate transpose or *Hermitian* operation. Here is a 3×3 example of a Hermitian operation on the matrix $\tilde{\mathbf{H}}$:
>
> $$\tilde{\mathbf{H}} = \begin{bmatrix} 5+6j & 2 & -4+9j \\ 7-2j & -1-5j & 3+3j \\ 6j & 9-2j & -8 \end{bmatrix} \qquad \tilde{\mathbf{H}}^{\dagger} = \begin{bmatrix} 5-6j & 7+2j & -6j \\ 2 & -1+5j & 9+2j \\ -4-9j & 3-3j & -8 \end{bmatrix}$$
>
> Do not confuse the Hermitian *operation* with the concept of a Hermitian *matrix* (a matrix that solves the equation $\tilde{\mathbf{A}} = \tilde{\mathbf{A}}^{\dagger}$). Generally, baseband MIMO channel matrices are *not* Hermitian matrices.

At the Receiver

Signal processing may also be performed at the receiver. After the channel operates on the transmitter output to produce $\tilde{\mathbf{H}}\tilde{\mathbf{V}}\tilde{\vec{\mathrm{x}}}(t)$, the receiver operates on the incoming signal with the matrix $\tilde{\mathbf{U}}^{\dagger}$. The final output signal is the vector $\tilde{\vec{\mathrm{z}}}$:

$$\tilde{\vec{\mathrm{z}}}(t) = \frac{1}{2}\overbrace{\tilde{\mathbf{U}}^{\dagger}\tilde{\mathbf{H}}\tilde{\mathbf{V}}}^{\tilde{\mathbf{D}}}\,\tilde{\vec{\mathrm{x}}}(t) \tag{9.2.5}$$

Let us assert that the receiver operation of $\tilde{\mathbf{U}}^{\dagger}$ is an N-by-N unitary matrix ($\tilde{\mathbf{U}}^{\dagger}\tilde{\mathbf{U}} = \tilde{\mathbf{I}}$). This condition assures that we have not added or subtracted any signal power - the receiver has simply rearranged the incoming signal with simple linear combining operations.

Of course, a wireless system has little control on the channel matrix operation, but it does have control of the $\tilde{\mathbf{U}}$ and $\tilde{\mathbf{V}}$ matrices. Let us impose an algorithm on the MIMO link that *diagonalizes* the entire net operation, $\tilde{\mathbf{D}}$, in Equation (9.2.5). Thus, we choose $\tilde{\mathbf{U}}$ and $\tilde{\mathbf{V}}$ so that end-to-end signal transmission operation resembles this (case shows $N > M$):

$$\tilde{\mathbf{D}} = \begin{bmatrix} \sqrt{\lambda_1} & 0 & 0 & \cdots & 0 \\ 0 & \sqrt{\lambda_2} & 0 & \cdots & 0 \\ 0 & 0 & \sqrt{\lambda_3} & \ddots & 0 \\ \vdots & \vdots & \ddots & \ddots & \vdots \\ 0 & 0 & 0 & \cdots & \sqrt{\lambda_M} \\ 0 & 0 & 0 & \cdots & 0 \\ \vdots & \vdots & \vdots & \ddots & \vdots \\ 0 & 0 & 0 & \cdots & 0 \end{bmatrix} \tag{9.2.6}$$

where the diagonal elements, $\{\sqrt{\lambda_i}\}$, are positive constants. The importance of performing this operation is the simplifying form of $\tilde{\vec{z}}(t)$

$$\tilde{\vec{z}}(t) = \frac{1}{2}\begin{bmatrix} \sqrt{\lambda_1}\tilde{x}_1(t) \\ \sqrt{\lambda_2}\tilde{x}_2(t) \\ \vdots \\ \sqrt{\lambda_M}\tilde{x}_M(t) \\ 0 \\ \vdots \end{bmatrix} \tag{9.2.7}$$

Mathematically, this shows that the MIMO channel may be viewed (for the case of $M \leq N$) as a set of M separate channels from transmitter to receiver.

For the case of more transmitter antennas than receiver antennas ($M \geq N$), the diagonal matrix of Equation (9.2.6) would support a maximum of N separate channels. Thus, the number of separate channels in a radio link with M transmitter antennas and N receiver antennas is $\min(M, N)$.

Singular Value Decomposition

Our signal processing steps in the previous sections had a distinct physical rationale. They rearrange the signals without adding or subtracting power (and changing the channel capacity through amplification). But what we have actually done is a *singular value decomposition* (SVD) on the channel matrix $\tilde{\mathbf{H}}$ [And00]. We can rearrange the equation

$$\tilde{\mathbf{D}} = \tilde{\mathbf{U}}^{\dagger}\tilde{\mathbf{H}}\tilde{\mathbf{V}} \tag{9.2.8}$$

to produce

$$\tilde{\mathbf{H}} = \tilde{\mathbf{U}}\tilde{\mathbf{D}}\tilde{\mathbf{V}}^{\dagger} \tag{9.2.9}$$

which is the definition for an SVD of the matrix $\tilde{\mathbf{H}}$. Furthermore, the SVD theorem states that those nonzero diagonal elements of $\tilde{\mathbf{D}}$ - the singular values of $\tilde{\mathbf{H}}$ - have a special and significant mathematical interpretation: When squared, they are actually the *eigenvalues* of the matrix $\tilde{\mathbf{H}}^{\dagger}\tilde{\mathbf{H}}$ for $N \geq M$ (or the eigenvalues of the matrix $\tilde{\mathbf{H}}\tilde{\mathbf{H}}^{\dagger}$ for $M \geq N$).

9.2.3 Separate Channels

If the same signals were transmitted on each channel in Equation (9.2.7), then we could coherently combine the signal branches in $\tilde{\vec{z}}(t)$ and achieve a capacity similar to Equation (9.1.6). If each signal in $\tilde{\vec{x}}(t)$ were a *different* signal, the individual channels as seen by $\tilde{\vec{z}}(t)$ would roughly have the capacity (for $N \geq M$)

$$C_{\text{single}} \approx B \log_2\left(1 + \frac{N}{M}\text{SINR}_0\right) \tag{9.2.10}$$

Yet there are M of these channels, so the total capacity of the system is given by

$$C \approx MB \log_2 \left(1 + \frac{N}{M}\mathrm{SINR_O}\right) \tag{9.2.11}$$

Notice how Equation (9.2.11) differs from the conventional multi-antenna channel capacity in Equation (9.1.6): The SINR on each channel diminishes, but the overall capacity multiplies due to the presence of M channels.

This is not a case of duplicitous accounting. Recall this basic property of logarithms:

$$x \log_n y = \log_n y^x \tag{9.2.12}$$

Multiplying a logarithm is equivalent to raising its argument to a power. Thus, dividing by M^2 in the argument of Equation (9.2.11) is usually overwhelmed by the simple multiplication of the logarithm by M:

$$\overbrace{MB \log_2 \left(1 + \frac{N}{M}\mathrm{SINR_O}\right)}^{\text{total capacity of separated channels}} > \overbrace{B \log_2 \left(1 + MN\mathrm{SINR_O}\right)}^{\text{capacity of a single channel}}$$

The key principle at work here is that it is more beneficial to transmit data using many different low-powered channels than using one single, high-powered channel.

> *Note: Know Thy Channel*
>
> If a system is to achieve the super-high capacity of Equation (9.2.11), it must be able to calculate the correct unitary matrices $\tilde{\mathbf{U}}$ and $\tilde{\mathbf{V}}$ for the SVD. Since this matrix operation depends on the channel matrix, $\tilde{\mathbf{H}}$, it becomes necessary to estimate the channel at both the transmitter and receiver. This leads to a lot of signal-processing overhead in MIMO systems. However, this overhead is a small price to pay for an order-of-magnitude increase in the bit rate.

It helps to have a physical interpretation of how a MIMO system operates rather than thinking in terms of abstract matrix operations. When the matrix $\tilde{\mathbf{V}}$ in Figure 9.6 operates on an input vector of symbols $\vec{\tilde{x}}(t)$, it is essentially providing a unique antenna pattern for each symbol in that vector. Each column of $\tilde{\mathbf{V}}$ contains the M amplitudes and phases for a symbol sent through the array of M transmitter antennas. Since a set of amplitudes and phases weightings for antenna elements defines a unique spatial radiation pattern for a radio signal, column 1 of $\tilde{\mathbf{V}}$ assigns a radiation pattern to $\tilde{x}_1(t)$, column 2 assigns a radiation pattern to $\tilde{x}_2(t)$, and so on.

The same may be said of the matrix $\tilde{\mathbf{U}}^\dagger$ as it operates on the received signals from the N receiver antenna elements. Each column of $\tilde{\mathbf{U}}$ (or, equivalently, each row of $\tilde{\mathbf{U}}^\dagger$) defines a radiation pattern that "picks out" one of the original symbols sent by the transmitter. This physical interpretation is shown in Figure 9.7. The weighting operations of $\tilde{\mathbf{V}}$ and $\tilde{\mathbf{U}}$ assign unique antenna patterns to each separate data stream, isolating individual data symbols from the others in the multipath channel.

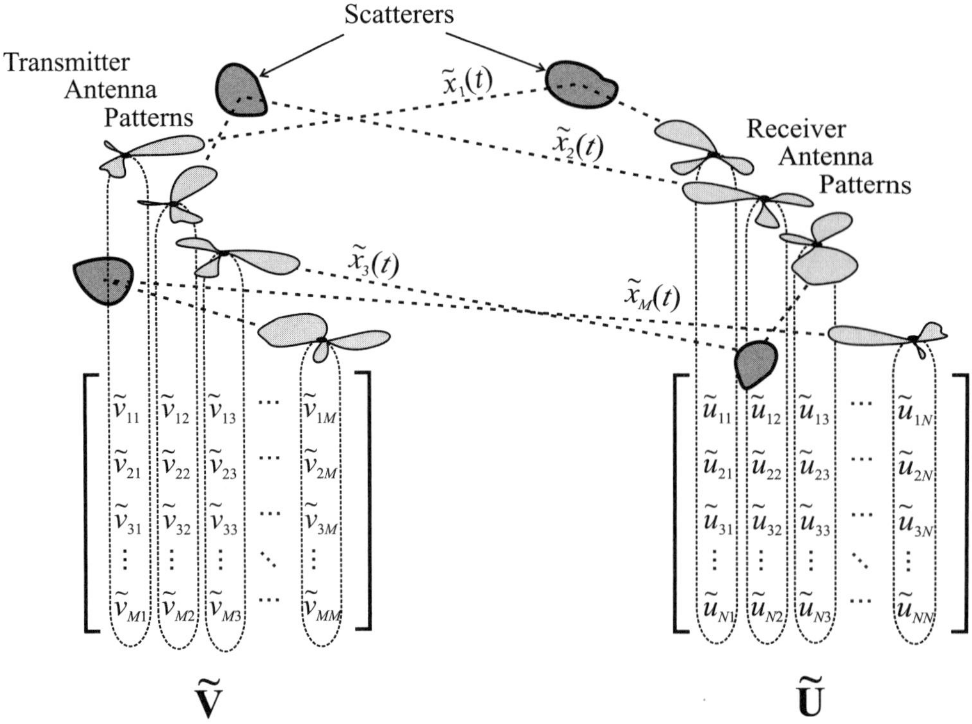

Figure 9.7 Physically, MIMO spatial coding sends different symbol streams to and from different directions in space.

> *Note: The Danger of Oversimplification*
>
> Remember, despite the simple sketch of Figure 9.7, an ideal MIMO system performs the mathematical SVD on the channel matrix to get perfect channel separation. The multipath channel is very complicated in real life, and antenna patterns have finite resolution, so the MIMO processing cannot simply "draw a bead" on a few dominant scatterers and get perfect separation of symbols. Still, the concept in Figure 9.7 is useful for a basic understanding.

9.2.4 Formal Capacity Expressions

To develop the conventional Shannon capacity equations for a flat-fading MIMO channel with bandwidth B, we start from the basic transmission equation of Equation (9.2.5) and add noise and interference to the received signal:

$$\tilde{\vec{z}}(t) = \frac{1}{2}\tilde{\mathbf{U}}^{\dagger}\tilde{\mathbf{H}}_0\tilde{\mathbf{V}}\tilde{\vec{x}}(t) + \tilde{\vec{n}}(t) \tag{9.2.13}$$

where $\tilde{\vec{n}}(t)$ is an N-element vector of noise+interference signals that have been received along with the desired signal $\tilde{\vec{x}}(t)$. In typical analysis, the MIMO channel matrix $\tilde{H}_0$ is normalized to the value $\tilde{\mathbf{H}}$:

$$\tilde{\mathbf{H}} = \frac{1}{H_0}\tilde{\mathbf{H}}_0 \quad \text{for } H_0 = \sqrt{\frac{1}{MN}\sum_{i=1}^{M}\sum_{j=1}^{N}\mathrm{E}\left\{|\tilde{h}_{ij}|^2\right\}} \tag{9.2.14}$$

Essentially, Equation (9.2.14) results in an $\tilde{\mathbf{H}}$ whose total channel components have, on average, unit power. This normalization separates the problems of calculating link-budget power and of studying the effects of channel fading introduced by $\tilde{\mathbf{H}}$.

> *Note: Dummy Signals*
>
> For the case of $M > N$, only N separate signals maybe sent through the channel. Since $\tilde{\vec{x}}(t)$ contains M signals, assume that the signals in the vector beyond the Nth position are set to zero. The tedious "*for* $N \geq M$ $\cdots$ *and for* $M \geq N$ $\cdots$" explanations can obscure the basic principles of the MIMO system, which are actually pretty simple.

Using the normalized channel matrix, $\tilde{\mathbf{H}}$, we can calculate the total theoretical channel capacity by summing the Shannon capacity for each individual separate channel:

$$C = \sum_{i=1}^{\min(M,N)} \log_2\left(1 + \lambda_i \mathrm{SINR}_i\right) \quad \text{for } \mathrm{SINR}_i = \frac{H_0^2\ \mathrm{E}\left\{|\tilde{x}_i(t)|^2\right\}}{4\ \ \mathrm{E}\left\{|\tilde{n}_i(t)|^2\right\}} \tag{9.2.15}$$

where λ_i are the singular values of the normalized channel matrix, $\tilde{\mathbf{H}}$. Often, the SINR's of the individual channels are identical, provided equal-powered symbols are transmitted and the average noise+intereference power on each signal branch is equal:

$$\mathrm{SINR}_0 = \frac{H_0^2\ \mathrm{E}\left\{|\tilde{x}_1(t)|^2\right\}}{4\ \ \mathrm{E}\left\{|\tilde{n}_1(t)|^2\right\}} = \frac{H_0^2\ \mathrm{E}\left\{|\tilde{x}_2(t)|^2\right\}}{4\ \ \mathrm{E}\left\{|\tilde{n}_2(t)|^2\right\}} = \cdots$$

For this case Equation (9.2.15) reduces to

$$C = \sum_{i=1}^{\min(M,N)} \log_2\left(1 + \lambda_i \mathrm{SINR}_0\right) \tag{9.2.16}$$

which assumes that the transmitter has excited each separate channel with equal power.

In truth, the capacity of Equation (9.2.16) can be increased by exciting unequal powers in the separate channels according to a *water-filling strategy* [Cov91]. In water-filling, more power is placed proportionally into stronger channels. Thus, if channel 1 was twice as strong as channel 2 ($\lambda_1 = 2\lambda_2$), then twice as much power would be sent through channel 1. The total channel capacity for this channel is given by

$$C = \sum_{i=1}^{\min(M,N)} \log_2\left(1 + \lambda_i \text{SINR}_i\right) \quad \text{for } \text{SINR}_i = \frac{\lambda_i}{\sum\limits_{i=1} \lambda_i} \text{SINR}_0 \tag{9.2.17}$$

Of course, the transmitter must know the SVD of the channel to allocate power in a water-filling strategy. Since feeding back channel information from receiver to transmitter is not desirable in wireless communications systems, Equation (9.2.16) is the implied practical limit for most MIMO analysis.

Rather than express capacity in terms of singular values, it is possible to develop an elegant channel capacity formula in terms of the normalized channel matrix, $\tilde{\mathbf{H}}$. The standard capacity formula for a MIMO system is usually given as

$$\begin{array}{cc} \text{Capacity for } M \leq N & \text{Capacity for } N \leq M \\ B\log_2\left(\det\left[\mathbf{I} + \text{SINR}_0\tilde{\mathbf{H}}^\dagger\tilde{\mathbf{H}}\right]\right) & B\log_2\left(\det\left[\mathbf{I} + \text{SINR}_0\tilde{\mathbf{H}}\tilde{\mathbf{H}}^\dagger\right]\right) \end{array} \tag{9.2.18}$$

which is derived as Theorem 9.1. Equation (9.2.18) is especially useful for calculating MIMO channel capacity in a mathematical software package.

Theorem 9.1: Matrix Channel Capacity

Statement: Prove the formulas for matrix channel capacity in Equation (9.2.18).

Proof: Starting from the capacity expression, we may write Equation (9.2.16) in an equivalent matrix notation

$$\begin{array}{cc} \text{Capacity for } M \leq N & \text{Capacity for } N \leq M \\ B\log_2\left(\det\left[\mathbf{I} + \text{SINR}_0\tilde{\mathbf{D}}^\dagger\tilde{\mathbf{D}}\right]\right) & B\log_2\left(\det\left[\mathbf{I} + \text{SINR}_0\tilde{\mathbf{D}}\tilde{\mathbf{D}}^\dagger\right]\right) \end{array}$$

where $\tilde{\mathbf{D}}$ is the diagonal matrix of singular values in Equation (9.2.6) and $\mathbf{I}$ is an appropriately sized identity matrix. (Recall that the determinant of a diagonal matrix is simply the product of the diagonal entries). Using the relationship $\tilde{\mathbf{D}} = \tilde{\mathbf{U}}^\dagger\tilde{\mathbf{H}}\tilde{\mathbf{V}}$, we can rewrite the equations as

$$B\log_2\left(\det\left[\mathbf{I} + \text{SINR}_0\tilde{\mathbf{V}}^\dagger\tilde{\mathbf{H}}^\dagger\tilde{\mathbf{H}}\tilde{\mathbf{V}}\right]\right) \quad B\log_2\left(\det\left[\mathbf{I} + \text{SINR}_0\tilde{\mathbf{U}}^\dagger\tilde{\mathbf{H}}\tilde{\mathbf{H}}^\dagger\tilde{\mathbf{U}}\right]\right)$$

Recall that $\tilde{\mathbf{V}}^\dagger\tilde{\mathbf{V}} = \mathbf{I}$ and $\tilde{\mathbf{U}}^\dagger\tilde{\mathbf{U}} = \mathbf{I}$, so it is possible to write these equations as

$$B\log_2\left(\det\left[\tilde{\mathbf{V}}^\dagger\left(\mathbf{I} + \text{SINR}_0\tilde{\mathbf{H}}^\dagger\tilde{\mathbf{H}}\right)\tilde{\mathbf{V}}\right]\right)$$

$$B \log_2 \left(\det \left[\tilde{\mathbf{U}}^\dagger \left(\mathbf{I} + \mathrm{SINR}_0 \tilde{\mathbf{H}} \tilde{\mathbf{H}}^\dagger \right) \tilde{\mathbf{U}} \right] \right)$$

Since $\tilde{\mathbf{V}}$ and $\tilde{\mathbf{U}}$ are unitary, they do not change the value of the determinant of the matrix that they multiply. Thus, they can be removed completely from the above equation to produce Equation (9.2.18).

9.2.5 MIMO Channels with Impaired Capacity

If you were to generate a channel matrix at random, filling $\tilde{\mathbf{H}}$ with random, independent Rayleigh-fading values, the capacity of Equation (9.2.18) would be quite high. Figure 9.8 illustrates these gains by plotting the CDF of MIMO channel capacity for various systems in an ideal Rayleigh-fading environment with $\mathrm{SINR}_0 = 10$ dB. For any MIMO system, there are small-scale fades in space that drop the capacity below average levels. Note, however, that deep fading is hardly a problem for MIMO links with larger numbers of transmitter and receiver antennas. Also note from Figure 9.8 the huge potential gains in average channel capacity, since C_0 is the capacity of the unfaded SISO channel. Thus, there is high gain due to the multiple spatial channels and there is also fade resistance, as if antenna diversity is also present.

The ideal Rayleigh channels are not always available to a MIMO radio link. There are several physical characteristics of the MIMO channel that may lead to channel capacity lower than the depiction in Figure 9.8.

Absence of Multipath

The channel capacity expression of Equation (9.2.18) is maximized when the signal power is evenly distributed among the singular values, λ_i, of the channel matrix. In the absence of multipath, the singular values become heavily skewed towards a single, dominant λ_i. In the extreme case of only one nonzero eigenvalue, it can be shown that channel capacity reverts to Equation (9.1.6) - the single-channel SINR-maximizing mode of operation. Multipath is necessary in order to break the MIMO channel into separate channels. As an example, contrast the graph of Figure 9.9, a graph of MIMO channel capacities in a Rician environment with $K = 20$ dB, with the graph of Figure 9.8. The concentration of power in the single specular multipath component reduces the capacity available in the aggregate channel.

Channel Correlation

Much like a simpler antenna diversity system, the MIMO system requires the individual channels in the channel matrix, $\tilde{\mathbf{H}}$, to be uncorrelated for high capacity. If the small-scale fading experienced by the receiver antennas is uncorrelated, then the singular values of $\tilde{\mathbf{H}}$ tend toward approximately equal values. If the small-scale fading of receiver antenna elements is correlated, then the singular values become skewed and one begins to dominate all the others. The same is also true for correlation between transmitter antenna elements.

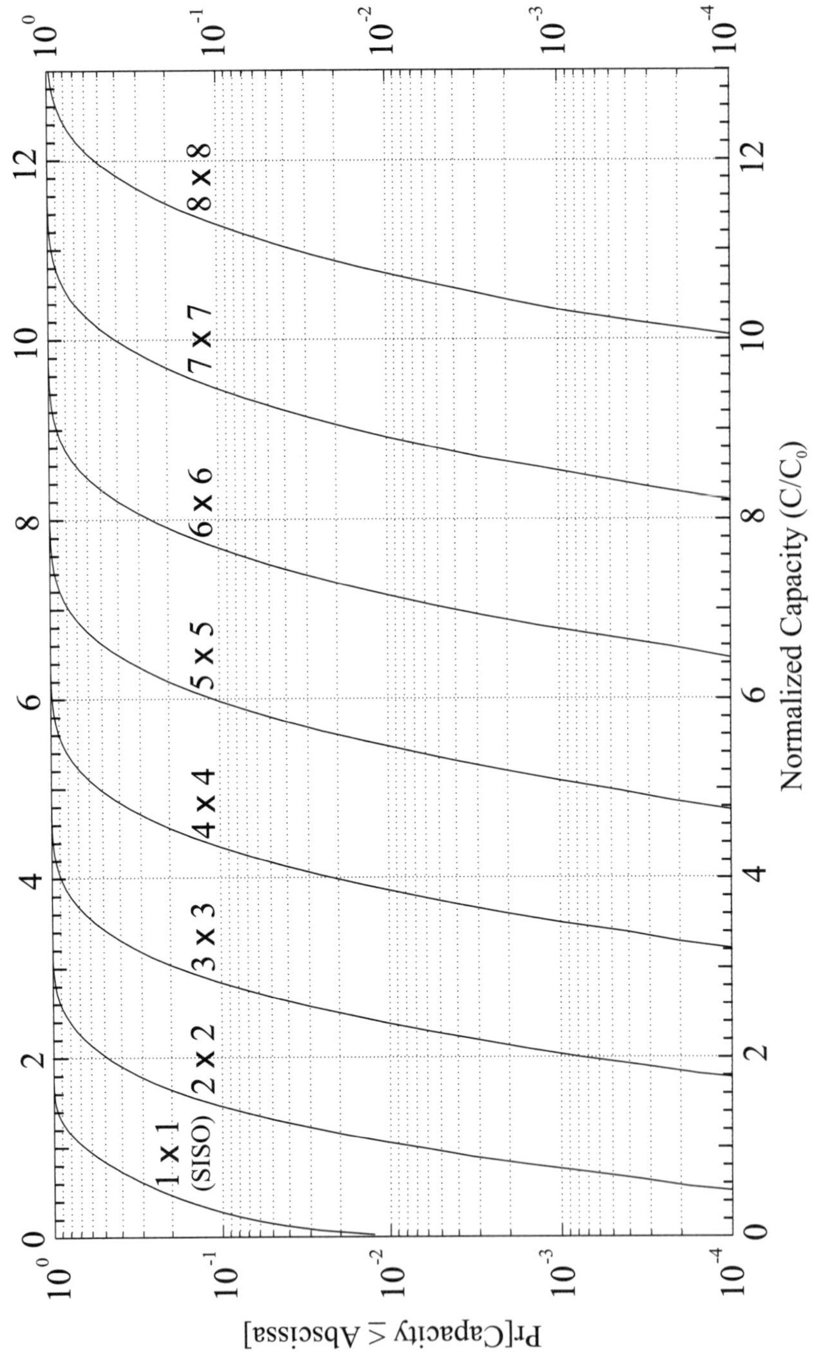

Figure 9.8 Channel capacity for ideal MIMO system in Rayleigh fading.

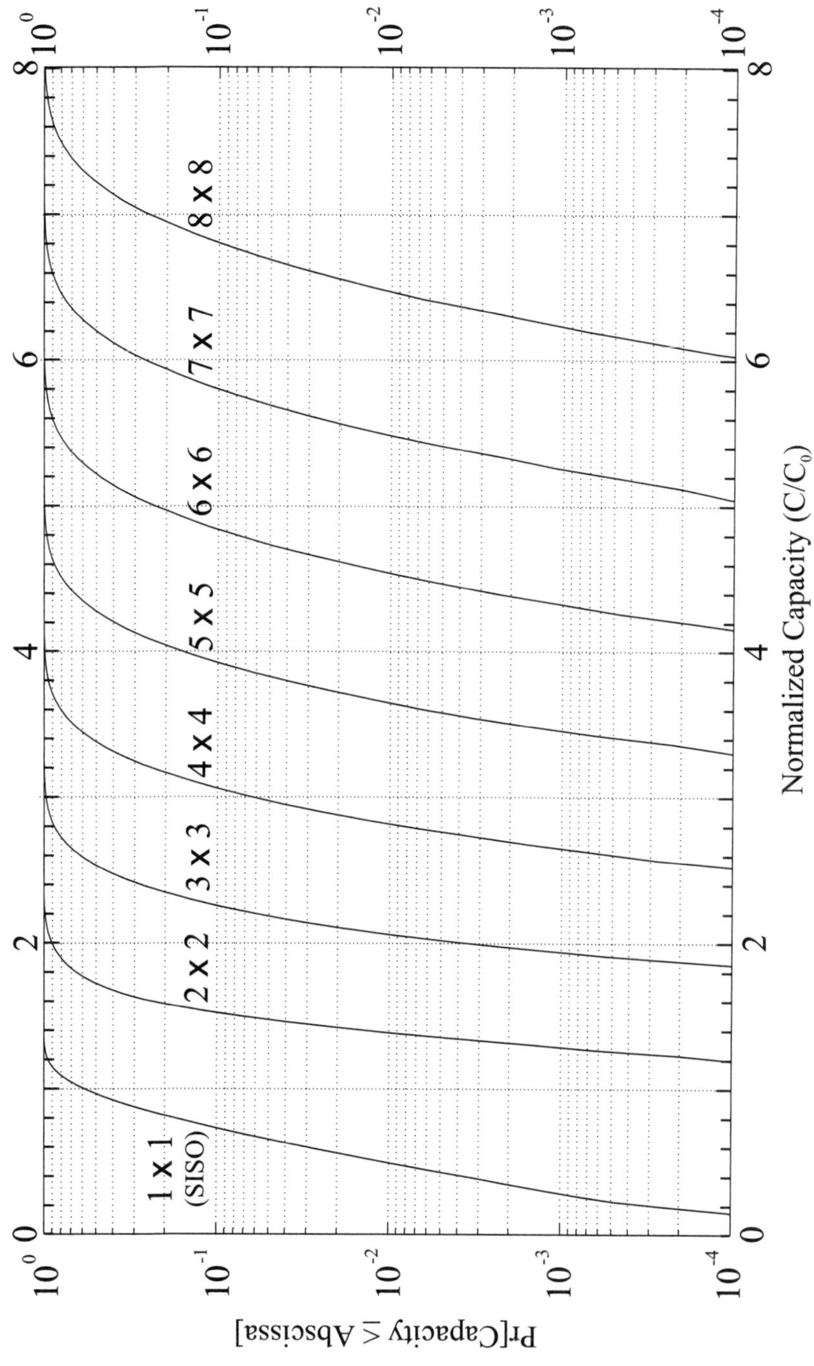

Figure 9.9 Channel capacity for an ideal MIMO system in Rician fading with $K = 20$ dB.

Unequal Average Branch Power

As discussed in the previous chapter, pattern and polarization diversity designs are not necessarily susceptible to channel correlation. These types of diversity were subject to unequal average branch powers. Antennas with unequal average branch powers are trouble for diversity systems, and MIMO systems are no different. If the average branch power on one antenna is higher than others, the signal power will once again gravitate towards a dominant single channel.

Keyhole Channel

In addition to the three previous channel impairments, there is a peculiar physical phenomenon called the *keyhole channel* that can impair a MIMO system [Chi00]. To understand the keyhole channel, first assume that a MIMO system is operating in a wireless channel that does not have any of the three previous impairments: It experiences a rich, Rayleigh multipath environment with equal branch power and plenty of decorrelation between both transmitter and receiver antenna elements. Now assume that in between transmitter and receiver is a spatial bottleneck - a tiny diffracting aperture (i.e., the "keyhole") - through which all propagation must pass. This scenario is illustrated in Figure 9.10.

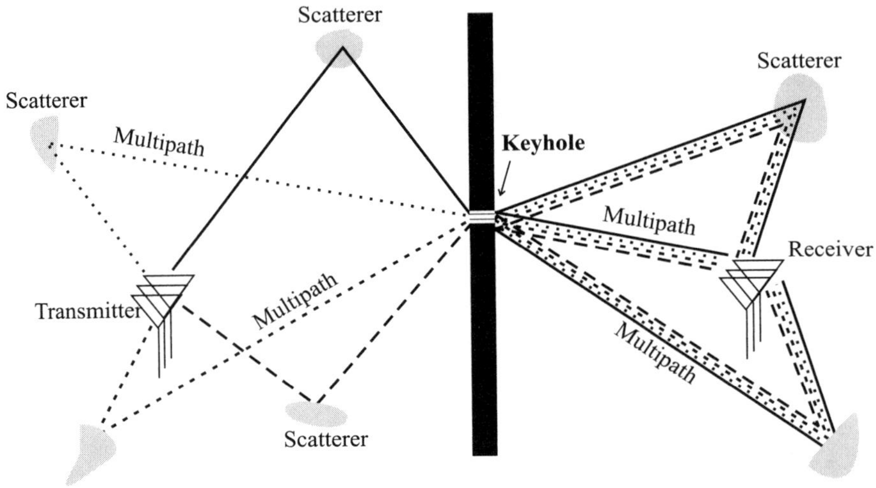

Figure 9.10 A keyhole channel creates a bottleneck for the MIMO system.

The spatial bottleneck in Figure 9.10 forces all multipath components on one side of the keyhole to propagate in conjunction on the other side of the keyhole. The end result is a MIMO channel that does not allow spatial separation at the receiver under any other favorable conditions; such a channel collapses to conventional Shannon capacity. The physical keyhole must be electromagnetically small (1 to 2 wavelengths or less) to create this type of channel. There has been speculation that doorways and hallways may act as keyholes for indoor MIMO systems, but there is little empirical evidence to show that keyholes are common in typical radio links.

9.3 Practical MIMO Signaling

This section presents an overview of MIMO implementation based on the space–time architecture by Foschini in [Fos96]. It exposes some practical design issues for MIMO systems.

9.3.1 Practical Signal Extraction

Like any Shannon channel-capacity expression, Equation (9.2.18) is the ideal limit. Practical systems will experience lower overall capacity. First of all, few wireless systems are going to perform the optimal SVD on the channel at both transmitter and receiver antennas, since this operation requires that *both* transmitter *and* receiver have reliable channel estimates. Instead - through the pretransmission of a known training sequence - it is the receiver only that knows the channel matrix components.

Even without knowledge of the channel at the transmitter, it is still possible to separate the channels. In this case, replace the receiver operating matrix, $\tilde{\mathbf{U}}^{\dagger}$, in Figure 9.6 with the inverse of the channel matrix, $\tilde{\mathbf{H}}^{-1}$. In the absence of any transmitter operations, the MIMO channels are easily (but not optimally) separated.

The mathematical operations of separating the incoming MIMO signals and detecting the original signal streams are best illustrated by this alternative representation of the signal vector $\tilde{\vec{y}}(t)$, the collection of received signals at the N receiver antennas:

$$\tilde{\vec{y}}(t) = \underbrace{\begin{bmatrix} \tilde{\vec{H}}_1 & \tilde{\vec{H}}_2 & \cdots & \tilde{\vec{H}}_M \end{bmatrix}}_{\tilde{\mathbf{H}}} \underbrace{\begin{bmatrix} \tilde{x}_1(t) \\ \tilde{x}_2(t) \\ \vdots \\ \tilde{x}_M(t) \end{bmatrix}}_{\tilde{\vec{x}}(t)} \tag{9.3.1}$$

(Note: The $\frac{1}{2}$ is omitted for clarity.) Each transmitted signal stream, $\tilde{x}_i(t)$, is mathematically multiplied by a column, $\tilde{\vec{H}}_i$, in the channel matrix, $\tilde{\mathbf{H}}$. Extracting the ith signal from $\tilde{\vec{y}}(t)$ is represented by the following linear combining operation:

$$\tilde{x}_i(t) = \tilde{\vec{u}}_i \cdot \tilde{\vec{y}}(t) \tag{9.3.2}$$

$$\text{where } \tilde{\mathbf{H}}^{-1} = \begin{bmatrix} \tilde{\vec{u}}_1 \; \tilde{\vec{u}}_2 \; \cdots \; \tilde{\vec{u}}_M \end{bmatrix}^{\mathrm{T}} \quad \text{and } \tilde{\vec{u}}_i \cdot \tilde{\vec{H}}_j = \begin{cases} 0 & \text{for } i \neq j \\ 1 & \text{for } i = j \end{cases}$$

The operation $\cdot$ is a scalar dot product. The dot product of $\tilde{\vec{u}}_i$ and any $\tilde{\vec{H}}_j$ will result in zero unless $i = j$, in which case the result is 1. In this way, we can see that any symbol stream $\tilde{x}_i(t)$ may be detected from the incoming MIMO signal by choosing the appropriate operation $\tilde{\vec{u}}_i$. This interpretation of signal detection is useful in later illustrations.

9.3.2 Transmission Technique

In our discussion of MIMO transmission and reception, we will use a 4×4 MIMO radio channel (i.e., four transmitter antennas and four receiver antennas) to illustrate concepts. Once this case is understood, it is possible to generalize the knowledge to other, more complicated MIMO systems. A diagram of our central example is shown in Figure 9.11, which follows a simplification of Foschini's famous layered space–time architecture in [Fos96].

In order to transmit a high-rate data stream through the 4×4 MIMO radio channel, the primary bit stream must first be demultiplexed into four slower data streams of equal bit rate, as shown at the top of Figure 9.11. These data streams are then arbitrarily channel-coded onto digital symbol streams. The diagram in Figure 9.11 shows 0s and 1s for the sake of simplicity, but any complex digital symbol is possible.

In some systems, these four data streams could be modulated directly to the carrier frequency and sent out through the four transmitter antennas. One problem with a direct transmission scheme, however, is the presence of low-powered channels through one or more of the transmitter antennas. The channels cannot be separated with equal SINR at the receiver by a simple inversion operation of $\tilde{\mathbf{H}}^{-1}$. Thus, some of the separable channels in the MIMO link may have very low SINR, possibly wiping out their data streams. This condition is problematic for a practical MIMO design; without channel estimation feedback, the transmitter has no way of knowing which of the separable channels cannot support the data rate of the transmitted streams.

To solve this problem, Foschini proposed a transmitter architecture that cycles the four demultiplexed streams [Fos96]. Each transmitted stream takes turns through the four transmitter antenna elements, spending a fixed duration (called a "time slot" in Figure 9.11) at each element and cycling back to its original element every 4th turn. Over a period of many time slots, each transmitted data stream experiences the same average channel so that some streams are not consistently wiped out on pathological channels.

9.3.3 Subtraction of Interference

Continuing with the example of a 4×4 system, each data stream is detected from a time slot according to Equation (9.3.2). For the first stream of data, we would write the operation as

$$\begin{aligned} \tilde{x}_1(t) &= \tilde{\vec{u}}_1 \cdot \tilde{\vec{y}}(t) \\ &= \tilde{\vec{u}}_1 \cdot \begin{bmatrix} \tilde{\vec{H}}_1 & \tilde{\vec{H}}_2 & \tilde{\vec{H}}_3 & \tilde{\vec{H}}_4 \end{bmatrix} \tilde{\vec{x}}(t) \\ &= \begin{bmatrix} 1 & 0 & 0 & 0 \end{bmatrix} \tilde{\vec{x}}(t) \end{aligned} \tag{9.3.3}$$

which extracts the first signal.

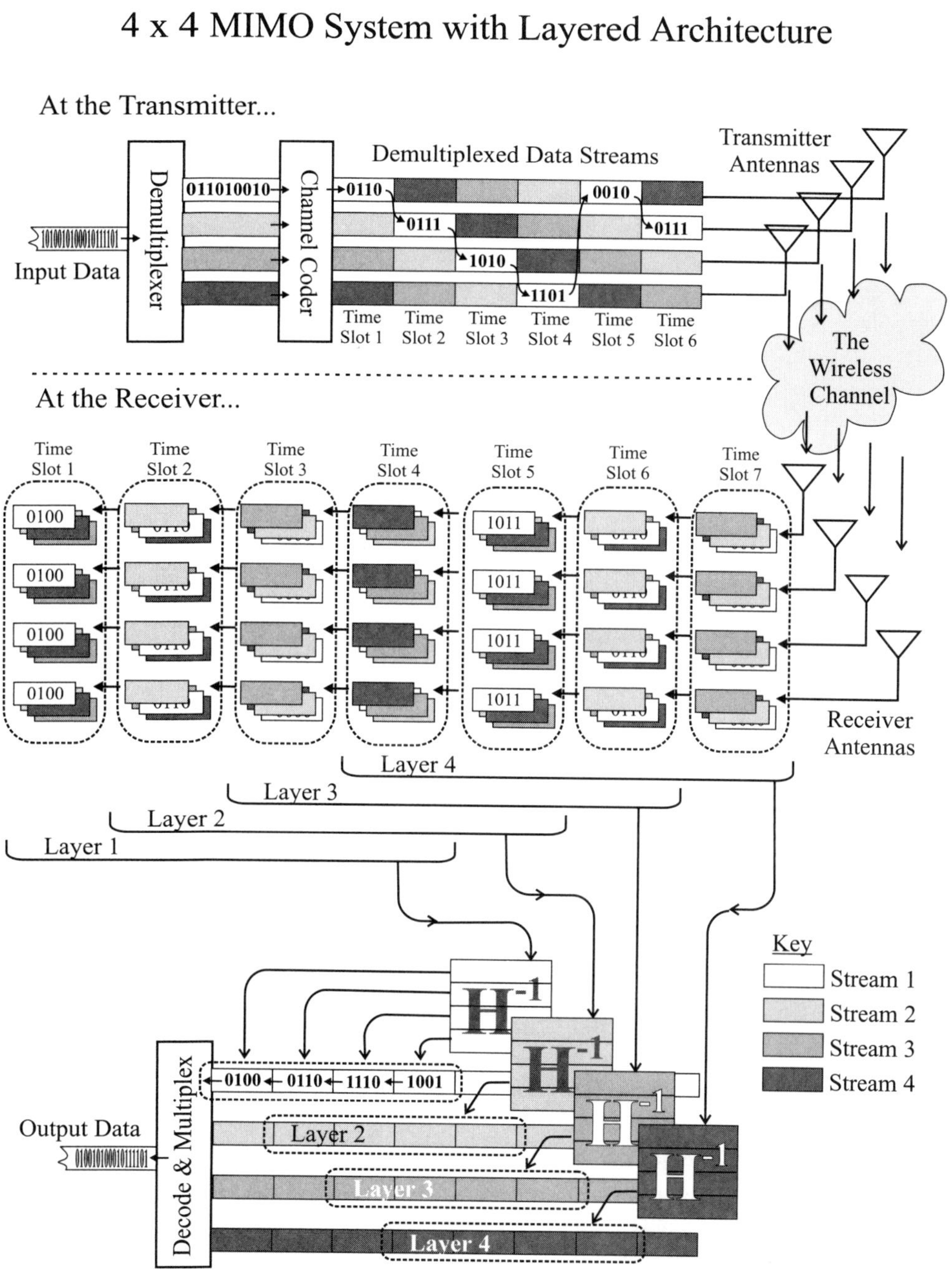

Figure 9.11 A practical $M = 4$, $N = 4$ MIMO transmission system [Fos96].

We could continue in this manner, extracting the other 3 streams with similar operations. However, there is an interesting opportunity to obtain signal gain through *interference cancellation.* Since $\tilde{x}_1(t)$ is now known, we may subtract its presence out of the signal vector, $\vec{\tilde{y}}(t)$. Thus, the second extraction is made in the following manner:

$$\begin{aligned} \tilde{x}_2(t) &= \vec{\tilde{u}}_2 \cdot \left(\vec{\tilde{y}}(t) - \tilde{x}_1(t)\vec{\tilde{H}}_1\right) \\ &= \vec{\tilde{u}}_2 \cdot \begin{bmatrix} 0 & \vec{\tilde{H}}_2 & \vec{\tilde{H}}_3 & \vec{\tilde{H}}_4 \end{bmatrix} \vec{\tilde{x}}(t) \\ &= \begin{bmatrix} 0 & 1 & 0 & 0 \end{bmatrix} \vec{\tilde{x}}(t) \end{aligned} \tag{9.3.4}$$

Note that, unlike Equation (9.3.3), there are only three defining criteria for $\vec{\tilde{u}}_2$:

$$\vec{\tilde{u}}_2 \cdot \vec{\tilde{H}}_2 = 1 \qquad \vec{\tilde{u}}_2 \cdot \vec{\tilde{H}}_3 = 0 \qquad \vec{\tilde{u}}_2 \cdot \vec{\tilde{H}}_4 = 0$$

If we view the four components of $\vec{\tilde{u}}_2$ as solutions to a linear system of equations, then the above set of conditions represents a system of three equations with four unknowns. Thus, removal of the first estimate, $\tilde{x}_1(t)$, provides an extra degree of freedom for constructing $\vec{\tilde{u}}_2$. Now there are *two* possible vectors $\vec{\tilde{u}}_2$ that null-out signals $\tilde{x}_3(t)$ and $\tilde{x}_4(t)$ while keeping signal $\tilde{x}_2(t)$. Since two independent estimates of $\tilde{x}_2(t)$ are available, the system can use combining algorithms to achieve diversity gain. The potential performance for this second data stream estimate is equivalent to MRC diversity using two branches.

Continuing with the interference cancellation, an estimate of $\tilde{x}_3(t)$ is made by subtracting the previous signal, $\tilde{x}_2(t)$, from the signal vector, $\vec{\tilde{y}}(t)$:

$$\begin{aligned} \tilde{x}_3(t) &= \vec{\tilde{u}}_3 \cdot \left(\vec{\tilde{y}}(t) - \tilde{x}_1(t)\vec{\tilde{H}}_1 - \tilde{x}_2(t)\vec{\tilde{H}}_2\right) \\ &= \vec{\tilde{u}}_3 \cdot \begin{bmatrix} 0 & 0 & \vec{\tilde{H}}_3 & \vec{\tilde{H}}_4 \end{bmatrix} \vec{\tilde{x}}(t) \\ &= \begin{bmatrix} 0 & 0 & 1 & 0 \end{bmatrix} \vec{\tilde{x}}(t) \end{aligned} \tag{9.3.5}$$

Now $\vec{\tilde{u}}_3$ has *three* degrees of freedom; following the model of Equation (9.3.5), we can obtain three independent estimates of $\tilde{x}_3(t)$. Diversity gain using MRC on these three signal branches is possible.

Upon reaching the last signal, $\tilde{x}_4(t)$, it is apparent that *four* independent estimates will exist. MRC diversity gain using four signal branches is possible. Much like peeling the layers of an onion, this interference cancellation algorithm strips away previously detected signals and in the process produces increasingly reliable estimates of the next detection with higher average SINR. In an ideal MIMO channel (uncorrelated Rayleigh fading with equal average power on all elements), the channel capacity of each data stream is expressed as the following (in bits/s/Hz):

$$\begin{aligned} C_{\tilde{x}1} &= \log_2(1 + \mathrm{SINR}_0\chi_2^2) \quad C_{\tilde{x}2} = \log_2(1 + \mathrm{SINR}_0\chi_4^2) \\ C_{\tilde{x}3} &= \log_2(1 + \mathrm{SINR}_0\chi_6^2) \quad C_{\tilde{x}4} = \log_2(1 + \mathrm{SINR}_0\chi_8^2) \end{aligned} \tag{9.3.6}$$

where χ_{2k} is a Chi-squared random variable with $2k$ degrees of freedom. (Recall from Section 8.2.2 that the output of k-branch MRC diversity follows the Chi-squared distribution). The total channel capacity for this reception using interference cancellation is the sum of the individual capacities of Equation (9.3.6). For a general $M \times M$ (square) MIMO system with this architecture, the channel capacity is given by

$$C = \sum_{k=1}^{M} \log_2(1 + \mathrm{SINR}_0 \chi^2_{2k}) \quad \text{bits/s/Hz} \tag{9.3.7}$$

Mathematically, Equation (9.3.7) is a lower bound of the full capacity expression in Equation (9.2.18) [Fos98].

> *Note: Serial Versus Parallel Cancellation*
>
> As described in this section, the detection employs *serial* interference cancellation, stripping estimates off one at a time. *Parallel* interference cancellation is also possible. In such a scheme, the algorithm iterates the serial interference rejection, repeating earlier signal estimates with a complete set of detected signals. Compared to its serial counterpart, parallel interference cancellation adds complexity to the receiver with modest improvement in overall bit rate.

9.3.4 Layered Reception Technique

The diagram of Figure 9.11 in our 4×4 example shows the flow of signal extraction with interference cancellation at the receiver. There are four raw data streams from the receiver antennas, down-converted to baseband, each carrying a different linear combination of the four original streams sent by the M transmitter antennas. Since the original streams were cycled over each transmitter antenna, as discussed in Section 9.3.2, different time slots contain a differently ordered arrangement of streams.

As presented in [Fos96], the receiver buffers the incoming streams of data and extracts streams in steps, each step lasting the duration of one time slot. At every step, the receiver detects a *layer* of data, consisting of a single stream spanning a block of four total time slots. For example, at time slot 1, the receiver detects four time slots' worth of data as layer 1, which spans time slots 1 through 4 in the buffer. At time slot 2, the receiver detects layer 2, which spans time slots 2 through 5 in the buffer. This continues for the other layers until time slot 5, when the receiver has returned to detect the layer corresponding to stream 1.

The layers are decoded and multiplexed onto a final stream of output data. Because the channel capacity of even a 4×4 MIMO link is much greater than a SISO link, huge end-to-end data rates are possible using a small amount of bandwidth. When the receiver detects layer 1, each time slot has had a different amount of interference removed from the data stream. In time slot 1, three other layers have been subtracted during previous detection operations; time slot 1 contains the most reliable data for layer 1. In time slot 4, *none* of the other layers have been subtracted;

time slot 4 contains the least reliable data. In this way, layer 1 is assembled from equal time slots that experience the four possible levels of interference. Every layer is constructed in this manner.

This symmetry of demodulation between all layers is the practical benefit of the layered space–time coding proposed by [Fos96]: Each data stream experiences all four channels in the MIMO link and all four levels of interference cancellation. Hence, their bit-error rates are nearly identical. Without cycled transmission and layered detection, the four different layers could have vastly different bit-error rates, producing bursts of errors and data loss in the final output signal. The system in Figure 9.11 is based on a "share the channel wealth" philosophy.

9.4 Space–Time Block Coding

Space–time block coding is another way to incorporate spatial and temporal signal processing for transmitters and receivers with multiple antennas.

9.4.1 MISO Revisited

The operation of a MISO channel that maximizes SINR at the receiver is problematic for implementation. In its most obvious conception, MISO signaling requires the transmitter to know the elements of the channel matrix, $\tilde{\mathbf{H}}$. To estimate a channel at the receiver, known training sequences must be sent by the transmitter so that the receiver can sound the weights in the channel matrix. These weights must then be sent *back* to the transmitter; the receiver must act as a transmitter and vice versa, complicating the link protocol.

To circumvent this problem, let us go back to a simple MISO case with two transmitter antennas and a single receiver antenna ($M = 2$, $N = 1$). The channel matrix for this case is given by

$$\tilde{\mathbf{H}} = \begin{bmatrix} \tilde{H}_{11} \\ \tilde{H}_{21} \end{bmatrix} \tag{9.4.1}$$

Since it is not possible to send more than one simultaneous signal, as outlined in Section 9.2.2, conventional thinking suggests that we should weight the two transmitter elements to maximize the signal power at the receiver. Let us divide the transmitted signal, $\tilde{x}(t)$, into its discrete digital symbols, $\{\tilde{x}_i\}$, where $\tilde{x}_1$ is the first symbol sent, $\tilde{x}_2$ is the second symbol sent, and so on. In the conventional MISO scheme, six digital symbols would be sent in the following manner:

$$\begin{matrix} \text{TX Antenna 1} \\ \text{TX Antenna 2} \end{matrix} \left[\begin{array}{c|c|c|c|c|c} \tilde{v}_{11}\tilde{x}_1 & \tilde{v}_{11}\tilde{x}_2 & \tilde{v}_{11}\tilde{x}_3 & \tilde{v}_{11}\tilde{x}_4 & \tilde{v}_{11}\tilde{x}_5 & \tilde{v}_{11}\tilde{x}_6 \\ \tilde{v}_{21}\tilde{x}_1 & \tilde{v}_{21}\tilde{x}_2 & \tilde{v}_{21}\tilde{x}_3 & \tilde{v}_{21}\tilde{x}_4 & \tilde{v}_{21}\tilde{x}_5 & \tilde{v}_{21}\tilde{x}_6 \end{array} \right]$$

where each column is a successive time slot. The weights $\tilde{v}_{11}$ and $\tilde{v}_{21}$ are chosen by the transmitter to maximize the power at the terminal of the single receiver antenna. But there is an even better way to accomplish this.

Even for this simple spatial signaling, the channel matrix coefficients must be fed back to the transmitter by the receiver; otherwise the transmitter cannot discern the optimum signal weights. Instead, consider this alternative signaling scheme:

$$\begin{array}{l} \text{TX Antenna 1} \\ \text{TX Antenna 2} \end{array} \left[\begin{array}{cc|cc|cc|c} \tilde{x}_1 & -\tilde{x}_2^* & \tilde{x}_3 & -\tilde{x}_4^* & \tilde{x}_5 & -\tilde{x}_6^* & \cdots \\ \tilde{x}_2 & \tilde{x}_1^* & \tilde{x}_4 & \tilde{x}_3^* & \tilde{x}_6 & \tilde{x}_5^* & \cdots \end{array} \right]$$

$$\text{Time} \longrightarrow$$

In the first time slot, symbol 1 is transmitted through the first antenna element and symbol 2 is transmitted through the second antenna element. The receiver sees the following symbol arrive during the first time slot:

$$\tilde{y}_1 = \frac{1}{2}\tilde{H}_{11}\tilde{x}_1 + \frac{1}{2}\tilde{H}_{21}\tilde{x}_2 + \tilde{n}_1 \tag{9.4.2}$$

which, alone, is a garbled combination of two data symbols. The symbol $\tilde{n}_1$ is any perturbation due to noise and interference received within the time slot. The received symbol for the second time slot is no better:

$$\tilde{y}_2 = -\frac{1}{2}\tilde{H}_{11}\tilde{x}_2^* + \frac{1}{2}\tilde{H}_{21}\tilde{x}_1^* + \tilde{n}_2 \tag{9.4.3}$$

Separately, $\tilde{y}_1$ and $\tilde{y}_2$ are useless. However, the two samples may be combined in the following manner:

$$\begin{aligned} \left[\frac{\tilde{H}_{11}^*}{\sqrt{|\tilde{H}_{11}|^2+|\tilde{H}_{21}|^2}}\right]\tilde{y}_1 + \left[\frac{\tilde{H}_{21}}{\sqrt{|\tilde{H}_{11}|^2+|\tilde{H}_{21}|^2}}\right]\tilde{y}_2^* &= \tfrac{1}{2}\tilde{x}_1\sqrt{|\tilde{H}_{11}|^2+|\tilde{H}_{21}|^2} + \tilde{n}_1' \\ \left[\frac{\tilde{H}_{21}^*}{\sqrt{|\tilde{H}_{11}|^2+|\tilde{H}_{21}|^2}}\right]\tilde{y}_1 - \left[\frac{\tilde{H}_{11}}{\sqrt{|\tilde{H}_{11}|^2+|\tilde{H}_{21}|^2}}\right]\tilde{y}_2^* &= \tfrac{1}{2}\tilde{x}_2\underbrace{\sqrt{|\tilde{H}_{11}|^2+|\tilde{H}_{21}|^2}}_{R} + \tilde{n}_2' \end{aligned}$$

The original digital symbols, $\tilde{x}_1$ and $\tilde{x}_2$, can be reconstructed for each pair of transmissions. But even more important, notice the real constant, R, that multiplies each symbol; this constant is equivalent to the output envelope of a two-branch diversity scheme with MRC. The single-antenna receiver has performed MRC on the transmitted symbols!

So, it is possible after all to use a MISO system without channel feedback to combat a fading channel, thanks to a cleverly constructed *space–time block code.* The receiver still needs to know the channel weights, but it does not have to send them back to the transmitter.

Example 9.1: Capacity of MISO Block Code

Problem: Compare the channel capacity of the 2×2 space–time block code to a conventional MISO channel that maximizes received power using channel state information (as discussed in Section 9.1.3). Assume an ideal Rayleigh, frequency-flat fading channel and equal average transmit power for each scenario.

Solution: Both the space–time block code and the optimum conventional MISO transmission produce the equivalent of two-element MRC diversity output. Assuming uncorrelated Rayleigh channels, the received power for both signals follows an identical Chi-squared PDF, as discussed in Section 8.2.2.

Space–time block code: The average input SINR is given by

$$\mathrm{SINR} = \frac{P_{\mathrm{dif}}}{\mathrm{E}\left\{|\tilde{n}|^2\right\}} = \mathrm{SINR}_0$$

where $\tilde{n}$ represents the random noise sample for each symbol estimate. We could then express channel capacity, C, in terms of random variables:

$$C = B\log_2(1 + P\,\mathrm{SINR}_0) \qquad f_P(p) = p\exp(-p)\,\mathrm{u}(p)$$

where B is bandwidth and P is a Chi-squared random variable.

Conventional with channel state: Assuming the same transmitted power is used by the conventional MISO technique, the transmitter antenna element weights $\tilde{v}_{11}$ and $\tilde{v}_{21}$ that maximize channel capacity are

$$\tilde{v}_{11} = \frac{\tilde{H}_{11}^*\sqrt{2}}{\sqrt{|\tilde{H}_{11}|^2 + |\tilde{H}_{21}|^2}} \qquad \tilde{v}_{21} = \frac{\tilde{H}_{21}^*\sqrt{2}}{\sqrt{|\tilde{H}_{11}|^2 + |\tilde{H}_{21}|^2}}$$

The space–time block-coding transmitter lacks these signal weights, effectively transmitting as $\tilde{v}_{11} = \tilde{v}_{21} = 1$. With the above weights, the average SINR at the receiver is actually *twice* as high as the space–time block-coding receiver. The capacity expression for this case is

$$C = B\log_2\left(1 + 2P\,\mathrm{SINR}_0\right) \qquad f_P(p) = p\exp(-p)\,\mathrm{u}(p)$$

Thus, the capacity is distributed similarly for each transmission technique, but the 2×2 space–time block code has less average capacity due to an effective -3 dB penalty in its overall SINR.

Note: Capacity for Nothing?

As Example 9.1 illustrates, the space–time block code does not give diversity gain for free. Under the same levels of output power, a MISO transmitter that uses channel information to perform MRC on the signal through the channel has higher capacity. Space–time block codes trade off some overall channel capacity for simplicity (no channel feedback).

9.4.2 Space–Time Block Codes

The name *space–time block code* refers to the joint coding of the input symbols $\{\tilde{x}_i\}$ in both space and time. The 2×2 space–time block code in the previous section was originally developed by Alamouti in [Ala98] and can be written in a compact matrix form:

$$\tilde{\mathbf{X}} = \begin{bmatrix} \tilde{x}_1 & -\tilde{x}_2^* \\ \tilde{x}_2 & \tilde{x}_1^* \end{bmatrix} \tag{9.4.4}$$

In this form, data is sent in blocks of symbols instead of individual, one-at-a-time symbols (hence the term *block*-coding). Although the receiver has a single antenna, Figure 9.12 illustrates how the space–time block codes construct an "array in time" at the receiver. During one time slot, the receiver experiences one type of spatial channel; in the successive time slot, the receiver experiences a different type of spatial channel. Signals in adjacent time slots are weighted and combined to produce the original symbols, just as in a two-element antenna array.

The result is somewhat surprising: The performance of the wireless link is *equivalent* to a radio signal sent from the receiver antenna to the two transmitter antennas, which then employ MRC diversity to construct a fade-resistant signal. This block diagram for this equivalent system is shown at the bottom of Figure 9.12. For this reason, the 2×2 space–time block code is vulnerable to the same potential channel conditions that degrade any antenna diversity - unequal average power or envelope correlation between each antenna element. Keep in mind, however, that it is the envelope correlation at the *transmitter antennas*, rather than at the receiver antennas, that corrupts link performance.

More than one antenna element may be used at the receiver; each element performs the same decoding operation using different channel weights. The outputs of two or more receiver antenna elements may then be combined for further diversity gain. Transmission through an arbitrary $M \times N$ MIMO channel using a space–time block code may be written as

$$\tilde{\mathbf{Y}} = \frac{1}{2}\tilde{\mathbf{H}}\tilde{\mathbf{X}} + \text{Interference} + \text{Noise} \tag{9.4.5}$$

where each column of $\tilde{\mathbf{X}}$ and $\tilde{\mathbf{Y}}$ represents a time slot and each row represents an antenna element. Despite the freedom to use multiple receiver antennas, it is the potential for diversity at the receiver with a single antenna element that makes the space–time block code so appealing. For mass-produced receiver units, a single antenna keeps the manufacturing costs low.

9.4.3 Other Codes

Higher level space–time block codes are possible, but the 2×2 code in Equation (9.4.4) is unique in its ability to transmit any complex baseband symbol at rate 1; that is, a one-to-one correspondence exists between the number of input and output symbols into the space–time coder ($\tilde{\mathbf{X}}$ is a square matrix). Space–time

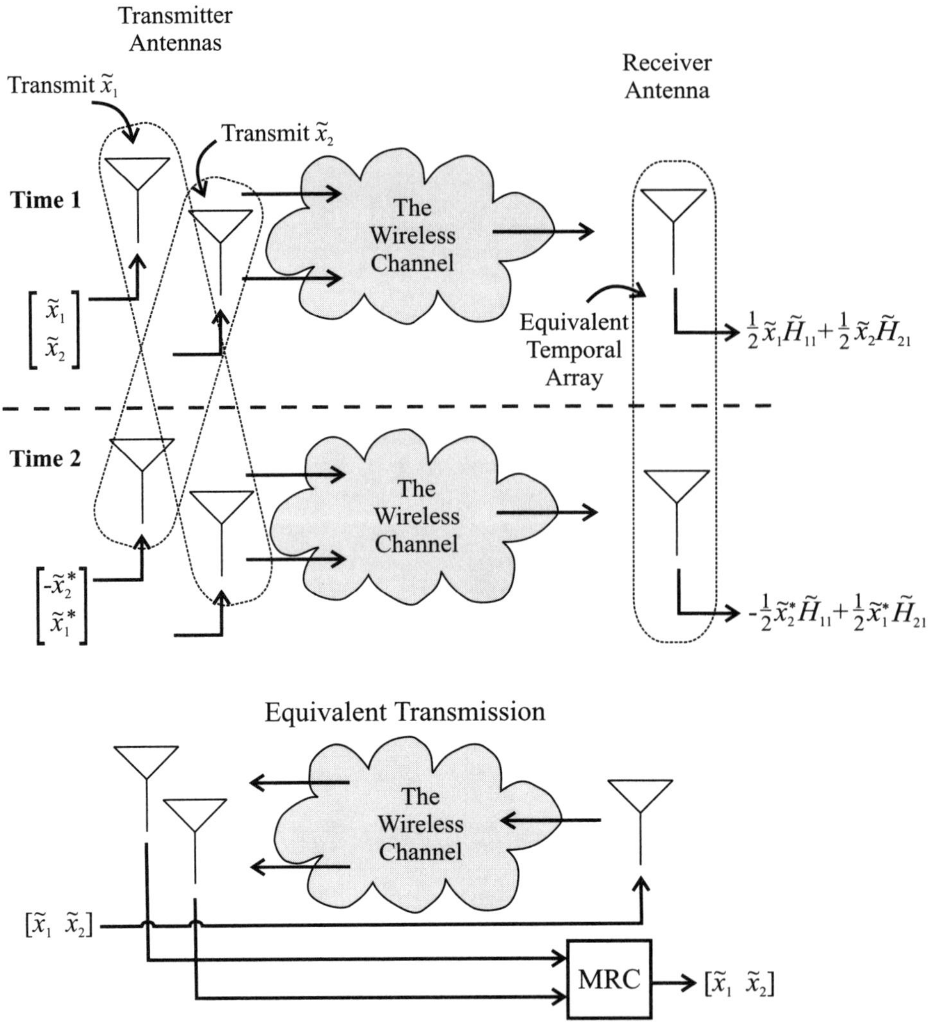

Figure 9.12 Space–time block codes allow the receiver antenna to operate like an array *in time*.

codes with more than two transmitter antennas often require a restricted type of signal input or a lower coding rate (more transmitted symbols than input symbols). Example 9.2 shows a space–time block code for four transmitter antenna elements.

Example 9.2: Large Space–Time Block Code

Problem: Describe how a receiver would extract the symbols from the following four transmit antenna space–time block code [Nag00]:

$$\tilde{\mathbf{X}} = \begin{bmatrix} \tilde{x}_1 & -\tilde{x}_2 & -\tilde{x}_3 & -\tilde{x}_4 & \tilde{x}_1^* & -\tilde{x}_2^* & -\tilde{x}_3^* & -\tilde{x}_4^* \\ \tilde{x}_2 & \tilde{x}_1 & \tilde{x}_4 & -\tilde{x}_3 & \tilde{x}_2^* & \tilde{x}_1^* & \tilde{x}_4^* & -\tilde{x}_3^* \\ \tilde{x}_3 & -\tilde{x}_4 & \tilde{x}_1 & \tilde{x}_2 & \tilde{x}_3^* & -\tilde{x}_4^* & \tilde{x}_1^* & \tilde{x}_2^* \\ \tilde{x}_4 & \tilde{x}_3 & -\tilde{x}_2 & \tilde{x}_1 & \tilde{x}_4^* & \tilde{x}_3^* & -\tilde{x}_2^* & \tilde{x}_1^* \end{bmatrix} = [\tilde{\mathbf{X}}_0 \ \tilde{\mathbf{X}}_0^*]$$

Solution:

First, we write the transmission equation, splitting the received block of symbols into a left portion, $\tilde{\mathbf{Y}}_L$, and a right portion, $\tilde{\mathbf{Y}}_R$:

$$\tilde{\mathbf{Y}} = [\tilde{\mathbf{Y}}_L \ \tilde{\mathbf{Y}}_R^*] = \frac{1}{2}\tilde{\mathbf{H}}[\tilde{\mathbf{X}}_0 \ \tilde{\mathbf{X}}_0^*] \longrightarrow \begin{matrix} \tilde{\mathbf{Y}}_L = \frac{1}{2}\tilde{\mathbf{H}}\tilde{\mathbf{X}}_0 \\ \tilde{\mathbf{Y}}_R = \frac{1}{2}\tilde{\mathbf{H}}^*\tilde{\mathbf{X}}_0 \end{matrix}$$

Let us assume a single receiver antenna, since the detection method for multiple receiver antennas follows easily from the single case. From the known or estimated channel coefficients, we construct the operation $\tilde{\mathbf{U}}$:

$$\tilde{\mathbf{U}} = \begin{bmatrix} \tilde{H}_{11}^* & \tilde{H}_{21}^* & \tilde{H}_{31}^* & \tilde{H}_{41}^* & \tilde{H}_{11} & \tilde{H}_{21} & \tilde{H}_{31} & \tilde{H}_{41} \\ \tilde{H}_{21}^* & -\tilde{H}_{11}^* & -\tilde{H}_{41}^* & \tilde{H}_{31}^* & \tilde{H}_{21} & -\tilde{H}_{11} & -\tilde{H}_{41} & \tilde{H}_{31} \\ \tilde{H}_{31}^* & \tilde{H}_{41}^* & -\tilde{H}_{11}^* & -\tilde{H}_{21}^* & \tilde{H}_{31} & \tilde{H}_{41} & -\tilde{H}_{11} & -\tilde{H}_{21} \\ \tilde{H}_{41}^* & -\tilde{H}_{31}^* & \tilde{H}_{21}^* & -\tilde{H}_{11}^* & \tilde{H}_{41} & -\tilde{H}_{31} & \tilde{H}_{21} & -\tilde{H}_{11} \end{bmatrix}$$

When $\tilde{\mathbf{U}}$ multiplies the received symbols,

$$\tilde{\mathbf{U}}[\tilde{\mathbf{Y}}_L \ \tilde{\mathbf{Y}}_R]^T = \left(\sum_{i=1}^{4} |\tilde{H}_{i1}|^2\right) \begin{bmatrix} \tilde{x}_1 & 0 & 0 & 0 \\ 0 & \tilde{x}_2 & 0 & 0 \\ 0 & 0 & \tilde{x}_3 & 0 \\ 0 & 0 & 0 & \tilde{x}_4 \end{bmatrix}$$

which separates the original symbols and produces a fourth-order diversity gain. Note that $\tilde{\mathbf{X}}$ is a rate 1/2 space–time block code; eight coded symbols in time are needed to send four uncoded data symbols.

Note: Space–Time Trellis Code Modulation

Space-time trellis code modulation (ST-TCM), first introduced by Tarokh et al. in [Tar98], is another form of space–time coding. Instead of a block-coding scheme, ST-TCM uses a space–time signaling lattice to encode the information. Digital constellation expansion and partitioning are applied in a manner similar to the conventional trellis code modulation originally proposed by Ungerboeck [Ung82], [Siw01].

9.5 Chapter Summary

No topic in wireless communications provides so elegant an application of space–time wireless channel modeling principles: The design of the antenna array in a MIMO system will be a function of the spatial selectivity in the channel; the frequency selectivity of the channel may lead to intersymbol interference; the temporal selectivity determines how often channel weights in the MIMO signaling schemes must be updated. There is truly a lot happening in the full MIMO channel.

A summary of the key points of MIMO channel modeling follows:

- In conventional multi-antenna transmission of single data streams, Shannon channel capacity is maximized when the received power (SINR) is maximized.
- In a MIMO channel, it is possible to transmit multiple data streams simultaneously through a multipath channel for high Shannon channel capacity.
  - ▷ The basic principle: Multiple low-powered channels have greater capacity than a single, high-powered channel.
  - ▷ Signal correlation, the absence of multipath, or unequal channel power degrade the data capacity of a MIMO channel.
- Practical MIMO signaling must address the following engineering challenges:
  - ▷ Geometry and integration of antenna arrays at transmitter and receiver.
  - ▷ Signal transmission without channel knowledge at the transmitter.
  - ▷ Signal extraction using imperfect channel estimates at the receiver.
- Space–time block codes use both the spatial and temporal dependencies of a channel to generate diversity gain, even for receivers with a single antenna.

The next chapter deals specifically with the geometrical design of antenna arrays for multipath environments. Chapter 10 blends the channel modeling theory of previous chapters with the applications of space diversity and MIMO signaling.

PROBLEMS

1. Here is a problem that illustrates the difference between coherent and incoherent combining. We desire to send a signal, $\tilde{x}_0 = 1$, through a radio channel. Upon reception, however, it is corrupted by additive Gaussian noise, $\tilde{n}$, with the property $\mathrm{E}\left\{|\tilde{n}|^2\right\} = 4$.

   a. If there is no other interference, what is the SINR for this signal?

   b. Because the channel is noisy, the transmitter repeatedly sends $\tilde{x}_0 = 1$ to the receiver. The ith reception of $\tilde{x}_0$ is described by $\tilde{x}_i = 1 + \tilde{n}_i$. The receiver estimates $\tilde{x}_0$ by summing N repeated transmissions. Derive an expression for the SINR of this accumulated signal as a function of N.

c. Repeat (b), but this time consider the case where $\tilde{x}_0$ has random phase upon every measurement, such that $\tilde{x}_0 = \exp(j\Phi)$, where Φ is a random phase variable uniformly distributed over the interval $[0, 2\pi)$.

2. Here are four different channel matrices that describe a 2×1 MISO radio link:

$$\tilde{\mathbf{H}} : [5+6j \quad 2-5j] \qquad [4 \quad 4j] \qquad [2+2j \quad 2-2j] \qquad [2+2j \quad 2+2j]$$

For each matrix, calculate the power gain in dB between the output of a receiver using space–time block codes and a conventional receiver.

3. A 5×5 MIMO radio channel is described by the channel matrix

$$\tilde{\mathbf{H}} = \begin{bmatrix} 3+7j & 8-j & 3 & 5+2j & -4+4j \\ 3+2j & -1-9j & 5+5j & 0 & -5+6j \\ -8+7j & -4+j & -2+4j & 5+2j & -6+j \\ 2j & 3-5j & 7-7j & 1+j & -2+5j \\ -9j & 0 & 2+9j & 10j & -3-3j \end{bmatrix}$$

Answer the following questions assuming the same average noise and interference power at each antenna element:

a. Which pair of antenna elements produce the best SISO channel?

b. Which of the five receiver antenna elements experiences the best MISO channel?

c. Which of the five transmitter antenna elements experiences the best SIMO channel?

d. If $\text{SINR}_0 = 3$ dB, what is the theoretical Shannon capacity for the full MIMO channel? (You may need a computer program to compute the determinant.)

4. A certain wideband MIMO channel takes the following form:

$$\tilde{\mathbf{H}}(\tau) = \sum_{i=1}^{3} \tilde{\mathbf{H}}_i \delta(\tau - i\Delta\tau)$$

$$\tilde{\mathbf{H}}_1 = \begin{bmatrix} 4-4j & 2-2j \\ 10 & 8j \end{bmatrix} \quad \tilde{\mathbf{H}}_2 = \begin{bmatrix} -1+j & 5 \\ 0 & 1+3j \end{bmatrix} \quad \tilde{\mathbf{H}}_2 = \begin{bmatrix} 1 & 2j \\ 1+j & 2j \end{bmatrix}$$

Now consider this same MIMO channel operating with *narrowband* radio signals. Based on the narrowband channel, answer the following questions assuming the same average noise and interference power at each antenna element:

a. Which transmitter and receiver antenna produces the best SISO channel?

b. Which of the two receiver antenna elements experiences the best MISO channel?

c. Which of the two transmitter antenna elements experiences the best SIMO channel?

d. If $\text{SINR}_0 = 6$ dB, what is the theoretical Shannon capacity for the full MIMO channel?

5. When we use a matrix to represent a complex channel for arrays of antennas, we are assuming that all multipath radio scatterers are in the *far field* (i.e., waves at the transmitter and receiver look like homogeneous plane waves). Using arguments from Chapter 4, describe why this is a valid representation for most wireless links.

6. Answer the following design questions based on an ideal MIMO radio link in Rayleigh fading with average SINR_0 of 10 dB. Refer to the chart in Figure 9.8 for help.

   a. If only 10 kHz of radio bandwidth is available, what is the minimum number of antenna elements needed to maintain a link at 50 kbit/s with less than 1% outage probability?

   b. What is the maximum data rate supported by a 4×4 MIMO system with 25 kHz of radio bandwidth at 0.5% outage probability?

   c. How much bandwidth is necessary to send 5 Mbit/s at 0.1% outage probability through a 6×6 MIMO system?

7. Generate a chart of Shannon channel capacity for an ideal MIMO radio link, similar to Figure 9.8, using the following scenarios:

   a. $N = 1, 2, 3, 4$, $\mathrm{SINR}_0 = 10$ dB

   b. $N = 4, 6, 8, 10$, $\mathrm{SINR}_0 = 5$ dB

   c. $N = 4, 6, 8, 10$, $\mathrm{SINR}_0 = 0$ dB

   d. $N = 1, 2, 4, 8$, $\mathrm{SINR}_0 = -3$ dB

8. Use simulation to generate a chart of Shannon channel capacity in a MIMO radio link, similar to Figure 9.8, for the following scenarios:

   a. Using a channel matrix consisting of independent Rayleigh components, calculate the PDF of capacity, $f_C(c)$, for the cases shown in Figure 9.8.

   b. Using a channel matrix consisting of independent *Rician* components, calculate the PDF of capacity, $f_C(c)$, for the 4×4 case at $\mathrm{SINR}_0 = 10$ dB. Make graphs for Rician K-factors of -3, 0, 3, and 10 dB.

   c. Repeat (b), this time calculating the CDF $F_C(c)$ of Rician channel capacity. Describe how capacity and outage probability depend on the Rician K-factor.

   d. Using a channel matrix consisting of independent *TWDP* components, calculate the PDF of capacity, $f_C(c)$, for the 4×4 case at $\mathrm{SINR}_0 = 10$ dB. Using a value of $K = 10$ dB, make graphs for $\Delta = 0.0$, 0.5, and 1.0.

   e. Repeat (d), this time calculating the CDF $F_C(c)$ of TWDP channel capacity. Describe how capacity and outage probability depend on the TWDP Δ-parameter.

9. In the practical space–time architecture of Section 9.3, the relative strengths of the separable channels in the random MIMO channel are unpredictable but have similar *average* power levels. Can you identify anything in the signal processing that might skew this symmetry as seen by the receiver?

10. Use the space–time block code in Example 9.2 to answer the following questions:

a. Prove the final result in the example that computes $\tilde{\mathbf{U}}[\tilde{\mathbf{Y}}_L\ \tilde{\mathbf{Y}}_R]^T$.

b. What are the trade-offs between using this space–time block code and the 2×2 code in Equation (9.4.4)?

Chapter 10

ARRAY DESIGN IN MULTIPATH

Wireless systems that require ultra-high data rates in interference-limited environments must use transmitters and receivers with multiple antennas to take advantage of spatial gains. There are numerous signal-processing schemes available for a multiple antenna system, whether it consists of a simple pair of diversity antennas or an array of antennas operating with sophisticated MIMO processing. Any discussion of these algorithms usually begins with a basic assumption: *uncorrelated Rayleigh fading on each antenna element.*

As we learned in Chapter 7, whether or not an antenna system has genuine uncorrelated Rayleigh fading on each transmitter or receiver element is heavily dependent upon the radio propagation environment and the multiple antenna element configuration. Engineers are at the mercy of the channel; they only have control over the configuration of antenna elements, so care must be taken to match array geometries and elements spacings according to the most likely propagation scenarios. This chapter discusses the following aspects of antenna geometry in the presence of multipath:

- Section 10.1: Use of shape factors to describe zones of decorrelation.
- Section 10.2: Descriptions of channels with double spatial dependencies.
- Section 10.3: Example of antenna geometry design in multipath.
- Section 10.4: Results from a space-time measurement campaign.
- Section 10.5: Summary of important concepts.

This chapter should provide the wireless engineer with an understanding of how to design antenna array geometries to match the typical multipath scenarios encountered in real life.

10.1 Rules of Spatial Decorrelation

The use of multipath shape factors greatly simplifies the design of antenna arrays in multipath environments [Dur01c]. This section discusses the geometrical guidelines for element decorrelation using shape factors.

10.1.1 Co-polar Versus Dissimilar Antenna Elements

Diversity and MIMO schemes require different antennas to experience unique radio channels. If the channels are correlated, the various antenna elements are not unique; the overall channel capacity drops, and the additional antenna elements become vestigial components of the radio link. Therefore, an antenna array operating in a multipath environment must avoid elements with correlated channels.

Identical, co-polar antenna elements are most sensitive to channel correlation. A common example of identical, co-polar elements is an array of vertical monopoles. Such antenna elements interact with the same sets of impinging multipath waves with the same gain pattern and polarization. Only their spatial displacement - and the associated phase-jumbling of multipath waves - is capable of producing unique radio channels on each element. Element correlation becomes a problem if the propagation conditions do not produce enough spatial phase-jumbling of multipath.

In order to avoid correlated channels, an engineer may use elements with different gain patterns or polarizations. Each antenna in such a design experiences different channels (due to the different spatial filtering of the propagation environment), so envelope correlation is minimal. It is a cruel twist of physics, however, that dissimilar antenna element systems are subject to an entirely different problem: channels that have unequal average power.

So, there is a fundamental trade-off in the design of multi-antenna schemes in multipath environments. Using identical, co-polar elements guarantees equal average power across the various channels; the drawback is a system vulnerable to channel correlation. Dissimilar antenna elements avoid the correlation problem; the drawback is a system vulnerable to a dominant channel. The rest of the section is devoted to avoiding both the correlation *and* dissimilar element problems by matching co-polar element geometry to typical multipath propagation conditions.

10.1.2 Approximate Autocovariance

The envelope autocovariance function, $\varrho_R(\Delta r, \theta_R)$, describes envelope correlation in space. For the general case of from-the-horizon propagation in a Rayleigh fading channel, Chapter 7 demonstrated that the spatial unit autocovariance for received envelope may be approximately expressed as a function of multipath shape factors as in Equation (7.2.10). Recall that the result is extremely accurate for the smaller spatial separations, Δr, that result in moderate-to-high correlation values - exactly the region that is critical to the performance of multiple antenna systems.

It should be noted that Equation (7.2.10) works equally well for describing correlation at a transmitter or receiver. If a unit acts as a receiver, then the shape factors are based on an azimuth spectrum that describes multipath angles-of-arrival. If a unit acts as a transmitter, then the shape factors are based on an azimuth spectrum that describes multipath angles-of-departure. Because a single radio link is reciprocal, the same azimuth spectrum describes a unit's angles-of-arrival or departure, as illustrated in Figure 10.1.

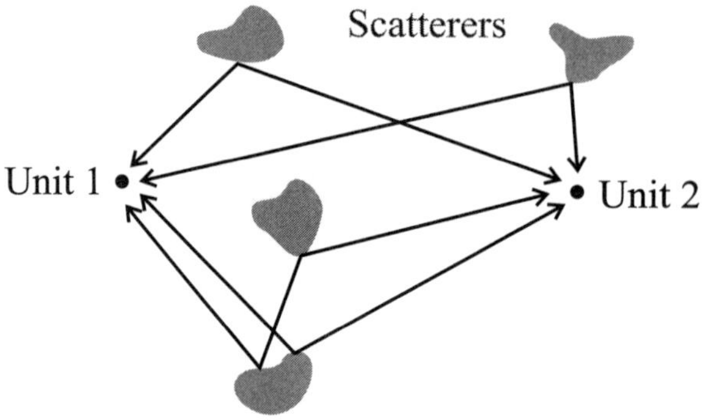

Figure 10.1 When the roles of transmitter and receiver are reversed, angles-of-arrival become angles-of-departure.

It also follows from Figure 10.1 that description of the radio link requires two sets of shape factors. One set, $\{\Lambda, \gamma, \theta_{\max}\}_1$, describes the azimuth spectrum (transmission or reception) for the first unit in the link. Another set, $\{\Lambda, \gamma, \theta_{\max}\}_2$, describes the azimuth spectrum for the second unit in the link. Depending on the environment, it is possible to have very different shape factors at each unit. The classical example of this is cellular mobile radio, where a tower-top base station experiences multipath propagation from a very small angular spread, but the mobile unit experiences multipath propagation with a high angular spread [Ert98].

A complex baseband channel with double spatial dependencies requires a mathematical description in the form of $\tilde{h}(\vec{r}_1, \vec{r}_2)$. From this channel, we may write an envelope process and derive envelope correlation statistics:

$$\varrho_R(\Delta\vec{r}_1, \Delta\vec{r}_2) \cong \varrho_{R1}(\Delta r_1, \theta_1)\varrho_{R2}(\Delta r_2, \theta_2) \tag{10.1.1}$$

$$\varrho_{R1}(\Delta r_1, \theta_1) \approx \exp\left(-23\Lambda_1^2\left[1 + \gamma_1\cos(2[\theta_1 - \theta_{\max,1}])\right]\left[\frac{\Delta r_1}{\lambda}\right]^2\right)$$

$$\varrho_{R2}(\Delta r_2, \theta_2) \approx \exp\left(-23\Lambda_2^2\left[1 + \gamma_2\cos(2[\theta_2 - \theta_{\max,2}])\right]\left[\frac{\Delta r_2}{\lambda}\right]^2\right)$$

Equation (10.1.1) is a valid description of random spatial fading at unit 1 and unit 2, provided the fading at each local area is independent and the incoming multipath waves lie in the azimuth plane. Equation (10.1.1) describes the correlation properties of identical, co-polar antennas at either unit 1 or unit 2, following the geometry of Figure 10.2.

Using shape factors and the assumption of separability, we may proceed to study correlation separately at transmitter arrays or receiver arrays. Hereafter, when we discuss correlation and antenna element separation, we will talk about only a single location; recognize that the principles apply equally to either transmitter or receiver antennas.

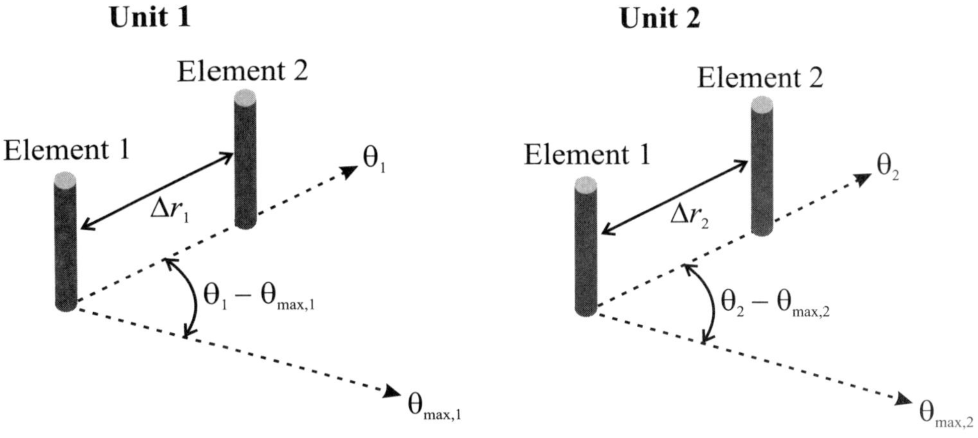

Figure 10.2 The geometry of element separation at unit 1 and unit 2.

> *Note: Reciprocity Theorem*
>
> The claim that angles-of-arrival become angles-of-departure and vice versa when transmitter and receiver antennas switch roles is just one consequence of the *reciprocity theorem* in electromagnetics. If the propagation medium is linear and time-invariant, then $\tilde{h}(f, \vec{r}_1, \vec{r}_2)$ describes the channel for transmission from unit 1 to unit 2 *or* from unit 2 to unit 1.

10.1.3 Forbidden Zones of Correlation

Before deciding the geometry of the antenna elements at either the transmitter or the receiver, we must first determine the maximum acceptable amount of correlation between two co-polar antenna elements. This critical correlation, ϱ_C, will depend on the system architecture, but usually falls between 0.4 and 0.7. Once known, we may set Equation (7.2.10) equal to the critical correlation, $\varrho(\Delta r, \theta_R) = \varrho_C$. This equation traces out an *isocorrelation* curve in polar coordinates, $(\Delta r, \theta_R)$, according to the following equation:

$$-\ln \varrho_C = 23\Lambda^2\left(1 + \gamma\cos[2(\theta_R - \theta_{\max})]\right)\left(\frac{\Delta r}{\lambda}\right)^2 \qquad (10.1.2)$$

Equation (10.1.2) defines an ellipse, with a minor axis in the azimuthal direction $\theta_{\max}$ and a major axis in the direction $\theta_{\max} \pm \frac{\pi}{2}$ [Dur01c], [Fle00]. A diagram of this ellipse is shown in Figure 10.3.

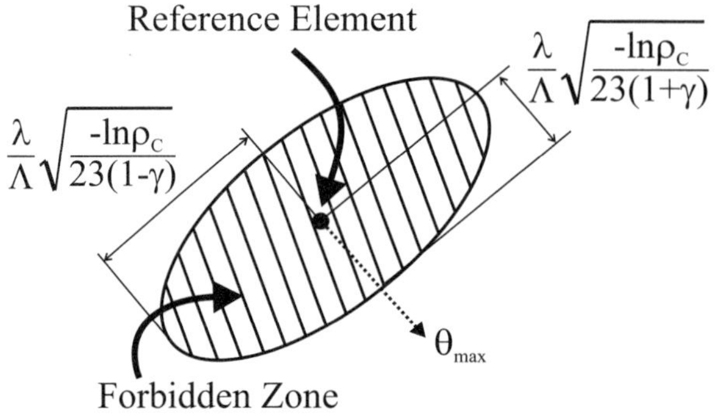

Figure 10.3 The correlation criterion, ρ_c, and the multipath shape factors determine the forbidden zone around an antenna element in azimuth [Dur01c].

This ellipse represents a *forbidden zone* of correlation. For the space inside the ellipse, the envelope correlation is higher than ϱ_C; outside the ellipse, the envelope correlation is lower than ϱ_C. The size and shape of the ellipse is a useful tool for spacing antenna elements. If elements are to produce decorrelated fading channels in space, then they must be placed mutually outside one another's forbidden zone. An example of such a configuration is illustrated in Figure 10.4.

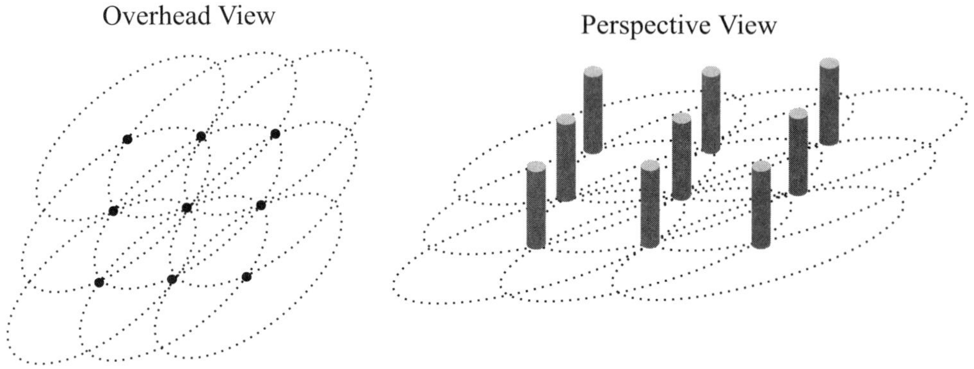

Figure 10.4 Antenna elements must be placed so that they do not intrude into the forbidden zone (dotted ellipse) of another element [Dur01c].

There are several important characteristics of the ellipse to note. First, the size of the ellipse becomes larger as the critical correlation level, ϱ_C, becomes smaller. This is intuitive, since systems that require low correlation must have antenna elements with greater separation.

The size of the elliptical forbidden zone is also inversely proportional to angular spread, Λ. As the angular spread drops, the size of the forbidden zone expands.

Finally, the elliptical eccentricity, ϵ, of the forbidden zone depends solely on the angular constriction, γ:

$$\epsilon = \sqrt{\frac{2\gamma}{1+\gamma}} \tag{10.1.3}$$

If angular constriction is 0, the forbidden zone is perfectly circular. If angular constriction is 1, the eccentricity becomes 1 and the forbidden zone stretches to infinity along its major axis.

10.1.4 Coupling Considerations

No discussion of co-polar antenna assemblies is complete without mentioning the effects of antenna coupling. When elements are placed very close to one another, their near electric and magnetic fields couple and cause changes in gain pattern, polarization, and impedance. It was first noted in [Ebi91] that coupling will actually cause *additional* decorrelation for some closely spaced antennas. This is not good news, however, because the decoupling usually occurs at the expense of overall gain on the individual elements. Therefore, coupling between elements should be avoided at both transmitter and receiver units.

Much like the case of correlation, we can draw a *coupling* forbidden zone around an antenna element. Within this forbidden zone, an object will couple with the antenna element. For example, the coupling forbidden zone of a monopole antenna is a circle centered on the element with $\frac{\lambda}{2}$ radius.

The coupling forbidden zone is superimposed atop the correlation forbidden zone to create a composite zone that describes an area that cannot tolerate other similar elements. There are three types of composite forbidden zones, as shown in Figure 10.5. First, there is the correlation-limited case where the correlation forbidden zone engulfs the smaller coupling forbidden zone. Second, there is the joint-limited case where the correlation and coupling forbidden zones are about the same size, creating an irregular composite forbidden zone. Finally, there is a coupling-limited case for a channel with a small correlation forbidden zone caused by highly spread, unconstricted multipath propagation. The latter case best represents the design process for classical beam-forming arrays, which are coupling-limited.

10.1.5 Random Orientation

A transmitter or receiver that operates with an unpredictable or random azimuthal orientation is a potential problem for element configuration design. This is especially true for heavily constricted propagation. For example, we may desire to place two identical, co-polar antennas on a commercial wireless handset. If the multipath has high angular constriction, the correlation zone becomes very eccentric and there is ambiguity as to which direction decorrelation will occur. After all, we cannot place constraints on how end users hold or orient their handsets and terminals in space.

There are two design philosophies for this case. Both are illustrated in Figure 10.6. First, the engineer may design the system according to the worst-case

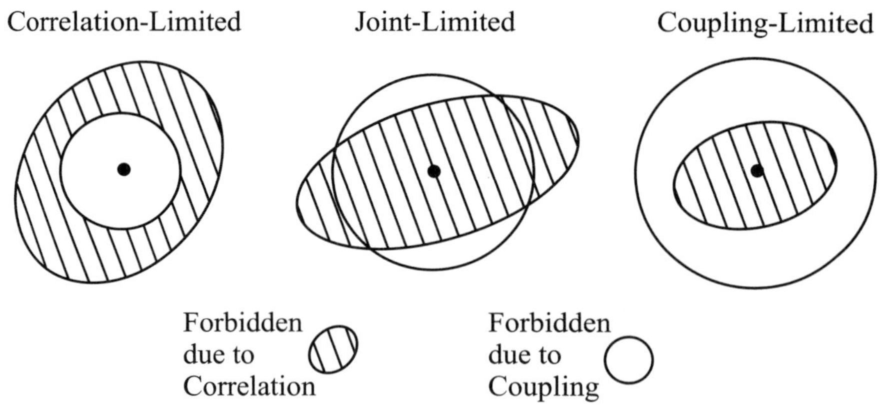

Figure 10.5 The three cases of forbidden zones when coupling is considered.

decorrelation distance, the major axis of the forbidden zone ellipse. This distance, R_c, depends only on angular spread, Λ, and angular constriction, γ:

$$R_c = \frac{\lambda}{\Lambda}\sqrt{\frac{-\ln \rho_c}{23(1-\gamma)}} \tag{10.1.4}$$

This distance ensures operation outside the forbidden zone for any orientation of two elements.

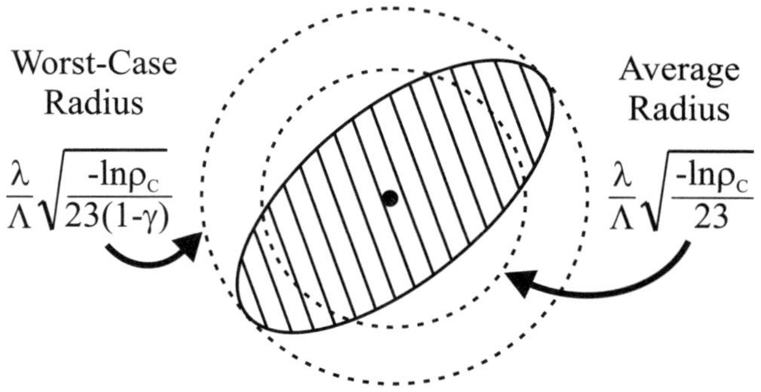

Figure 10.6 Two design philosophies when designing systems in which the orientation of propagation with respect to transmitter or receiver is unknown.

Second, the engineer may design the system to the average-case decorrelation distance. This distance depends only on angular spread, Λ [Dur99a]:

$$R_c = \frac{\lambda}{\Lambda}\sqrt{\frac{-\ln \rho_c}{23}} \tag{10.1.5}$$

Two elements operating with this separation distance risk spending some time within the correlation forbidden zone in certain orientations. This distance is a convenient alternative for multipath with high angular constriction and a worst-case distance that is physically unworkable for the transmitter or receiver.

10.2 Modeling Double Spatial Dependencies

The techniques we used to describe spatial complex baseband channels in a SLAC model are easily extended to cover channels that vary with two spatial dependencies: transmitter position and receiver position. The modeling technique allows us to construct realistic MIMO channel matrices, $\tilde{\mathbf{H}}$, and to reproduce realistic correlation behavior.

10.2.1 SLAC Modeling Technique for MIMO Channels

Model development of the baseband channel matrix, $\tilde{\mathbf{H}}$, is challenging because the individual elements must obey the laws of free-space wave propagation. The technique in this section applies the local area assumption to create a realistic MIMO channel model. The approach is to represent the channel matrix as the sum of L individual channel matrices, $\tilde{\mathbf{H}}_l$, which are due to single multipath waves in the environment:

$$\tilde{\mathbf{H}} = \sum_{l=1}^{L} \tilde{\mathbf{H}}_l \tag{10.2.1}$$

Once we apply Equation (10.2.1), we can construct the individual elements of $\tilde{\mathbf{H}}_l$ based on wave propagation in Figure 10.7. The channel between the ith transmitter element and the jth receiver element due to the lth multipath wave is written as [Dur01b]:

$$\tilde{h}_{ijl} = V_l \exp\left(j\left[\Phi_l - \frac{2\pi}{\lambda}\left(\hat{\mathrm{k}}_{Al} \cdot \vec{\mathrm{r}}_{Rj} + \hat{\mathrm{k}}_{Dl} \cdot \vec{\mathrm{r}}_{Ti}\right)\right]\right) \tag{10.2.2}$$

A summary of the variables in Equation (10.2.2) is given below:

V_l	–	amplitude of lth multipath wave
Φ_l	–	phase change between local areas
$\vec{\mathrm{r}}_{Rj}$	–	position of ith transmit element
$\vec{\mathrm{r}}_{Ti}$	–	position of jth receive element
$\hat{\mathrm{k}}_{Al}$	–	unit vector pointing in the direction of the lth wave arrival
$\hat{\mathrm{k}}_{Dl}$	–	unit vector pointing in the direction of the lth wave departure

Using this systematic approach, a complete MIMO channel matrix may be constructed that contains the correct correlation behavior for the various configurations of antenna elements.

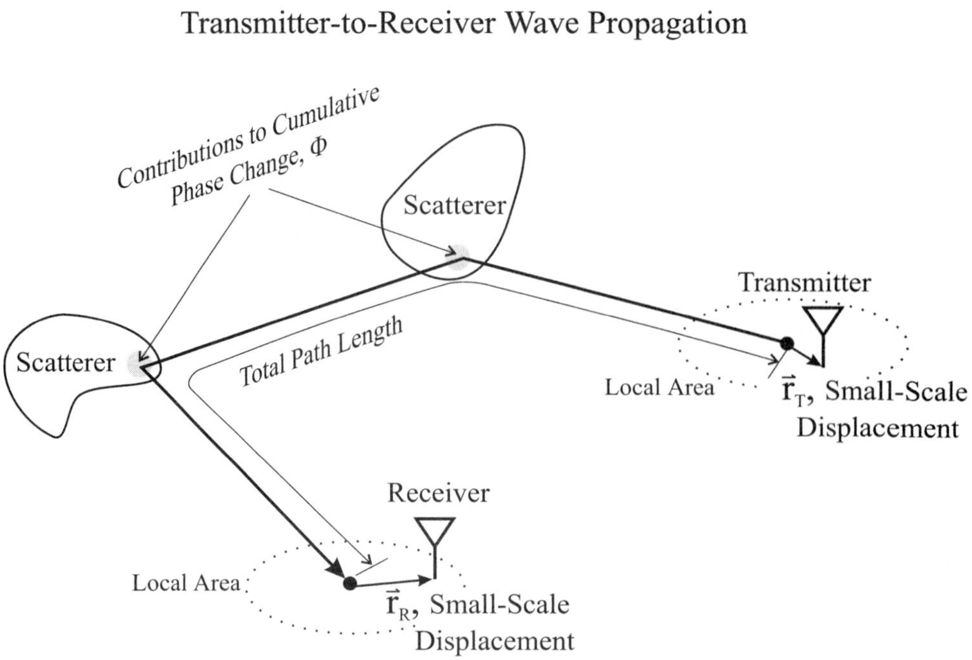

Figure 10.7 The physics of local area propagation for transmitter and receiver antennas [Dur01b].

10.2.2 Double Spatial Channel Correlation

Correlation statistics are based on the *envelope* of each element in the channel matrix, $\tilde{\mathbf{H}}$. From the complex baseband channel representation, $\tilde{\mathbf{H}}$ of Equation (9.2.3), we define the envelope of a single channel, R_{ij}, to be the magnitude of $\tilde{H}_{ij}$. Since an envelope, R_{ij}, is a positive quantity, it is typical to study the correlation statistics of envelope with its mean removed ($R_{ij} - \langle R_{ij} \rangle$, where the operation $\langle \cdot \rangle$ can be an ensemble average or spatial average operation).

As an example, consider the correlation between the ith and jth received element channels, R_{mi} and R_{mj}, both originating from the mth transmitter element. The correlation coefficient for these two channels is mathematically defined as

$$\rho^{\mathrm{R}}_{mij} = \frac{\langle R_{mi} R_{mj} \rangle - \langle R_{mi} \rangle \langle R_{mj} \rangle}{\sqrt{\langle R^2_{mi} \rangle \langle R^2_{mj} \rangle} - \langle R_{mi} \rangle \langle R_{mj} \rangle} \tag{10.2.3}$$

Equation (10.2.3) is normalized so that $|\rho^{\mathrm{R}}_{mij}|$ is always between 1 (perfectly correlated) and 0 (completely uncorrelated).

A similar definition exists for correlation between the ith and jth transmitted element channels, R_{in} and R_{jn}, both arriving at the nth receiver element. The correlation coefficient for these two channels is defined as

$$\rho^{\mathrm{T}}_{ijn} = \frac{\langle R_{in} R_{jn} \rangle - \langle R_{in} \rangle \langle R_{jn} \rangle}{\sqrt{\langle R^2_{in} \rangle \langle R^2_{jn} \rangle - \langle R_{in} \rangle \langle R_{jn} \rangle}} \tag{10.2.4}$$

If it were necessary to tabulate all of the possible correlation coefficients in an M transmitter, N receiver system in this manner, we would require $(M + N) \times M \times N$ coefficients!

Fortunately, we may make several simplifying assumptions in a co-polar, similar element system by invoking the *local area* assumption. First, we assume that the averages of signal envelope or signal strength among similar elements are identical: $\langle R_{mi} \rangle = \langle R_{mj} \rangle$ and $\langle R^2_{mi} \rangle = \langle R^2_{mj} \rangle$. The local area assumption also implies that correlation between two receiver elements is identical, regardless of the transmitter element the channel originated:

$$\rho^{\mathrm{R}}_{1ij} = \rho^{\mathrm{R}}_{2ij} = \cdots \rho^{\mathrm{R}}_{Mij} \tag{10.2.5}$$

This principle also holds for transmitter correlation:

$$\rho^{\mathrm{T}}_{ij1} = \rho^{\mathrm{T}}_{ij2} = \cdots \rho^{\mathrm{T}}_{ijN} \tag{10.2.6}$$

Channel correlation between two transmitter elements is independent of which receiver acts as an endpoint in the link.

The local area assumption conveniently reduces Equation (10.2.3) and Equation (10.2.4) to the following basic set of correlation values:

$$\rho^{\mathrm{R}}_{ij} = \frac{\langle R_{mi} R_{mj} \rangle - \langle R_{mi} \rangle^2}{\langle R^2_{mi} \rangle - \langle R_{mi} \rangle^2} \tag{10.2.7}$$

$$\rho^{\mathrm{T}}_{ij} = \frac{\langle R_{in} R_{jn} \rangle - \langle R_{in} \rangle^2}{\langle R^2_{in} \rangle - \langle R_{in} \rangle^2} \tag{10.2.8}$$

where m and n are arbitrary. These envelope statistics produce a number with magnitude between 0 and 1 that measures correlation between transmitter or receiver antenna elements.

10.2.3 Example Model

Here is an intuitive illustration of how a random MIMO channel is modeled for the linear arrangement of transmitter and receiver antennas in Figure 10.8. In this scenario, multipath angles-of-*departure* are spread evenly over a 20° sector pointing transverse to the line of transmitter antennas; multipath angles-of-*arrival* are spread evenly over the entire azimuth. To simulate this situation, let us choose 256 multipath components - L in Equation (10.2.1) - with equal amplitudes and random angles-of-arrival/departure according to the distributions in Figure 10.8. There are 128 transmitter antennas, M, and 128 receiver antennas, N, both sets uniformly spaced with 0.05λ between each element. Although the separation distance is small and the quantity of elements is large compared to any useful wireless system, this dense configuration clearly illustrates the spatial correlation properties that we are trying to reproduce.

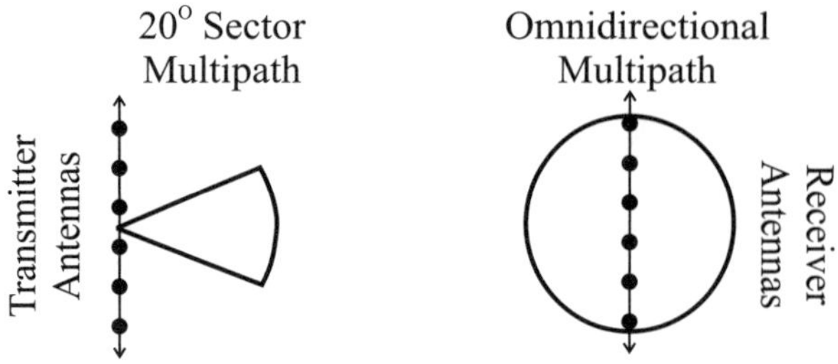

Figure 10.8 Top view of the antenna configurations and multipath angles-of-departure (from transmitter) and angles-of-arrival (to receiver) [Dur01b].

Figure 10.9 shows the envelope unit autocovariance of the channel matrix, $\tilde{\mathbf{H}}$, for one realization of the propagation in Figure 10.8. The top graph of Figure 10.9 demonstrates how the envelope of individual channels, $\{|\tilde{H}_{ij}|\}$, are correlated to one another. The bottom graph of Figure 10.9 shows two correlation graphs - one comparing transmit envelopes, $|\tilde{H}_{xj}|$, with the same j and the other comparing receive antenna envelopes, $|\tilde{H}_{ix}|$, with the same i. The omnidirectional multipath causes the receiver envelopes to decorrelate rapidly (about 0.2λ), while the sector multipath causes the transmitter envelopes to decorrelate much more slowly - like the conventional behavior predicted for single-excitation spatial channels [Dur00a].

10.3 Example System

The following is an example of multiple antenna design in a neighborhood *wireless local area network* (WLAN) deployed using the unlicensed 5 GHz to 6 GHz band (U-NII or HIPERLAN) [Dur98b]. The hypothetical system and propagation scenarios illustrate the concepts in this chapter.

10.3.1 Problem Statement

A WLAN is to be deployed in a neighborhood environment, where base-station access points are placed atop outdoor utility poles and user terminals are located in homes. Such a system is illustrated in Figure 10.10 and could be used to deliver Internet, digital video, or other wireless multimedia content to everyone in the residential neighborhood. The WLAN operates at a carrier frequency of 5.3 GHz; the wavelength is 0.057 m.

The goal is to design the simplest possible array of monopole antennas at both the base station and the user terminals. The element geometry must be matched to

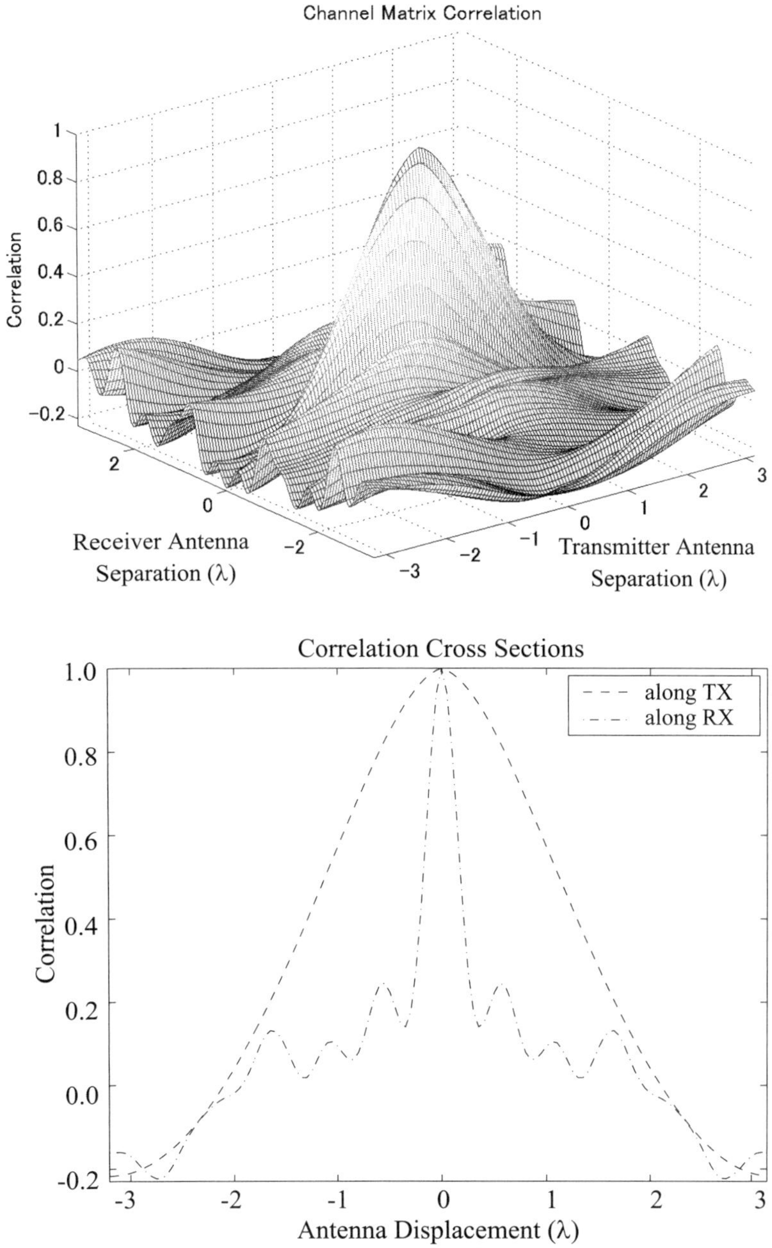

Figure 10.9 Unit autocovariance of the channel matrix envelope [Dur01b].

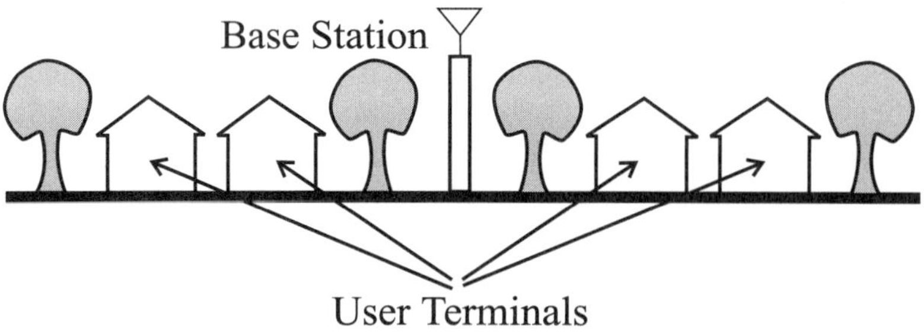

Figure 10.10 A neighborhood WLAN deployment.

the multipath conditions since the link will use a MIMO signal-processing algorithm to combat spatial selectivity and boost data capacity through the channel. To ensure that the array will work properly, we will impose a condition that there must be at least two elements with correlation of $\exp(-1)$ or less. All space diversity and MIMO processing algorithms work at near-optimum levels with this degree of correlation.

10.3.2 Angle-of-Arrival Models

We will use simple angle-of-arrival models to describe typical cases of multipath propagation at the base station and user terminals.

At the Base Station: Multipath signals arriving at an outdoor tower-top antenna from a user terminal are usually spread over a narrow angle. We will use a sectored multipath model, illustrated in Figure 10.11, choosing a slender sector width, α, of $10°$ or $\frac{\pi}{18}$ radians. This spread is slightly larger than a PCS or cellular base station, since these systems operate on taller towers and across longer average distances - factors that lead to a smaller spread.

At the User Terminal: Indoor terminals typically experience a larger multipath azimuth spread than their outdoor counterparts due to the dense concentration of potential scatterers about the horizon; multipath also tends to resonate within a room, causing a $180°$ bifurcation in the arriving propagation [Dur01a]. Thus, we will use a double-sectored multipath model, illustrated in Figure 10.11, choosing a moderate sector width, α, of $20°$ or $\frac{\pi}{9}$ radians.

10.3.3 Base-Station Design

The base-station antenna configuration is challenging because of the random orientation of its radio links with user terminals. Ideally, one base station serves a neighborhood of users that exist in many spatial directions. Ignoring the orientation issue for a moment, we can at least calculate two shape factors from the sector model. For the case of $\alpha = 10°$, the angular spread is $\Lambda = 0.0504$ and the angular constriction is $\gamma = 0.9990$.

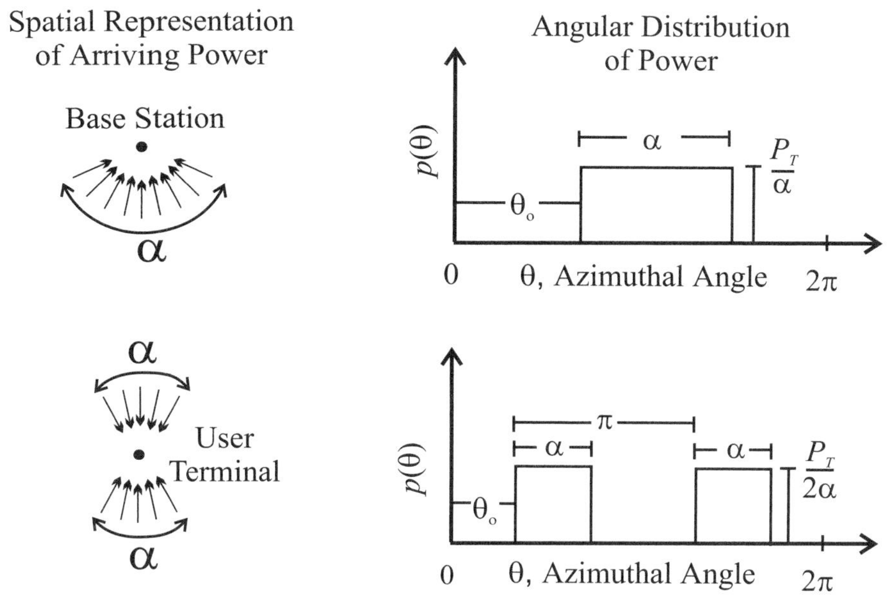

Figure 10.11 Multipath angle spectrum models for the base station and user terminals.

A channel with such a small angular spread and such a large angular constriction will likely require more than two elements. A design with two elements in the worst-case scenario would require over 7m of separation - impractical for a utility pole top. A design with two elements in the average-case scenario would create some inadequate, high-correlation links in the neighborhood. One solution is to design *three elements* to the average case, arranging them into the configuration of Figure 10.12. The average-case spatial separation is 0.24m, calculated from Equation (10.1.5), and would guarantee at least one pair of antenna would have sufficient decorrelation.

10.3.4 User Terminal Design

The user terminal also suffers from the random orientation problem. We cannot impose a user to adjust a desktop transceiver unit in a way that optimizes propagation conditions. Instead, we must design to the worst-case scenario using the double-sector shape factors. The double-sector azimuth spectrum always has an angle spread of $\Lambda = 1$. For the case of $\alpha = 20°$, we get an angular constriction of $\gamma = 0.9798$. Inserting this result into Equation (10.1.4) produces the minimum separation distance of 8.3 cm - a reasonable amount for a desktop unit that might plug into a desktop computer.

This distance could be lowered further in two ways. We could raise the critical correlation level, ρ_C, which may degrade some radio links. We could also incorporate

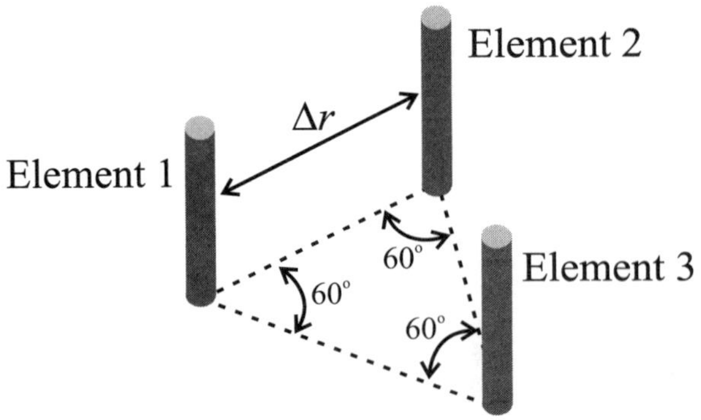

Figure 10.12 Three elements in an equilateral triangle configuration.

a third element, similar to the base-station antenna design. The disadvantage of a third element is increased cost and complexity, which is less attractive for a mass-produced unit such as a user terminal.

10.4 Peer-to-Peer Space–Time Measurements

This section presents space–time measurements for the *peer-to-peer* radio channel at a center frequency of 1920 MHz with 140 MHz of radio-frequency bandwidth. The data illustrate typical values for spectral spreads and shape factors in realistic channels and are useful for antenna designs as discussed in the previous sections.

10.4.1 The Peer-to-Peer Channel

In the peer-to-peer configuration, both transmitter and receiver are positioned at a height of roughly 1.5m above ground (head-level). This position is common for portable, handheld communication devices that may operate in mobile ad hoc networks. Because both transmitter and receiver are buried in clutter, the peer-to-peer radio link is quite lossy and rife with multipath. This is particularly true for longer distanced links; unlike a high-tier system such as a cellular network, the transmitter does not have the benefit of a tall radio tower to clear the tops of obstructions. Future applications for these peer-to-peer, mobile ad hoc networks include communications between such users as students, soldiers, businessmen, and emergency rescue workers.

10.4.2 Description of Peer-to-Peer Measurement Technique

The measurement campaign was performed during November 1999 on the campus of Virginia Tech. Twelve locations were measured during this campaign, using the peer-to-peer configuration of transmitter and receiver antennas. The measurement technique uses a unique combination of directional antennas and mechanical positioning (similar to [Egg95], [Spe00]), as opposed to a system based on large antenna arrays.

The first set of six measurements were outdoor-to-outdoor locations, which emphasized long-distance, obstructed links with transmitter-receiver (TR) separation distances between 240m and 910m. The signal had to propagate over irregular campus terrain, which included hills, multistory buildings, and leafless trees.

Then, three indoor-to-indoor locations were measured inside a four-story modern office building (Durham Hall on the Virginia Tech campus) with interior walls made from either cinderblocks or drywall. The same transmitter - placed inside a fourth-floor office - was used for all three indoor-to-indoor measurements. One receiver location was measured on the same floor as the transmitter, on the opposite side of the building. Another receiver location was measured on the ground floor, on the same side of the building as the transmitter. A third location was measured on a different floor than the transmitter *and* on the opposite side of the building.

Finally, three outdoor-to-indoor locations were measured using a transmitter placed 330m away from the four-story office building. The exterior of this office building is a combination of glass, concrete, and stone masonry. One receiver location was measured on the ground floor, on the same side of the building as the transmitter. Another receiver location was measured on the fourth floor, also on the same side of the building as the transmitter. A third location on the back side of the building was measured as well. A graphical summary of the six indoor local area configurations may be found in Figure 10.13.

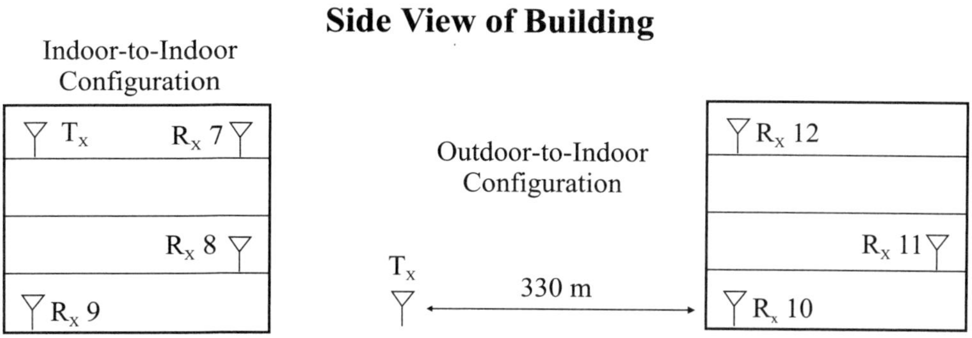

Figure 10.13 The transmitter-receiver configurations for the six local areas measured indoors [Dur03].

A precise automated positioning system was used to place the receiver antenna along a linear track and, for a directional antenna, to orient the antenna with respect to azimuth. The antenna platform is positioned using stepper motors that drive a rotating table and a long serpentine-drive track. The positioning error for placement along the track is $\pm 10\mu$m and for rotation about an axis is $\pm 0.01°$. In this measurement campaign, there were two types of measurement sequences: *track measurements* and *rotational measurements*, shown in Figure 10.14.

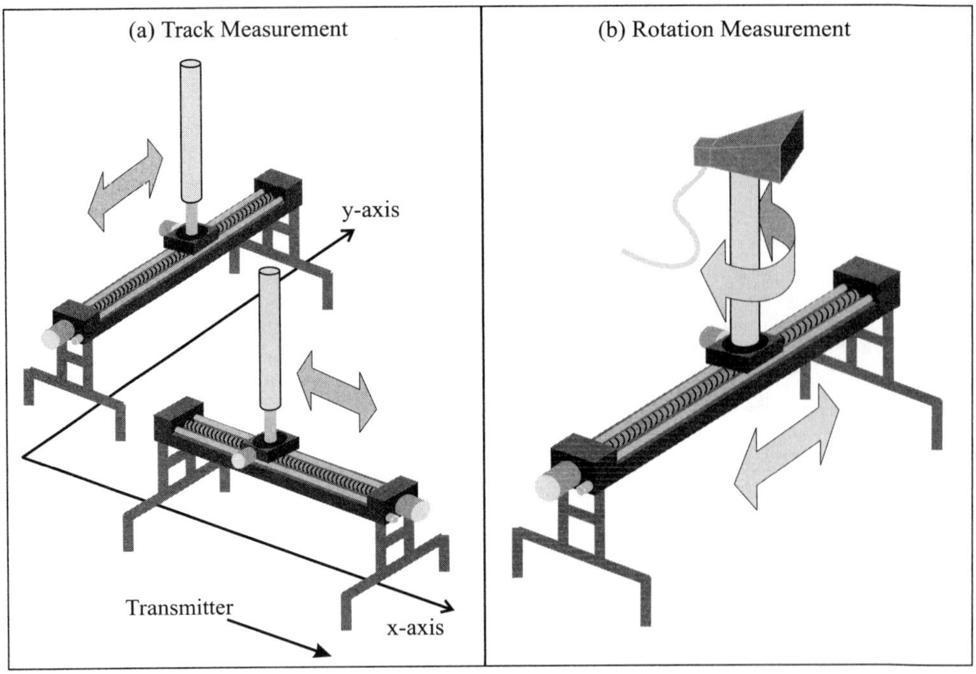

Figure 10.14 In a local area, power delay profiles are measured along two orthogonal linear tracks using an omnidirectional antenna (a) and also by spatially averaging angular sweeps with a directional antenna (b) [Dur03].

Track Measurements: For the track measurement sequence, an omnidirectional antenna is mounted atop the positioning table. Two linear track measurements are performed, each using different track orientations. The first orientation aligns the track along an axis (referred to as the x-axis) directed along the line connecting the transmitter to the receiver. The second orientation aligns the track along an axis (referred to as the y-axis) transverse to the direction of the transmitter location. The orientations are depicted in Figure 10.14. For each linear track measurement, snapshots of the channel PDP (power delay profile) are taken along the length of the measurement track (about 9 wavelengths at 1920 MHz). Each PDP snapshot is spaced 0.25 wavelengths apart, producing a total of 36 snapshots per linear track measurement. Thus, a measurement along two orthogonal tracks produces a total of 72 PDP snapshots. An example of a linear track measurement is photographed in Figure 10.15, and an example of collected PDPs as a function of position is shown in Figure 10.16.

Rotational Measurements: The second measurement sequence at a location is a rotational measurement using a directional antenna. PDP snapshots are recorded from the channel sounder as the test antenna is rotated in steps across the entire

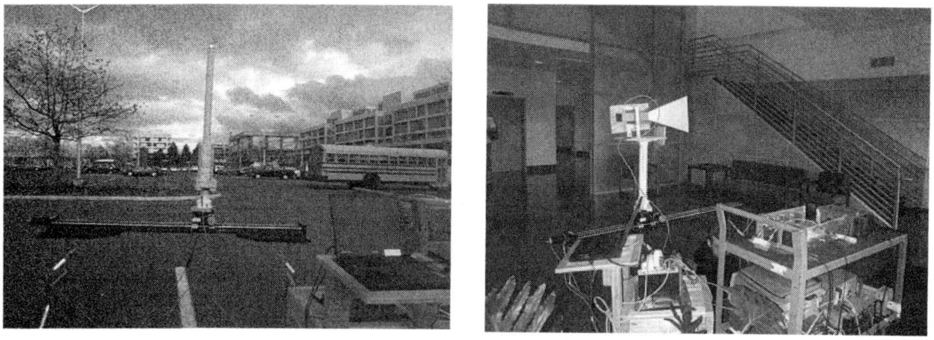

Figure 10.15 A track measurement (left) and a rotational measurement (right) are taken with the mechanical positioning system.

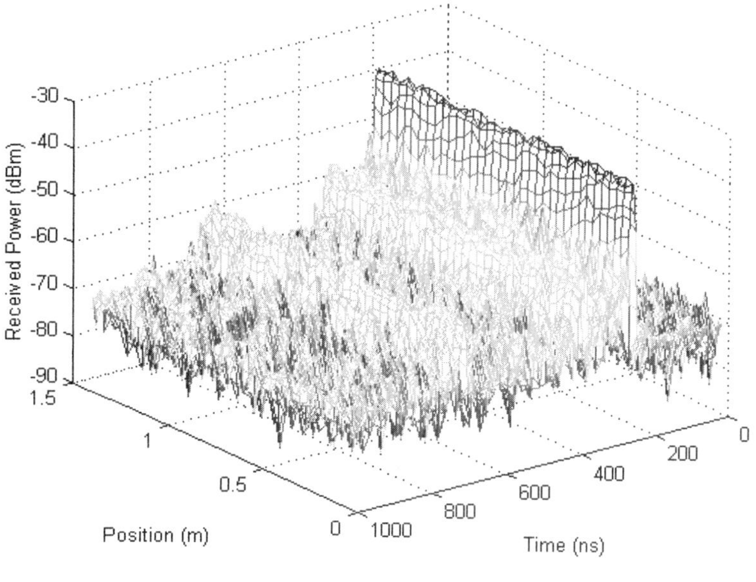

Figure 10.16 A series of PDP snapshots along a track, measured with an omnidirectional receiver antenna [Dur03].

horizon in evenly spaced 10° increments. Thus, a single sweep in the rotational measurement results in a total of 36 PDP snapshots. The antenna platform is then moved along the track by 2.67 wavelengths (0.42 meters), and another series of 36 rotational PDP snapshots are recorded. This procedure is repeated until a fourth rotational measurement is made. Figure 10.14 illustrates this sequence of measurements. In all, the rotational measurement sequence results in a total of 144 PDP snapshots. An example of a linear track measurement is photographed in

Figure 10.15, and an example of collected PDPs as a function of azimuth orientation is shown in Figure 10.17.

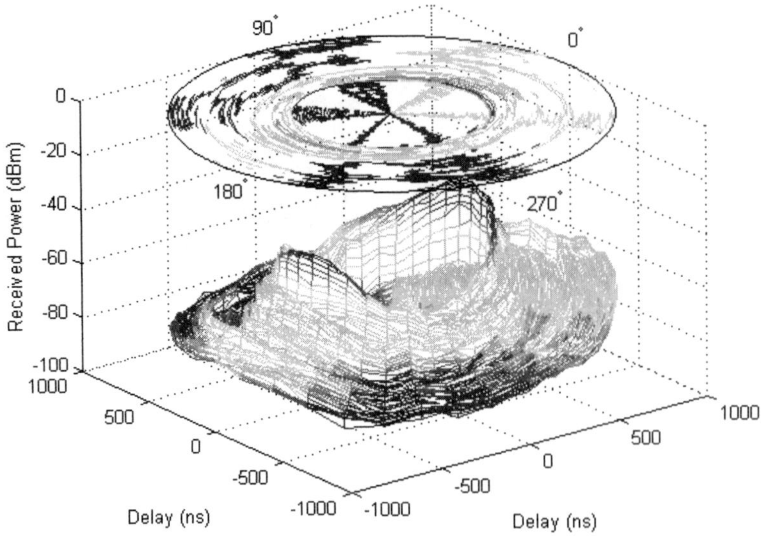

Figure 10.17 A local area angle–delay spectrum as measured from a set of rotational measurements [Dur03].

10.4.3 Delay Dispersion Results

Table 10.1 records all of the dispersion results for the 12 measured locations. This table records delay spread, centroid jitter, centroid standard deviation, timing jitter, and timing standard deviation for PDPs measured along orthogonal tracks. (See Appendix 10.A for a discussion on how temporal and spatial parameters are calculated from measured data.)

For outdoor-to-outdoor links, the delay spreads in Table 10.1 range as low as 17 ns to as high as 219 ns. Low values for delay spread are found at location 4, the only line-of-sight link in the outdoor-to-outdoor measurements, as well as at locations 5 and 6. It should be noted that the values in Table 10.1 are comparable to delay spreads measured by Patwari in [Pat99] (from 25 ns to 333 ns), despite the increased average link distances. In fact, the correlation between delay spread and TR separation is weak. The longest obstructed link - location 5 with 910m of TR separation - has one of the smallest delay spreads at an average of 46 ns.

Delay spreads for indoor receivers demonstrate much more homogeneous behavior, independent of whether the transmitter is indoors or outdoors. The 12 indoor delay spreads (two for each track for locations 7–12) fall within a 27 ns to 45 ns range. Thus, a much simpler equalizer may be used if the radio is guaranteed to be operating indoors.

Table 10.1 Dispersion Statistics Calculated from Track Measurements [Dur03]

	X-AXIS			Y-AXIS		
Location	Delay Jitter	Centroid Jitter	Timing Std. Dev.	Delay Spread	Centroid Jitter	Timing Jitter
1 OO	164 ns	118 ns	128 ns	138 ns	95 ns	146 ns
2 OO	196	89	90	148	101	101
3 OO	185	115	53	219	84	71
4 OO	55	17	24	22	15	20
5 OO	51	38	51	40	25	35
6 OO	25	8	16	17	6	10
7 II	30	33	13	29	26	18
8 II	45	29	24	45	29	20
9 II	44	40	25	42	31	25
10 OI	28	10	25	31	47	39
11 OI	35	37	46	44	46	44
12 OI	43	39	40	27	31	31
Average	75	48	44	67	45	47
Std. Dev.	62	36	32	62	30	39

OO – outdoor-to-outdoor II – indoor-to-indoor OI – outdoor-to-indoor

10.4.4 Angle Dispersion Results

A number of other multipath parameters may also be calculated from the measured track and rotational data. Table 10.2 records TR separation distance, path loss with respect to 1m free space, the angular spread, and the peak multipath direction of arrival.

One trend that is apparent from Table 10.2 is in the peak direction of multipath arrival. This parameter measures the direction in azimuth that the horn antenna was pointing when the maximum total power was received. According to Table 10.2, the peak direction of multipath arrival is almost always in the direction of the transmitter (corresponding to 0°). This is true even for obstructed receiver locations. The one exception in Table 10.2 is location 9, where the peak direction of multipath arrival is 140°. However, location 9 is an indoor location corresponding to an indoor transmitter that is nearly directly above the receiver, with two floors in between. Given this unique location, the deviation from the trend is understandable. This property of the peak multipath arrival angle implies that an array scanning for peak power results in an effective direction-finding algorithm, even in heavily obstructed channels.

Another trend may be observed in the angular spread data. For indoor receivers

Table 10.2 Spatial Multipath Parameters Calculated from Spatially Averaged Azimuthal Sweeps of a Horn Antenna [Dur03]

Location	Path Loss w.r.t. 1m FS	TR Sep Dist.	Λ	γ	θ_{max}	Peak AOA
1 OO	90 dB	770 m	0.82	0.69	44°	0°
2 OO	51 dB	550 m	0.65	0.44	−63°	0°
3 OO	39 dB	240 m	0.91	0.51	36°	20°
4 OO	81 dB	585 m	0.52	0.77	89°	0°
5 OO	72 dB	910 m	0.46	0.70	76°	−10°
6 OO	83 dB	410 m	0.36	0.76	85°	0°
7 II	70 dB	29 m	0.73	0.52	84°	−10°
8 II	82 dB	33 m	0.74	0.50	30°	30°
9 II	76 dB	15 m	0.90	0.33	−74°	140°
10 OI	65 dB	340 m	0.78	0.72	85°	0°
11 OI	84 dB	365 m	0.86	0.19	−46°	10°
12 OI	85 dB	340 m	0.98	0.42	−38°	−50°
Average	73 dB	–	0.73	0.55	–	11°
Std. Dev.	15 dB	–	0.19	0.19	–	43°

OO – outdoor-to-outdoor II – indoor-to-indoor OI – outdoor-to-indoor

at locations 7 through 12, the angular spread falls within the range 0.73 to 0.98. Thus, indoor angular spread values are almost always near the maximum value of 1.00. For outdoor receivers at locations 1 through 6, the angular spread falls, on average, within a range of lower values, 0.36 to 0.91. Thus, an omnidirectional fading model for narrowband fading, such as the Clarke model in [Cla68], is not accurate for long-distance peer-to-peer links. The increase in angular spread indoors as opposed to outdoors may be explained by the increased density of scatterers (doors, walls, shelves, etc.) in all directions in an indoor environment.

Still another trend may be observed in the angular constriction data. From Table 10.2, we see that the average value for angular constriction, γ, is 0.55. This fairly large value indicates that multipath power is clustering about a few directions instead of being uniformly spread out. This is another indication that idealized, uniform multipath models may not characterize the spatial fading behavior for the peer-to-peer channel.

10.4.5 Joint Angle-Delay Statistics

The graph in Figure 10.18 shows angular spread versus delay spread for the 12 measured local areas. We would expect that higher delay spreads indicate more multipath components from a larger variety of scattering mechanisms; under these circumstances, the angular spread should increase as well. While counterexamples certainly exist, most of the measured points in Figure 10.18 follow this basic trend.

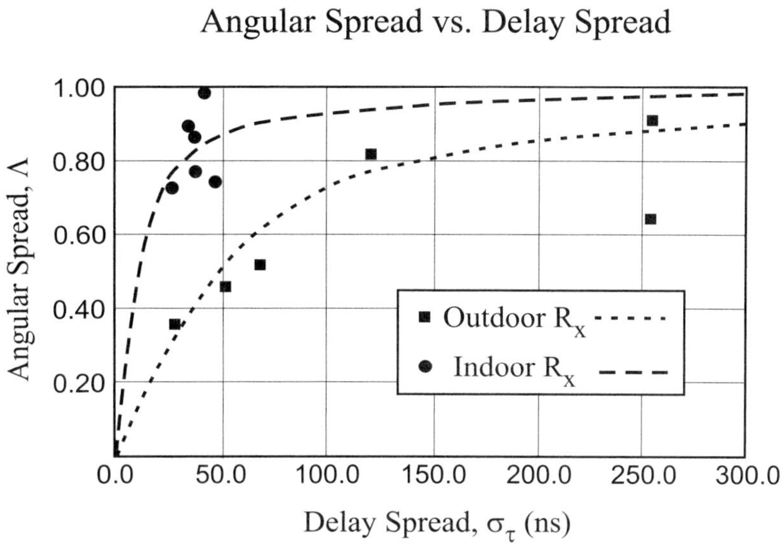

Figure 10.18 The trend between multipath angular spread, Λ, and RMS delay spread, σ_T, for indoor and outdoor receiver locations [Dur03].

Note that the *rate* at which angular spread increases as a function of delay spread depends heavily on whether the receiver is indoors or outdoors. To study this effect quantitatively, we propose the following empirical guideline for angular spread, Λ, as a function of delay spread, σ_T:

$$\Lambda \approx \exp\left(-\frac{\sigma_c}{\sigma_T}\right) \tag{10.4.1}$$

The critical delay spread, σ_c, is the key parameter in Equation (10.4.1) for determining the rate of angular spread increase. The critical delay spread may be calculated using linear regression on a set of measurement points, $(-\ln\Lambda, \frac{1}{\sigma_T})$. The six indoor points produce a critical delay spread value of 7.4 ns, while the six outdoor points produce a much larger σ_c of 32.5 ns. Plots of Equation (10.4.1) for these two critical delay spreads are shown in Figure 10.18.

10.5 Chapter Summary

This chapter presented a methodology for designing radio links that employ multiple antennas at either transmitter or receiver, and provided some insight into design and measurement.

- There are simple geometrical rules for constructing antenna arrays in multipath environments that minimize correlation for diversity and MIMO systems.
  - ▷ Forbidden zones of correlation are constructed from multipath shape factors and the desired correlation levels.
  - ▷ Forbidden zones of coupling may be superimposed as well.
  - ▷ The effects of random orientation are often important for user terminals.
- Modeling a MIMO channel matrix, $\tilde{\mathbf{H}}$, with realistic correlation properties requires a modification of the SLAC model of Chapter 4 that had only a single spatial dependency.
- The concept of forbidden zones works for the design of antenna arrays in multipath environments at either the transmitter *or* the receiver.
- Spatial probing with a wideband channel sounder is a powerful technique for measuring the joint space–time parameters of a baseband wireless channel.

This chapter helps to complete our understanding of the multiple antenna systems introduced in Chapter 8 and Chapter 9.

PROBLEMS

1. Directional antenna elements are often characterized by their *front-to-back ratio* (FBR), the ratio (often reported in dB) between the gain in the peak direction and the gain in the opposite direction. Below is a simple gain pattern for an electrically small antenna with FBR:

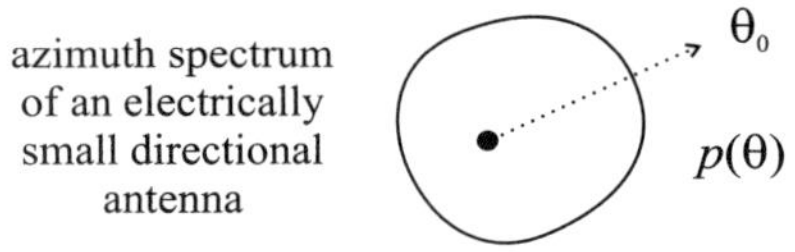

$$p(\theta) = 1 + \frac{\text{FBR-1}}{\text{FBR+1}} \cos(\theta - \theta_0)$$

Answer the following questions based on this antenna gain pattern:

a. If this antenna is operated in a multipath environment described by an omnidirectional azimuth spectrum, what are the shape factors that describe the spatial fading for this antenna element?

b. Two Yagi-Uda antennas are used in a diversity array where each is oriented with peak gain pointing transverse to their separation. If FBR = 3 dB and critical correlation is 0.3, what should the separation distance be in wavelengths? (Ignore coupling effects.)

c. If the Yagi-Uda antennas are replaced by co-polar antennas with FBR $= -\infty$ dB, what should the separation distance be for a critical correlation of 0.3?

d. Based on your answer in (c), might element coupling be a problem for this design? Explain two alternative ways to use these antennas for diversity that avoid any coupling problems.

2. Consider a four-element array of co-polar monopole antenna elements used in a multipath environment. Below are top-view sketches of three possible array geometries, with r_L being the longest separation distance between two elements and r_S being the shortest separation distance between two elements.

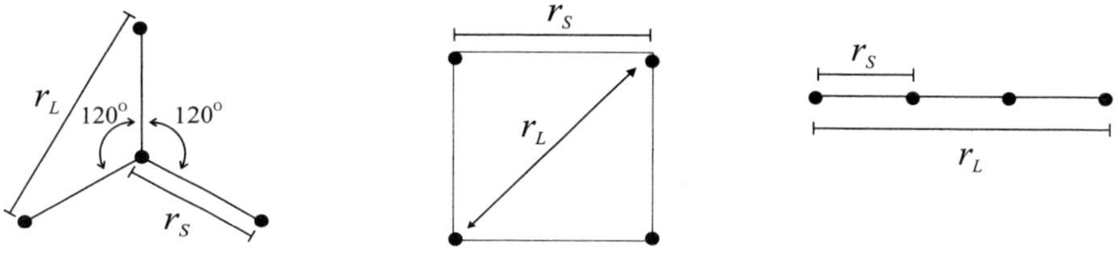

Answer the following questions based on these three geometries, assuming a critical correlation of 0.6:

a. For each geometry, calculate the relationship between r_L and r_S.

b. These arrays are used in a multipath environment described by a sector azimuth spectrum with width 35°. Calculate the shape factors for this channel and assume this model for the remaining questions.

c. Calculate r_L and r_S (in wavelengths) if the goal of each design is to guarantee decorrelation between all elements, regardless of orientation.

d. Calculate r_L and r_S if the goal of each design is to guarantee decorrelation between at least two elements, regardless or orientation.

e. Calculate r_L and r_S if the goal of each design is to guarantee average-case decorrelation between all elements.

3. Consider an N-element circular array of co-polar monopole antenna elements. Below are top-view sketches of this geometry and the double "bent" sector azimuth spectrum that describes the multipath propagation.

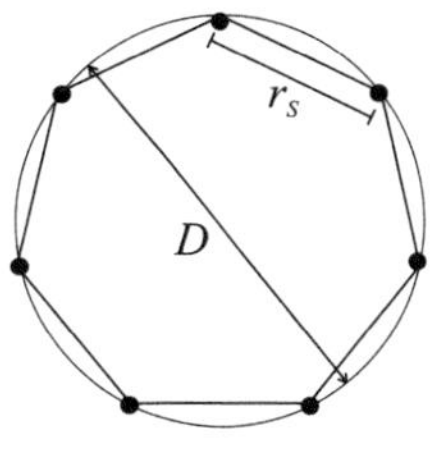

N-element circular array

Azimuth Spectrum

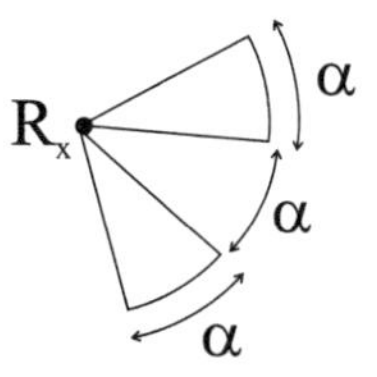

Answer the following questions based on these three geometries, assuming a critical correlation of $\exp(-1)$:

a. For a given number N, calculate the relationship between r_S and D.

b. Calculate the shape factors for this channel as a function of α.

c. For $N = 6$ and $\alpha = 45°$, what are the smallest dimensions of the array that guarantee decorrelation for at least two elements?

d. For $N = 5$ and $\alpha = 15°$, what are the smallest dimensions of the array that guarantee decorrelation for all elements?

e. For $N = 8$ and $\alpha = 10°$, what are the smallest dimensions of the array that guarantee at least four of the elements have mutually decorrelated channels?

f. Derive an expression for N as a function of α resulting in an array geometry that sets r_S equal to the best-case decorrelation distance and D to the worst-case decorrelation distance.

4. Let us define a *perfect linear array* (PLA) to be the configuration of co-polar monopole antennas with the following three criteria: (1) the element separation is uniform, (2) the length of the linear array guarantees decorrelation between the two elements with the largest separation distance regardless of orientation in azimuth, and (3) the element separation distance is set for decorrelation in the "best-case" azimuth orientation. Using a critical correlation of $\exp(-1)$ and ignoring the effects of element coupling, find the optimum number of elements, N, for the PLA operating in the following multipath channels:

N-element linear array

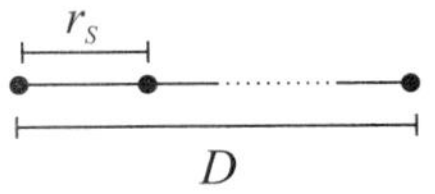

(a) azimuth spectrum

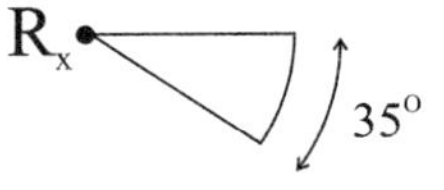

(b) azimuth spectrum

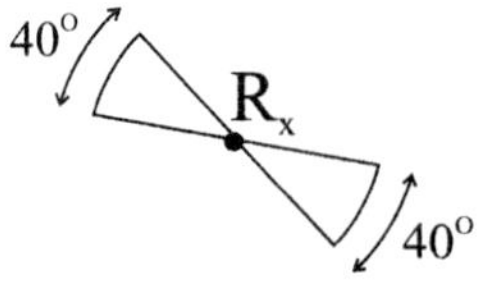

(c) azimuth spectrum

$p(\theta) = \cos(4\theta) \quad |\theta| \leq \frac{\pi}{8}$

(d) azimuth spectrum

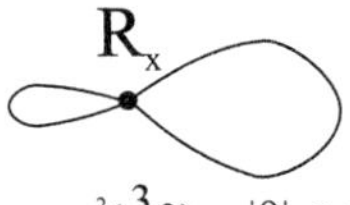

$p(\theta) = \cos^2(\frac{3}{4}\theta) \quad |\theta| \leq \pi$

(e) azimuth spectrum

5. Repeat the previous exercise, now considering the effects of coupling for separation distances less than $\frac{\lambda}{2}$. Which of your answers have changed from the previous calculation?

6. Use the unit autocovariance expression in Equation (7.2.10) to verify the accuracy of the transmitter and receiver correlation properties in Figure 10.9.

7. Simulate the correlation properties of a channel with a double spatial dependency using the modeling technique discussed in Section 10.2. Use the following multipath azimuth spectra for transmitter angles-of-departure and receiver angles-of-arrival.

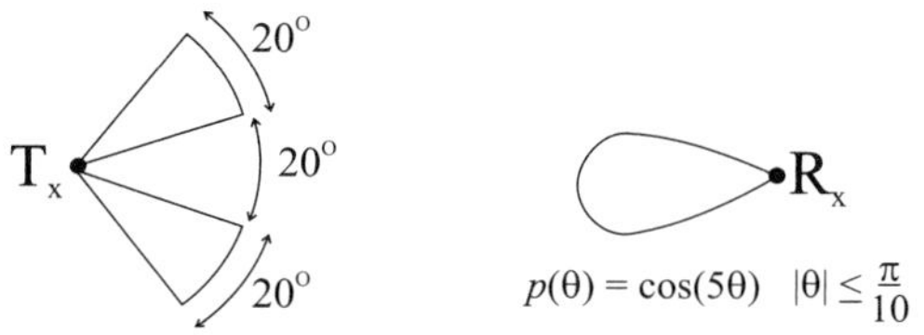

8. A pair of peer-to-peer MIMO transceivers are communicating between one another in an indoor office environment with an extremely high bit rate. However, when the units are moved outdoors, the overall data rate plummets. Use the data in Table 10.2 and the concept of forbidden zones to explain why this happened.

9. You wish to deploy an indoor wireless LAN system that uses a compact two-element co-polar space diversity at the receiver. Unfortunately, you have no spatial measurements to confirm whether there is enough multipath in the indoor environment to produce optimum diversity gain at the receiver. All you have is a set of wideband measurements showing that most delay spreads in the typical environment will be between 40 and 60 ns. Will the system work as desired? Explain your answer.

10. Below is a table of spatial and temporal parameters taken from a hypothetical factory environment.

Location	1	2	3	4	5	6	7	8	9	10
Delay Spread (ns)	46	66	23	55	19	90	61	45	33	81
Angular Spread (%)	73	47	49	77	34	99	92	81	64	89

From this data, calculate the critical delay spread, as specified in Section 10.4.4. Plot the data with the critical delay spread trend-line on the same graph.

11. As an exercise in creativity, invent an analytical expression that describes any observable trend between path loss in dB and angular constriction in Table 10.2. Fit this data to your expression. Can you physically explain any observed trends? How reliable is this trend?

10.A Description of Measured Parameters

This appendix defines the terminology and parameters used in the analysis of the wideband, space-varying channels.

10.A.1 Noncoherent Channel Measurements

The wideband radio channel is measured as a function of space in the November 1999 measurement campaign at Virginia Tech using a *single-channel, noncoherent receiver.* Below is a list of terminology used to describe the types of channels measured during this campaign.

Power Delay Profile (PDP): Noncoherent channel-sounding hardware measures a PDP instead of a CIR. The PDP has units of *power* (related to physical units of *Watts* by a constant of proportionality) and is defined as the magnitude-squared of the CIR:

$$P(\tau, \vec{r}) = |\tilde{H}(\tau, \vec{r})|^2 \tag{10.A.1}$$

Written without a θ-dependence, it may be assumed that Equation (10.A.1) represents a measurement with an omnidirectional antenna.

Power Angle Profile (PAP): The PAP is the spatial equivalent of a PDP. The PAP has units of *power* and is defined as

$$P(\theta, \vec{r}) = |\tilde{H}(\theta, \vec{r})|^2 \tag{10.A.2}$$

Written without a τ-dependence, Equation (10.A.2) represents the angle-of-arrival characteristics of a narrowband channel.

Angle-Delay Profile (ADP): When a directional antenna at position, $\vec{r}$, and azimuthal orientation, θ, is connected to a wideband noncoherent channel sounder, an ADP is measured. The ADP has units of *power* and is defined as

$$P(\tau, \theta, \vec{r}) = |\tilde{H}(\tau, \theta, \vec{r})|^2 \tag{10.A.3}$$

10.A.2 Power Spectra

When power delay or angle profiles are linearly averaged in space, an estimate of a *power spectrum* (and hence PSD) is produced. Averaging various power profiles will produce the following power spectra:

Delay Spectrum: Spatially averaging a collection of PDPs measured within the same local area produces an estimate for the delay spectrum, $S_{\tilde{h}}(\tau)$, of the channel:

$$S_{\tilde{h}}(\tau) \cong \frac{\sum_{i=1}^{N} P(\tau, \vec{r}_i)}{\int_{-\infty}^{+\infty} \left[\sum_{i=1}^{N} P(\tau, \vec{r}_i) \right] d\tau} \tag{10.A.4}$$

where $\{\vec{r}_i\}$ is the set of measurement positions.

Angle Spectrum: Spatially averaging a collection of PAPs measured within the same local area produces an estimate for the azimuth spectrum, $p(\theta)$, of the channel:

$$p(\theta) \cong \frac{\sum_{i=1}^{N} P(\theta, \vec{r}_i)}{\int_{-\infty}^{+\infty} \left[\sum_{i=1}^{N} P(\theta, \vec{r}_i) \right] d\theta} \tag{10.A.5}$$

Angle-Delay Spectrum: Spatially averaging a set of full ADPs within the same local area produces an estimate for the joint angle-delay spectrum, $S_{\tilde{h}}(\tau, \theta)$:

$$S_{\tilde{h}}(\tau, \theta) \cong \frac{\sum_{i=1}^{N} P(\tau, \theta, \vec{r}_i)}{\int_{-\infty}^{+\infty} \int_{-\infty}^{+\infty} \left[\sum_{i=1}^{N} P(\tau, \theta, \vec{r}_i) \right] d\theta \, d\tau} \tag{10.A.6}$$

Note that the use of power spectra in channel modeling assumes radio channels that are wide-sense stationary in space and frequency over basic intervals of interest. The frequency interval of interest is the *RF bandwidth* of the transmitted signal and the spatial interval of interest is the *local area.*

10.A.3 Time Delay Parameters

Delay dispersion statistics can be measured from a collection of local area PDPs. Note that before each statistic is calculated, all power at or below the noise+interference floor of the received profile is zeroed.

Centroid Jitter: If the centroid were calculated from a single PDP instead of the delay spectrum, then each local area would have a set of centroids as a function of space, $\{\overline{\tau}(\vec{r}_i)\}$. The *centroid jitter* is the maximum centroid value minus the minimum centroid value measured in the same local area. This value measures range of possible centroid fluctuations within a local area.

Timing Jitter: This parameter, defined by Devasirvatham in [Dev87], is based on the set of instantaneous RMS delay spreads, $\{\sigma_\tau(\vec{r}_i)\}$, calculated from PDPs in a local area. The timing jitter is the maximum delay spread minus the minimum delay spread measured in the same local area. This value measures the range of possible RMS delay spread fluctuations within a local area.

10.A.4 Angle-of-Arrival Parameters

There are also several angular dispersion parameters that are measured in this campaign and reported in this paper. Each parameter is also calculated for a measured local area and defined below:

Peak Angle-of-Arrival, θ_{peak}: This parameter, calculated from the estimate of angle spectrum $p(\theta)$, is the azimuthal angle in which the largest average multipath power is received.

Shape Factors: These parameters (angular spread, angular constriction, and direction of maximum fading) are calculated from the angle spectrum estimate, consistent with their definitions in Chapter 6.

Appendix A

SPECIAL FUNCTIONS

During the course of analysis, a number of special functions arise that cannot easily be described in terms of elementary transcendental functions (i.e., sines, exponents, logarithms, and so on) This appendix defines the special functions used in this book, providing graphs and calculation aides whenever appropriate.

A.1 Singularity Functions

Singularity functions arise whenever descriptions of discontinuous or piecewise continuous functions are studied. The first singularity function we must define is the *unit step function* (also called the Heaviside unit step function):

$$\mathrm{u}(t) = \begin{cases} 0, & t < 0 \\ 1, & t \geq 0 \end{cases} \tag{A.1.1}$$

The unit step function is zero for all negative values of t and is one for all positive values of t.

The delta function is closely related to the unit step function. Often called the Dirac delta function or the impulse function, the delta function is the derivative of the unit step function.

$$\delta(t) = \frac{d\mathrm{u}(t)}{dt} \tag{A.1.2}$$

Since the unit step function has no slope for nonzero t, the delta function is 0 for all values $t \neq 0$. The discontinuity at $t = 0$ implies that the delta function has an infinite value at this point. The delta function represents an ideal pulse, a concentration of finite energy at a single point in time or other dependency. Table A.1 records some of the most important properties of the delta function.

In graphs, the delta function is represented as a line segment terminating in an arrow; the length of the segment denotes amplitude. Delta functions and unit step functions may be combined in many ways to construct interesting signals and mathematical function shapes. Just a few examples of these combinations are shown in Table A.2.

Table A.1 Properties of the Delta Function, $\delta(t)$

Unit Step Relation	$\delta(t) = \frac{du(t)}{dt} \qquad u(t) = \int_{-\infty}^{t} \delta(t_0)\,dt_0$
Pulse Definitions	$\delta(t) = \lim_{B\to\infty} \text{sn}(Bt) = \lim_{T\to 0} \frac{1}{T\sqrt{2\pi}} \exp\left(\frac{-t^2}{2T^2}\right)$
Reversal/Scaling	$\delta(-t) = \delta(t) \qquad \delta(at) = \frac{1}{\lvert a\rvert}\delta(t)$
Integral Relation	$\delta(t) = \lim_{B\to\infty} \int_{-B/2}^{B/2} \exp(j2\pi ft)\,df$
Sifting Property	$f(t)\delta(t-t_0) = f(t_0)\delta(t-t_0)$
Sifting Integral	$f(t_0)u(t_1-t_0) = \int_{-\infty}^{t_1} f(t)\delta(t-t_0)\,dt$
Integral Identity	$\delta(t) = \lim_{B\to\infty} \int_{-B/2}^{B/2} \exp(j2\pi ft)df \cong \int_{-\infty}^{+\infty} \exp(j2\pi ft)df$
Function Argument	$\delta(y-f(t)) = \sum_i \left\lvert \frac{df(t_i)}{dt} \right\rvert^{-1} \delta(t-t_i)$ where t_i are the zeros of $f(t_i) = y$
Vector Notation	$\delta(\vec{r}) = \delta(x)\delta(y)\delta(z)$ where $\vec{r} = x\hat{x} + y\hat{y} + z\hat{z}$
Spherical Expansion	$\delta(\vec{r} - \vec{r}_0) = \frac{\delta(r-r_0)\delta(\theta-\theta_0)\delta(\varphi-\varphi_0)}{r_0^2 \cos(\varphi_0)}$

A.2 Sinc Function

A function that appears time and time again in basic communication theory is the *sinc* function, $\text{sn}(t)$. Mathematically, the sinc function is defined to be

$$\text{sn}(t) = \frac{\sin \pi t}{\pi t} \tag{A.2.1}$$

which is simple enough to evaluate, provided $t \neq 0$. In this case, we must take the limit:

$$\text{sn}(0) = \lim_{t\to 0} \frac{\sin \pi t}{\pi t} = 1 \tag{A.2.2}$$

Thus, the continuous sinc function is defined for all values of t. Figure A.1 shows a graph of the sinc function.

Table A.2 Common Examples of Singularity Functions

FUNCTION TYPE	MATHEMATICAL FORM	FUNCTION SHAPE
Delta Function	$\delta(t)$	
Impulse Train	$\sum_i A_i \delta(t - t_i)$	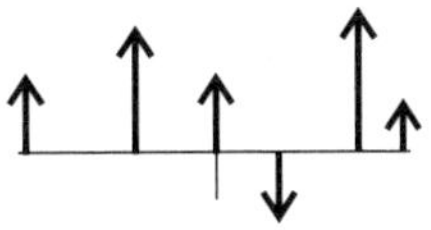
Unit Step Function	$\mathrm{u}(t)$	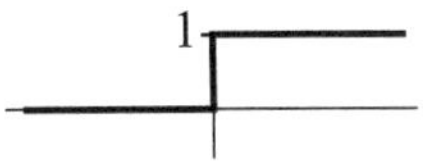
Box Function	$\mathrm{u}(\frac{T}{2} - \lvert t \rvert)$	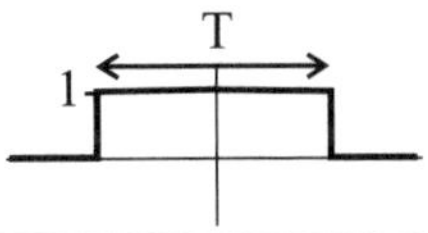
Staircase Function	$\sum_i A_i \mathrm{u}(t - t_i)$	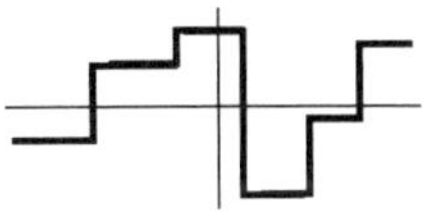
Ramp Function	$t\mathrm{u}(t)$	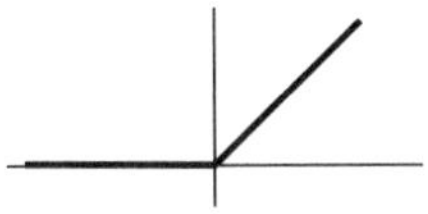
Clipped Function	$f(t)\mathrm{u}(t)$	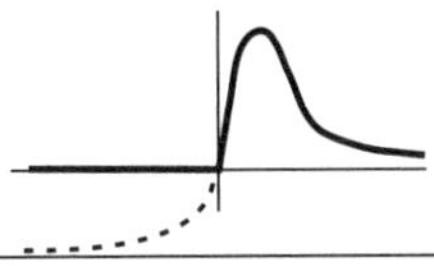
Framed Function	$f(t)\mathrm{u}(T - \lvert t \rvert)$	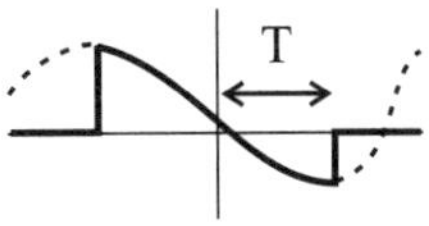
Crossing Counter	$\left\lvert \frac{df(t)}{dt} \right\rvert \delta(f(t) - A)$	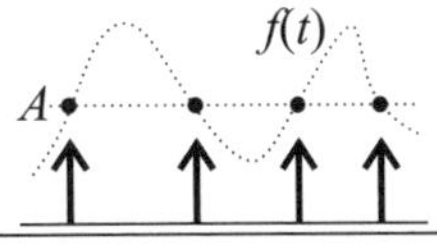

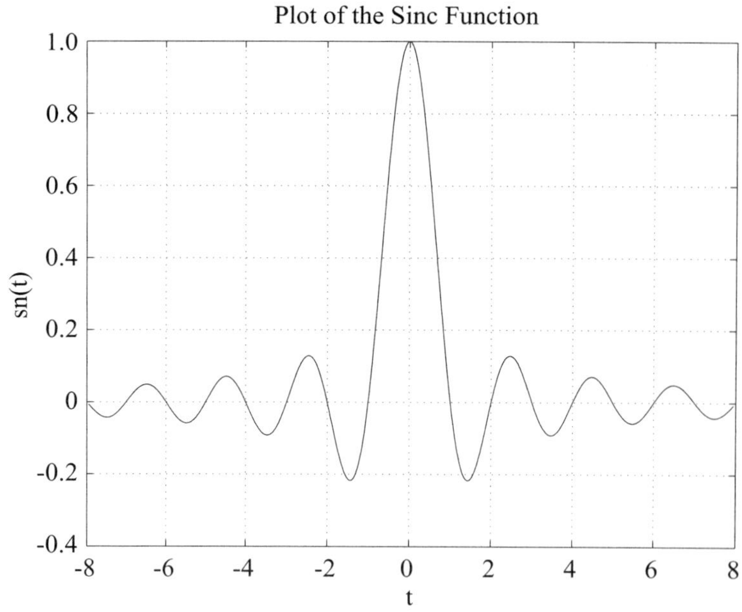

Figure A.1 A graph of the communications sinc function.

> *Note: The Sa Function*
>
> Occasionally, you will find an Sa function (pronounced like the word *essay*). The Sa function is defined as $\mathrm{Sa}(x) = \frac{\sin x}{x}$ and is used instead of the sinc function in some textbooks. The Sa function is mostly used in controls engineering, while the sinc function is mostly used in communications engineering. The reason: Controls engineers find it most convenient to operate in a radian-frequency domain (like our Doppler domain) where the π's of the sinc function are already included in the frequency domain variable.

A.3 Gamma Function

The gamma function is defined using the following definite integral:

$$\Gamma = \int_0^{\infty} t^{x-1} \exp(-t)\, dt \tag{A.3.1}$$

Common values of the gamma function include:

$$\Gamma(0) = \infty \qquad \Gamma(1/2) = \sqrt{\pi} \qquad \Gamma(1) = 1 \qquad \Gamma(3/2) = \frac{\sqrt{\pi}}{2}$$

The most common form of the gamma function takes an integer argument, n, and satisfies the factorial relationship

$$\Gamma(1+n) \;=\; n! \;=\; n \cdot (n-1) \cdot (n-2) \cdots 3 \cdot 2 \cdot 1 \qquad \text{(A.3.2)}$$

You will often find mathematical texts using the Gamma function to express series and definite integrals.

A.4 Bessel Functions

Two types of Bessel functions are used in this book. The ordinary Bessel function (often called a Bessel function of the first kind) is denoted by $\mathrm{J}_\nu(x)$, where ν is the *order*. This special function is a solution to Bessel's differential equation, which is listed in Table A.3 with other analytical relationships. Although ν may take on any real value, integer orders are most common. As is the convention in mathematics, relationships for integer-order Bessel functions set $\nu = n$. A plot of several Bessel functions with integer order is shown in Figure A.2. These functions decay with respect to increasing argument and generally exhibit aperiodic oscillation.

The modified Bessel function, $\mathrm{I}_\nu(x)$, is similar in definition to the ordinary Bessel function. This special function is a solution to Bessel's modified differential equation, also shown in Table A.3 along with useful analytical relationships. Figure A.2 records a plot of modified Bessel functions for integer orders 0 to 4. Unlike the ordinary Bessel function, $\mathrm{I}_\nu(x)$ is a monotonically increasing function. Values of both types of 0-order Bessel functions are collected in Table A.4.

A.5 Complete Elliptic Integral Functions

There are two kinds of complete elliptic integral functions. They are defined using the following definite integrals:

$$\text{First Kind:} \quad K(m) = \int_0^{\frac{\pi}{2}} \frac{d\theta}{\sqrt{1+m\sin^2\theta}} \qquad \text{for } 0 \le m \le 1 \qquad \text{(A.5.1)}$$

$$\text{Second Kind:} \quad E(m) = \int_0^{\frac{\pi}{2}} \sqrt{1+m\sin^2\theta}\, d\theta \qquad \text{for } 0 \le m \le 1 \qquad \text{(A.5.2)}$$

The solutions to these integrals may be expressed as infinite series:

$$K(m) = \frac{\pi}{2}\left[1 - \left(\frac{1}{2}\right)^2 m + \left(\frac{1\cdot 3}{2\cdot 4}\right)^2 m^2 - \left(\frac{1\cdot 3\cdot 5}{2\cdot 4\cdot 6}\right)^2 m^3 + \cdots\right] \qquad \text{(A.5.3)}$$

$$E(m) = \frac{\pi}{2}\left[1 - \left(\frac{1}{2}\right)^2 m - \left(\frac{1\cdot 3}{2\cdot 4}\right)^2 m^2 - \left(\frac{1\cdot 3\cdot 5}{2\cdot 4\cdot 6}\right)^2 m^3 - \cdots\right] \qquad \text{(A.5.4)}$$

Table A.3 Useful Bessel Function Properties

BESSEL FUNCTION, J	MODIFIED BESSEL FUNCTION, I
Defining Differential Equation	
$x^2\frac{d^2y}{dx^2} + x\frac{dy}{dx} + (x^2 - \nu^2)y = 0$	$x^2\frac{d^2y}{dx^2} + x\frac{dy}{dx} - (x^2 + \nu^2)y = 0$
Defining Integral Equation	
$\mathrm{J}_n(x) = \frac{1}{j^n\pi}\int_0^\pi \exp(jx\cos\theta)\cos(n\theta)\,d\theta$	$\mathrm{I}_n(x) = \frac{1}{\pi}\int_0^\pi \exp(x\cos\theta)\cos(n\theta)\,d\theta$
$\mathrm{J}_\nu(x) = \frac{1}{\pi}\int_0^\pi \cos(\nu\theta - x\sin\theta)\,d\theta$	
Series Expansion	
$\mathrm{J}_\nu(x) = \left(\frac{x}{2}\right)^\nu \sum_{i=0}^{\infty} \frac{(-1)^i x^{2i}}{4^i\Gamma(1+\nu+i)i!}$	$\mathrm{I}_\nu(x) = \left(\frac{x}{2}\right)^\nu \sum_{i=0}^{\infty} \frac{x^{2i}}{4^i\Gamma(1+\nu+i)i!}$
$\mathrm{J}_n(x) = \left(\frac{x}{2}\right)^n \sum_{i=0}^{\infty} \frac{(-1)^i x^{2i}}{4^i(n+i)!i!}$	$\mathrm{I}_n(x) = \left(\frac{x}{2}\right)^n \sum_{i=0}^{\infty} \frac{x^{2i}}{4^i(n+i)!i!}$
Recurrence Relationships	
$\mathrm{J}_{\nu+1}(x) - \mathrm{J}_{\nu-1}(x) = -2\frac{d\mathrm{J}_\nu(x)}{dx}$	$\mathrm{I}_{\nu+1}(x) - \mathrm{I}_{\nu-1}(x) = -2\frac{d\mathrm{I}_\nu(x)}{dx}$
$\mathrm{J}_{\nu+1}(x) + \mathrm{J}_{\nu-1}(x) = \frac{2\nu}{x}\mathrm{J}_\nu(x)$	$\mathrm{I}_{\nu+1}(x) + \mathrm{I}_{\nu-1}(x) = \frac{2\nu}{x}\mathrm{I}_\nu(x)$
Symmetry Properties	
$\mathrm{J}_n(-x) = (-1)^n\mathrm{J}_n(x)$	$\mathrm{I}_n(-x) = (-1)^n\mathrm{I}_n(x)$
Large Argument Approximations $x \gg 1$	
$\mathrm{J}_\nu(x) \approx \sqrt{\frac{2}{x\pi}}\cos\left(x - \frac{\nu\pi}{2} - \frac{\pi}{4}\right)$	$\mathrm{I}_\nu(x) \approx \frac{1}{\sqrt{2\pi x}}\exp\left(x - \frac{4\nu^2-1}{8x}\right)$
Other Useful Relationships	
$\int_0^\infty \mathrm{J}_\nu(x)\,dx = 1$	$j^\nu\mathrm{J}_\nu(x) = \mathrm{I}_\nu(jx)$
$\mathrm{J}_{\frac{1}{2}}(x) = \sqrt{\frac{2}{x\pi}}\sin(x)$	$\mathrm{J}_{-\frac{1}{2}}(x) = \sqrt{\frac{2}{x\pi}}\cos(x)$
$\int_0^\infty \exp(-c^2x^2)J_\nu(ax)J_\nu(bx)x\,dx = \frac{1}{2c^2}\exp\left(-\frac{a^2+b^2}{4c^2}\right)\mathrm{I}_\nu\left(\frac{ab}{2c^2}\right)$	

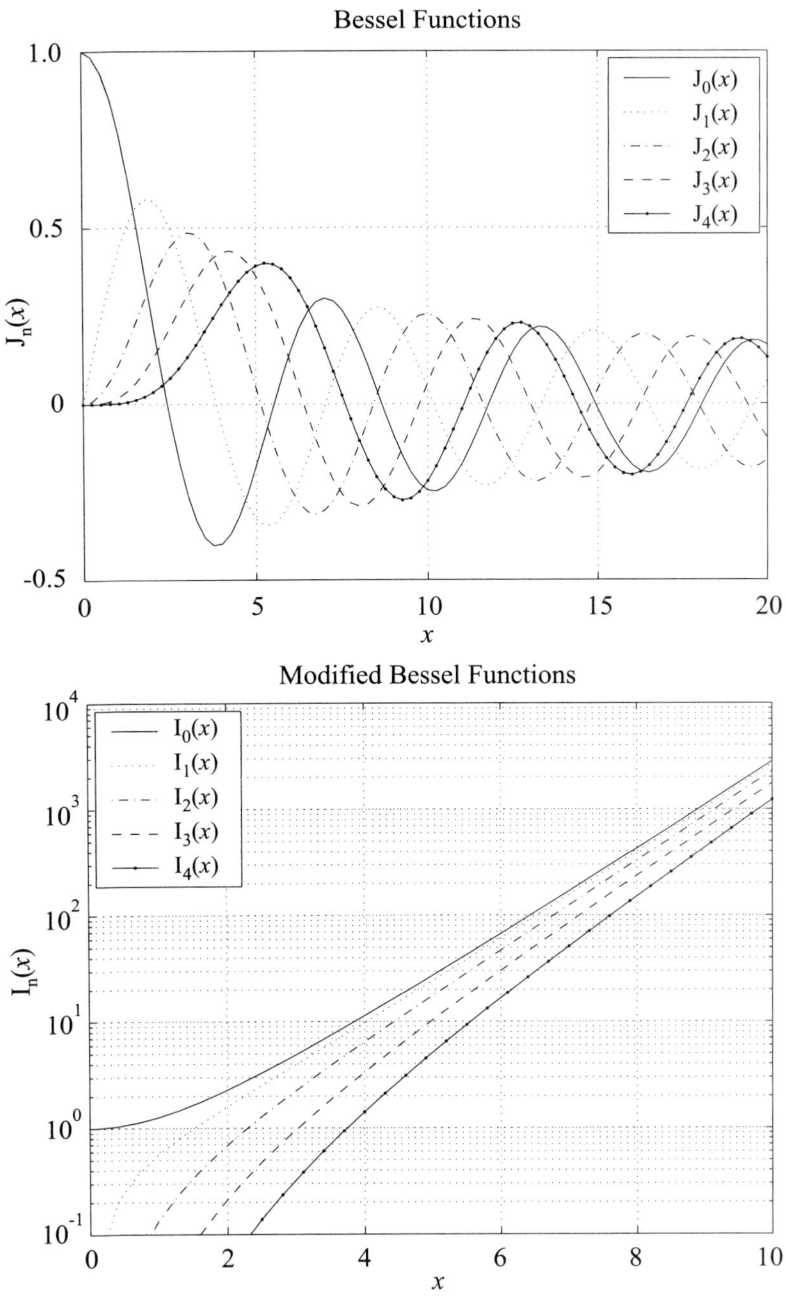

Figure A.2 A graph of Bessel functions and modified Bessel functions (orders 0–4).

Table A.4 Tabulated Values for the Bessel Function, $J_0(x)$, and Modified Bessel Function, $I_0(x)$.

x	$J_0(x)$	x	$J_0(x)$	x	$J_0(x)$	x	$J_0(x)$	x	$J_0(x)$	x	$J_0(x)$
0.0	1.0000	**4.0**	-0.3971	**8.0**	0.1717	**12.0**	0.0477	**16.0**	-0.1749	**20.0**	0.1670
0.2	0.9900	**4.2**	-0.3766	**8.2**	0.1222	**12.2**	0.0908	**16.2**	-0.1893	**20.2**	0.1505
0.4	0.9604	**4.4**	-0.3423	**8.4**	0.0692	**12.4**	0.1296	**16.4**	-0.1960	**20.4**	0.1282
0.6	0.9120	**4.6**	-0.2961	**8.6**	0.0146	**12.6**	0.1626	**16.6**	-0.1948	**20.6**	0.1010
0.8	0.8463	**4.8**	-0.2404	**8.8**	-0.0392	**12.8**	0.1887	**16.8**	-0.1860	**20.8**	0.0700
1.0	0.7652	**5.0**	-0.1776	**9.0**	-0.0903	**13.0**	0.2069	**17.0**	-0.1699	**21.0**	0.0366
1.2	0.6711	**5.2**	-0.1103	**9.2**	-0.1367	**13.2**	0.2167	**17.2**	-0.1472	**21.2**	0.0020
1.4	0.5669	**5.4**	-0.0412	**9.4**	-0.1768	**13.4**	0.2177	**17.4**	-0.1190	**21.4**	-0.0323
1.6	0.4554	**5.6**	0.0270	**9.6**	-0.2090	**13.6**	0.2101	**17.6**	-0.0863	**21.6**	-0.0650
1.8	0.3400	**5.8**	0.0917	**9.8**	-0.2323	**13.8**	0.1943	**17.8**	-0.0506	**21.8**	-0.0949
2.0	0.2239	**6.0**	0.1506	**10.0**	-0.2459	**14.0**	0.1711	**18.0**	-0.0134	**22.0**	-0.1207
2.2	0.1104	**6.2**	0.2017	**10.2**	-0.2496	**14.2**	0.1414	**18.2**	0.0241	**22.2**	-0.1414
2.4	0.0025	**6.4**	0.2433	**10.4**	-0.2434	**14.4**	0.1065	**18.4**	0.0601	**22.4**	-0.1564
2.6	-0.0968	**6.6**	0.2740	**10.6**	-0.2276	**14.6**	0.0679	**18.6**	0.0934	**22.6**	-0.1650
2.8	-0.1850	**6.8**	0.2931	**10.8**	-0.2032	**14.8**	0.0271	**18.8**	0.1226	**22.8**	-0.1671
3.0	-0.2601	**7.0**	0.3001	**11.0**	-0.1712	**15.0**	-0.0142	**19.0**	0.1466	**23.0**	-0.1624
3.2	-0.3202	**7.2**	0.2951	**11.2**	-0.1330	**15.2**	-0.0544	**19.2**	0.1646	**23.2**	-0.1514
3.4	-0.3643	**7.4**	0.2786	**11.4**	-0.0902	**15.4**	-0.0919	**19.4**	0.1759	**23.4**	-0.1344
3.6	-0.3918	**7.6**	0.2516	**11.6**	-0.0446	**15.6**	-0.1253	**19.6**	0.1800	**23.6**	-0.1123
3.8	-0.4026	**7.8**	0.2154	**11.8**	0.0020	**15.8**	-0.1533	**19.8**	0.1770	**23.8**	-0.0858
4.0	-0.3971	**8.0**	0.1717	**12.0**	0.0477	**16.0**	-0.1749	**20.0**	0.1670	**24.0**	-0.0562

x	$I_0(x)$	x	$I_0(x)$	x	$I_0(x)$	x	$I_0(x)$	x	$I_0(x)$	x	$I_0(x)$
0.0	1.0000	**2.0**	2.280	**4.0**	11.302	**6.0**	67.23	**8.0**	427.6	**10.0**	2816
0.1	1.0025	**2.1**	2.446	**4.1**	12.324	**6.1**	73.66	**8.1**	469.5	**10.1**	3096
0.2	1.0100	**2.2**	2.629	**4.2**	13.442	**6.2**	80.72	**8.2**	515.6	**10.2**	3404
0.3	1.0226	**2.3**	2.830	**4.3**	14.668	**6.3**	88.46	**8.3**	566.3	**10.3**	3744
0.4	1.0404	**2.4**	3.049	**4.4**	16.010	**6.4**	96.96	**8.4**	621.9	**10.4**	4117
0.5	1.0635	**2.5**	3.290	**4.5**	17.481	**6.5**	106.29	**8.5**	683.2	**10.5**	4527
0.6	1.0920	**2.6**	3.553	**4.6**	19.093	**6.6**	116.54	**8.6**	750.5	**10.6**	4979
0.7	1.1263	**2.7**	3.842	**4.7**	20.86	**6.7**	127.79	**8.7**	824.4	**10.7**	5477
0.8	1.1665	**2.8**	4.157	**4.8**	22.79	**6.8**	140.14	**8.8**	905.8	**10.8**	6024
0.9	1.2130	**2.9**	4.503	**4.9**	24.91	**6.9**	153.70	**8.9**	995.2	**10.9**	6626
1.0	1.2661	**3.0**	4.881	**5.0**	27.24	**7.0**	168.59	**9.0**	1093.6	**11.0**	7288
1.1	1.3262	**3.1**	5.294	**5.1**	29.79	**7.1**	184.95	**9.1**	1201.7	**11.1**	8018
1.2	1.3937	**3.2**	5.747	**5.2**	32.58	**7.2**	202.9	**9.2**	1320.7	**11.2**	8820
1.3	1.4693	**3.3**	6.243	**5.3**	35.65	**7.3**	222.7	**9.3**	1451.4	**11.3**	9704
1.4	1.5534	**3.4**	6.785	**5.4**	39.01	**7.4**	244.3	**9.4**	1595.3	**11.4**	10676
1.5	1.6467	**3.5**	7.378	**5.5**	42.69	**7.5**	268.2	**9.5**	1753.5	**11.5**	11746
1.6	1.7500	**3.6**	8.028	**5.6**	46.74	**7.6**	294.3	**9.6**	1927.5	**11.6**	12924
1.7	1.8640	**3.7**	8.739	**5.7**	51.17	**7.7**	323.1	**9.7**	2119	**11.7**	14221
1.8	1.9896	**3.8**	9.517	**5.8**	56.04	**7.8**	354.7	**9.8**	2329	**11.8**	15648
1.9	2.128	**3.9**	10.369	**5.9**	61.38	**7.9**	389.4	**9.9**	2561	**11.9**	17219
2.0	2.280	**4.0**	11.302	**6.0**	67.23	**8.0**	427.6	**10.0**	2816	**12.0**	18949

A graph of both elliptic integrals is shown in Figure A.3, and their values are recorded in Table A.5.

> *Note: Sometimes k Instead of m*
>
> In some books and papers, the complete elliptic integrals are defined with the modulus k instead of m, to make $K(k)$ and $E(k)$. If this is the case, an alternate definition of the functions is being used where $m = k^2$.

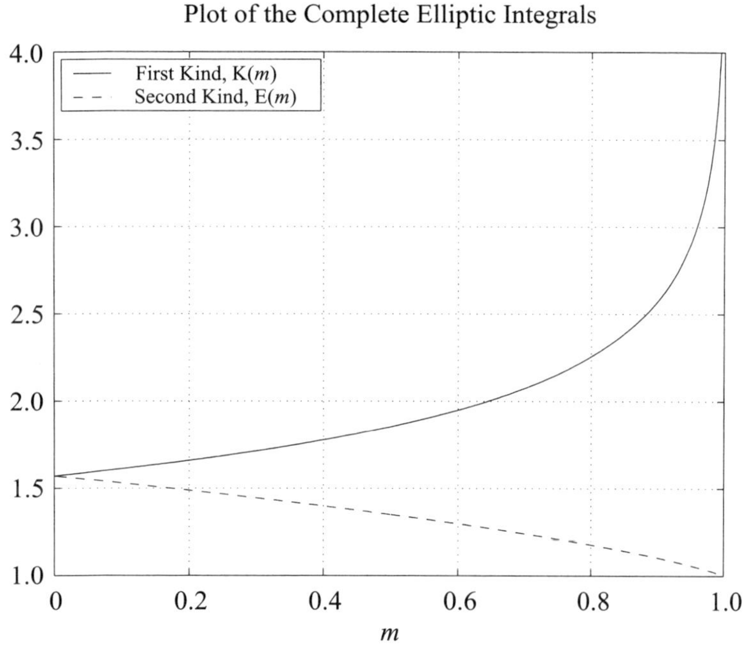

Figure A.3 A graph of complete elliptic integral functions of the first and second kind.

A.6 Q-function

The Q-function is used by engineers for probability analysis involving Gaussian PDFs. The mathematical definition of a Q-function is given by the following definite integral:

$$\mathrm{Q}(x) = \frac{1}{\sqrt{2\pi}} \int_{x}^{\infty} \exp\left(-\frac{z^2}{2}\right) dz \tag{A.6.1}$$

Although the Gaussian CDF cannot be expressed in terms of elementary functions, it may be expressed in terms of a Q-function. For a Gaussian random variable, X,

Table A.5 Tabulated Values for the Complete Elliptic Integrals

m	K(m)	m	K(m)
0.00	1.5708	**0.50**	1.8541
0.02	1.5787	**0.52**	1.8714
0.04	1.5869	**0.54**	1.8895
0.06	1.5952	**0.56**	1.9085
0.08	1.6037	**0.58**	1.9285
0.10	1.6124	**0.60**	1.9496
0.12	1.6214	**0.62**	1.9718
0.14	1.6306	**0.64**	1.9953
0.16	1.6400	**0.66**	2.0203
0.18	1.6497	**0.68**	2.0469
0.20	1.6596	**0.70**	2.0754
0.22	1.6699	**0.72**	2.1059
0.24	1.6804	**0.74**	2.1390
0.26	1.6912	**0.76**	2.1748
0.28	1.7024	**0.78**	2.2140
0.30	1.7139	**0.80**	2.2572
0.32	1.7258	**0.82**	2.3052
0.34	1.7381	**0.84**	2.3593
0.36	1.7508	**0.86**	2.4209
0.38	1.7639	**0.88**	2.4926
0.40	1.7775	**0.90**	2.5781
0.42	1.7917	**0.92**	2.6836
0.44	1.8063	**0.94**	2.8208
0.46	1.8216	**0.96**	3.0161
0.48	1.8375	**0.98**	3.3541
0.50	1.8541	**1.00**	∞

m	E(m)	m	E(m)
0.00	1.5708	**0.50**	1.3506
0.02	1.5629	**0.52**	1.3405
0.04	1.5550	**0.54**	1.3302
0.06	1.5470	**0.56**	1.3198
0.08	1.5389	**0.58**	1.3092
0.10	1.5308	**0.60**	1.2984
0.12	1.5226	**0.62**	1.2875
0.14	1.5143	**0.64**	1.2763
0.16	1.5059	**0.66**	1.2650
0.18	1.4975	**0.68**	1.2535
0.20	1.4890	**0.70**	1.2417
0.22	1.4805	**0.72**	1.2296
0.24	1.4718	**0.74**	1.2173
0.26	1.4631	**0.76**	1.2047
0.28	1.4543	**0.78**	1.1918
0.30	1.4454	**0.80**	1.1785
0.32	1.4364	**0.82**	1.1648
0.34	1.4273	**0.84**	1.1507
0.36	1.4181	**0.86**	1.1360
0.38	1.4088	**0.88**	1.1207
0.40	1.3994	**0.90**	1.1048
0.42	1.3899	**0.92**	1.0879
0.44	1.3803	**0.94**	1.0700
0.46	1.3705	**0.96**	1.0505
0.48	1.3606	**0.98**	1.0286
0.50	1.3506	**1.00**	1.0000

with mean μ and standard deviation σ, the CDF is given by

$$F_X(x) = \Pr\left[X \leq x\right] = 1 - \mathrm{Q}\left(\frac{x-\mu}{\sigma}\right) \tag{A.6.2}$$

A plot of the Q-function is shown in Figure A.4. A Q-function may be evaluated manually using the values in Table A.6. Note that $\mathrm{Q}(-x) = 1 - \mathrm{Q}(x)$.

Note: Error Function Versus Q-Function

The Q-function is used extensively in communications engineering, but is really just another form of the mathematical error function, $\text{erf}(x)$, and complementary error function, $\text{erfc}(x)$. These functions are related by

$$Q(x) = \frac{1}{2} - \frac{1}{2}\text{erf}\left(\frac{x}{\sqrt{2}}\right) = \frac{1}{2}\text{erfc}\left(\frac{x}{\sqrt{2}}\right)$$

As a rule, engineers enjoy the convenient form of the Q-function, while some mathematicians scoff at the use of an infinite limit of integration used in the definition of Equation (A.6.1).

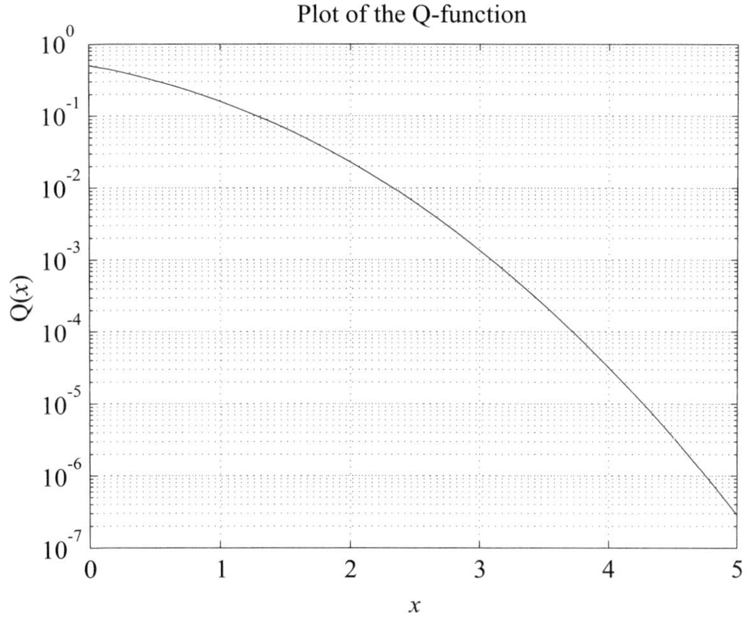

Figure A.4 A graph of the Q-function.

There are two useful approximations for evaluating Q-functions with large arguments. If $x > 3$, the following simple approximation may be used to evaluate $Q(x)$ [Skl01]:

$$Q(x) \approx \frac{1}{x\sqrt{2\pi}} \exp\left(-\frac{x^2}{2}\right) \tag{A.6.3}$$

This approximation is very accurate for the region outside of Table A.6.

Table A.6 Tabulated Values for the Q-function, $Q(x)$

x	$Q(x)$	x	$Q(x)$	x	$Q(x)$	x	$Q(x)$	x	$Q(x)$
0.00	0.5000	**1.00**	0.1587	**2.00**	0.02275	**3.00**	0.001350	**4.00**	0.00003167
0.05	0.4801	**1.05**	0.1469	**2.05**	0.02018	**3.05**	0.001144	**4.05**	0.00002561
0.10	0.4602	**1.10**	0.1357	**2.10**	0.01786	**3.10**	0.0009676	**4.10**	0.00002066
0.15	0.4404	**1.15**	0.1251	**2.15**	0.01578	**3.15**	0.0008164	**4.15**	0.00001662
0.20	0.4207	**1.20**	0.1151	**2.20**	0.01390	**3.20**	0.0006871	**4.20**	0.00001335
0.25	0.4013	**1.25**	0.1056	**2.25**	0.01222	**3.25**	0.0005770	**4.25**	0.00001069
0.30	0.3821	**1.30**	0.09680	**2.30**	0.01072	**3.30**	0.0004834	**4.30**	0.000008540
0.35	0.3632	**1.35**	0.08851	**2.35**	0.009387	**3.35**	0.0004041	**4.35**	0.000006807
0.40	0.3446	**1.40**	0.08076	**2.40**	0.008198	**3.40**	0.0003369	**4.40**	0.000005413
0.45	0.3264	**1.45**	0.07353	**2.45**	0.007143	**3.45**	0.0002803	**4.45**	0.000004294
0.50	0.3085	**1.50**	0.06681	**2.50**	0.006210	**3.50**	0.0002326	**4.50**	0.000003398
0.55	0.2912	**1.55**	0.06057	**2.55**	0.005386	**3.55**	0.0001926	**4.55**	0.000002682
0.60	0.2743	**1.60**	0.05480	**2.60**	0.004661	**3.60**	0.0001591	**4.60**	0.000002112
0.65	0.2578	**1.65**	0.04947	**2.65**	0.004025	**3.65**	0.0001311	**4.65**	0.000001660
0.70	0.2420	**1.70**	0.04457	**2.70**	0.003467	**3.70**	0.0001078	**4.70**	0.000001301
0.75	0.2266	**1.75**	0.04006	**2.75**	0.002980	**3.75**	0.00008842	**4.75**	0.000001017
0.80	0.2119	**1.80**	0.03593	**2.80**	0.002555	**3.80**	0.00007235	**4.80**	0.0000007933
0.85	0.1977	**1.85**	0.03216	**2.85**	0.002186	**3.85**	0.00005906	**4.85**	0.0000006173
0.90	0.1841	**1.90**	0.02872	**2.90**	0.001866	**3.90**	0.00004810	**4.90**	0.0000004792
0.95	0.1711	**1.95**	0.02559	**2.95**	0.001589	**3.95**	0.00003908	**4.95**	0.0000003711
1.00	0.1587	**2.00**	0.02275	**3.00**	0.001350	**4.00**	0.00003167	**5.00**	0.0000002867

Appendix B

FOURIER ANALYSIS

Basic Fourier concepts may not be difficult for a seasoned engineer, but hard-core analysis and calculation is rife with difficulties due to the numerous transform conventions that exist in the literature. In the field of space–time channel modeling, this problem is exacerbated by joint analysis of two or more pairs of spectral domains. Conventions, however, are purely mathematical and can be overcome by keeping a few simple rules in mind. This section is intended to be a recipe book for defining, calculating, and manipulating Fourier transform pairs *regardless of their defined conventions.* The results will assist analysis in spectral domains such as the time–frequency, time–Doppler, frequency–delay, position–wavenumber, and others used in this book or in other science and engineering texts.

B.1 Basic Fourier Transform Definitions

All analytical communication theory begins with the basic time–frequency Fourier transform definitions. A time domain function, $s(t)$, has a unique transform pair, $S(f)$, in the frequency domain. The standard definitions for the forward transform (from time to frequency) and the inverse transform (from frequency to time) are given below:

$$\text{Forward Transform:} \quad S(f) = \int_{-\infty}^{+\infty} s(t) \exp(-j2\pi f t)\, dt \tag{B.1.1}$$

$$\text{Inverse Transform:} \quad s(t) = \int_{-\infty}^{+\infty} S(f) \exp(j2\pi f t)\, df \tag{B.1.2}$$

These definitions are general for complex-valued $s(t)$ and $S(f)$. Table B.1 and Table B.2 show a list of time-frequency Fourier transform pairs that have a closed analytical form.

In addition to the basic definitions, it is also useful to understand the many mathematical manipulations that can be made on Fourier transform pairs (linear addition, scaling, differentiation, and time-shifting, to name just a few). If a manipulation is performed in one domain, a corresponding operation occurs on the transform pair. The basic relationships between time and frequency domain operations are summarized in Table B.3. By using Table B.3 in conjunction with Table B.1 and Table B.2, an endless number of mathematical manipulations can be made on the base and spectral domains of analytical transform pairs without resorting to the basic transform integrations.

Example B.1: Frequency Domain to Time Domain

Problem: Find the time-domain function corresponding to this "truncated ramp" spectrum:

$$S(f) = f\mathrm{u}\left(\frac{W}{2} - |f|\right)$$

Solution: This problem becomes much easier if we work with the *derivative* of $S(f)$:

$$\frac{dS(f)}{df} = -\frac{W}{2}\delta\left(f + \frac{W}{2}\right) + \mathrm{u}\left(\frac{W}{2} - |f|\right) - \frac{W}{2}\delta\left(f - \frac{W}{2}\right)$$

Now we can apply linearity and solve for the Fourier transform in parts. From Table B.1 we recognize the cosine/even-delta transform pair:

$$\cos(\pi W t) \longleftrightarrow \frac{1}{2}\left[\delta\left(f + \frac{W}{2}\right) + \delta\left(f - \frac{W}{2}\right)\right]$$

Using the duality operation, we also recognize the sinc/box transform pair:

$$W\mathrm{sn}(Wt) \longleftrightarrow \mathrm{u}\left(\frac{W}{2} - |f|\right)$$

Thus, the transform pair for the derivative is given by

$$\frac{dS(f)}{df} \longleftrightarrow W\left[\mathrm{sn}(Wt) - \cos(\pi W t)\right]$$

The operation of integration in the frequency domain, according to Table B.3, is equivalent to multiplication by $\frac{-1}{j2\pi t}$ in the time domain (the negative sign is due to duality). The final answer is given by

$$S(f) \longleftrightarrow \frac{jW}{2\pi t}\left[\mathrm{sn}(Wt) - \cos(\pi W t)\right]$$

Note: The Physics Convention

In physics literature, there is often a different convention for defining Fourier transform pairs. Besides using $i^2 = -1$ instead of $j^2 = -1$ to denote imaginary components, the conjugates of the complex exponential functions are used:

$$S(f) = \int_{-\infty}^{+\infty} s(t) \exp(i2\pi f t)\, dt \qquad s(t) = \int_{-\infty}^{+\infty} S(f) \exp(-i2\pi f t)\, df$$

The physics Fourier conventions, like the engineering conventions used in this book, are entirely self-consistent. The only problems arise when mixing the two conventions in the same set of calculations.

B.2 Time–Doppler Transforms

The cut-and-dry analysis of time–frequency transforms becomes more difficult when we perform the time–Doppler/frequency–delay split in channel modeling or add a space–wavenumber dependency. Mathematically, however, the differences between defining time-frequency transform pairs and other-domain transform pairs is merely cosmetic. For the case of time–Doppler transforms, $s(t) \longleftrightarrow S(\omega)$, the only difference is that Doppler domain is in units of radian frequency. Thus, the standard time–frequency relationships can be used, provided a substitution $\omega = 2\pi f$ is made. The basic steps for this manipulation is summarized in Table B.4.

Example B.2: Doppler Domain to Time Domain

Problem: Find the time domain representation of the Doppler domain function $S(\omega) = \frac{7}{2+\omega}$.

Solution: Note from Table B.4 that we must first substitute $\omega = 2\pi f$:

$$S(f) = \frac{7}{2 + 2\pi f}$$

Now the transform tables can be used. The corresponding time-domain function is the exponential function for $a = 2$:

$$s(t) = 7 \exp(-2t)\mathrm{u}(t)$$

According to Table B.4, this is the final answer. No more modification is needed.

B.3 Frequency–Delay Transforms

Frequency–delay Fourier transform pairs are also easy to adapt for use with the time–frequency tables. The units of frequency (Hz) and delay (seconds) do not constitute a radian transform pair, so this simplifies the application. Realize, however,

Table B.1 Elementary Fourier Transform Pairs

Time Domain		Frequency Domain	
Constant	A	$A\delta(f)$	Impulse
Complex Exponent	$\exp(j2\pi f_0 t)$	$\delta(f - f_0)$	Shifted Impulse (f_0)
Cosine	$\cos(2\pi f_0 t)$	$\frac{1}{2}[\delta(f-f_0) + \delta(f+f_0)]$	Even Deltas
Sine	$\sin(2\pi f_0 t)$	$\frac{1}{2j}[\delta(f-f_0) - \delta(f+f_0)]$	Odd Deltas (Im)
Unit Step (1)	$\mathrm{u}(t)$	$\frac{1}{2}\delta(f) + \frac{1}{j2\pi f}$	|S(f)|
Signum (1, -1)	$2\mathrm{u}(t) - 1$	$\frac{1}{j\pi f}$	|S(f)|
Box (T, 1)	$\mathrm{u}(\frac{T}{2} - \|t\|)$	$T\mathrm{sn}(Tf)$	Sinc
Triangle (T, 1)	$(1 - \frac{\|t\|}{T})\,\mathrm{u}(T - \|t\|)$	$T\mathrm{sn}^2(Tf)$	Squared Sinc
Exponential	$\exp(-at)\,\mathrm{u}(t)$	$\frac{1}{a+j2\pi f}$ for $a > 0$	|S(f)|
Ramped Exp.	$t^\nu \exp(-at)\,\mathrm{u}(t)$	$\frac{\Gamma(\nu+1)}{(a+j2\pi f)^{\nu+1}}$ for $a > 0$	|S(f)|
Damped Cosine	$\exp(-at)\cos(\omega_0 t)\,\mathrm{u}(t)$	$\frac{a+j2\pi f}{(a+j2\pi f)^2+\omega_0^2}$ for $a \geq 0$	|S(f)|
Damped Sine	$\exp(-at)\sin(\omega_0 t)\,\mathrm{u}(t)$	$\frac{\omega_0 - ja}{(a+j2\pi f)^2+\omega_0^2}$ for $a \geq 0$	|S(f)|
Damp + Ramp Cos	$t^\nu \exp(-at)\cos(\omega_0 t)\,\mathrm{u}(t)$	$\frac{\Gamma(\nu+1)\left[(a+j2\pi f-j\omega_0)^{\nu+1}+(a+j2\pi f+j\omega_0)^{\nu+1}\right]}{2\left[(a+j2\pi f)^2+\omega_0^2\right]^{\nu+1}}$	
Damp + Ramp Sin	$t^\nu \exp(-at)\sin(\omega_0 t)\,\mathrm{u}(t)$	$\frac{\Gamma(\nu+1)\left[(a+j2\pi f+j\omega_0)^{\nu+1}-(a+j2\pi f-j\omega_0)^{\nu+1}\right]}{2j\left[(a+j2\pi f)^2+\omega_0^2\right]^{\nu+1}}$	

Table B.2 Advanced Fourier Transform Pairs

TIME DOMAIN		FREQUENCY DOMAIN	
Double Exp. 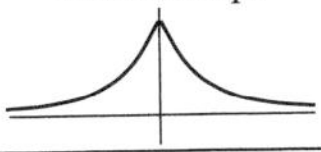	$\exp(-a\|t\|)$	$\frac{2a}{a^2+4\pi^2 f^2}$ for $a > 0$	Reciprocal Parabola 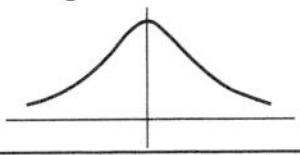
Gaussian 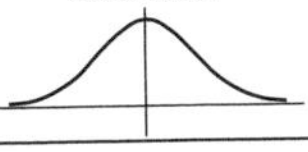	$\exp(-at^2)$	$\sqrt{\frac{\pi}{a}}\exp\left(-\frac{\pi^2 f^2}{a}\right)$ for $a > 0$	Gaussian 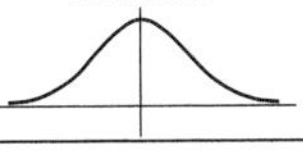
Chirp 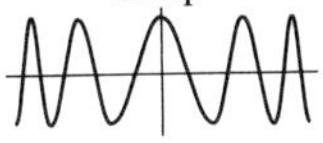	$\cos(\omega_0^2 t^2)$ $\sin(\omega_0^2 t^2)$	$\frac{\sqrt{\pi}}{\|\omega_0\|}\cos\left(\frac{f^2\pi^2}{\omega_0^2}-\frac{\pi}{4}\right)$ $\frac{\sqrt{\pi}}{\|\omega_0\|}\sin\left(\frac{f^2\pi^2}{\omega_0^2}-\frac{\pi}{4}\right)$	Chirp 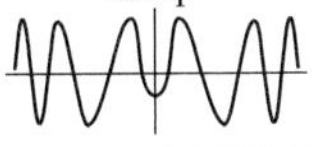
Gaussian Chirp 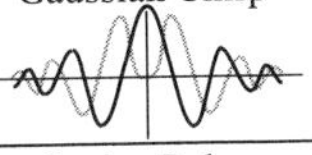	$\exp(-\kappa\exp(j\theta)t^2)$	$\sqrt{\frac{\pi}{\kappa}}\exp\left(-\frac{\pi^2 f^2}{\kappa\exp(j\theta)}-\frac{j\theta}{2}\right)$ $-\frac{\pi}{2}\le\theta\le\frac{\pi}{2}$ and $\kappa\ge 0$	Gaussian Chirp 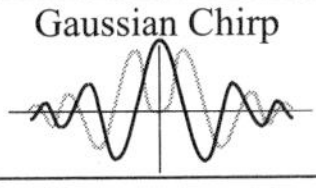
Cosine Pulse 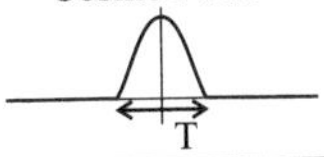	$\cos\left(\frac{\pi t}{T}\right)\mathrm{u}\left(\frac{T}{2}-\|t\|\right)$	$\frac{T}{2}\left[\mathrm{sn}(Tf+\frac{1}{2})+\mathrm{sn}(Tf-\frac{1}{2})\right]$	Offset Sincs 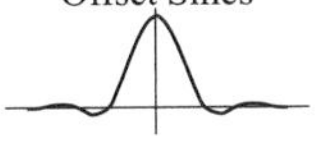
U-Shape 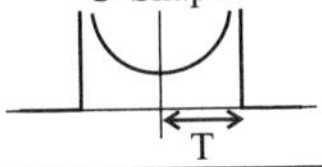	$\frac{\mathrm{u}(T-\|t\|)}{\sqrt{1-\left(\frac{t}{T}\right)^2}}$	$\pi T\mathrm{J}_0(2\pi Tf)$	Bessel 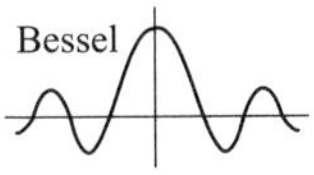
Half Circle 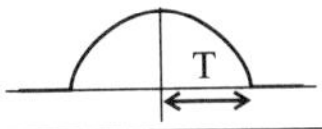	$\sqrt{1-\left(\frac{t}{T}\right)^2}\mathrm{u}(T-\|t\|)$	$\frac{1}{2f}\mathrm{J}_1(2\pi Tf)$	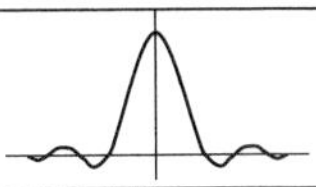
Generic Pulse 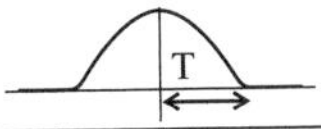	$\frac{\mathrm{u}(T-\|t\|)}{\left[1-\left(\frac{t}{T}\right)^2\right]^{\frac{1}{2}-\nu}}$	$\frac{2\pi T\Gamma(2\nu)}{(4\pi Tf)^{\nu}\Gamma(\nu)}\mathrm{J}_\nu(2\pi Tf)$	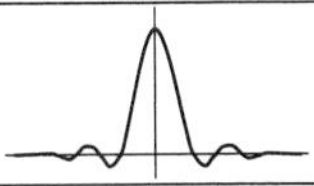
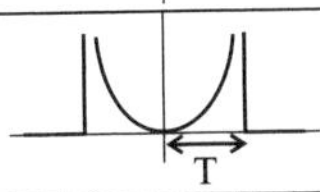	$\frac{\left(\frac{t}{T}\right)^2\mathrm{u}(T-\|t\|)}{\sqrt{1-\left(\frac{t}{T}\right)^2}}$	$\frac{\pi T}{2}\left[\mathrm{J}_0(2\pi Tf)-\mathrm{J}_2(2\pi Tf)\right]$	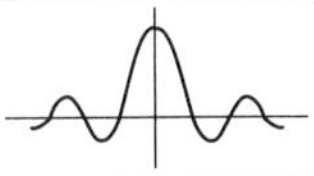
Rippled Bowl 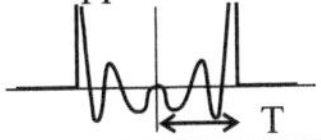	$\frac{\cos\left(n\cos^{-1}\frac{t}{T}\right)\mathrm{u}(T-\|t\|)}{\sqrt{1-\left(\frac{t}{T}\right)^2}}$	$\frac{\pi T}{j^n}\mathrm{J}_n(2\pi Tf)$	Bessel 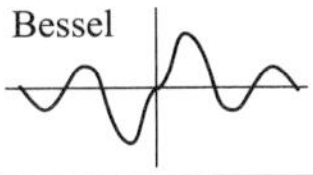
Elliptic K 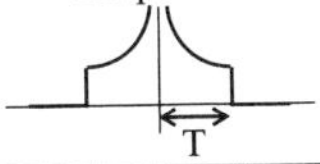	$\mathrm{K}\left(\frac{\sqrt{T^2-t^2}}{T}\right)\mathrm{u}(T-\|t\|)$	$\frac{\pi^2 T}{2}\mathrm{J}_0^2(\pi Tf)$	Squared Bessel 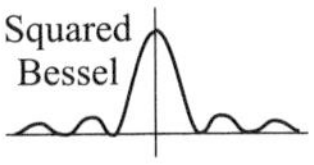
Impulse Train 	$\sum_{n=-\infty}^{+\infty}\delta(t-nT)$	$\frac{1}{T}\sum_{n=-\infty}^{+\infty}\delta\left(f-\frac{n}{T}\right)$	Impulse Train 
Raised Cosine Pulse 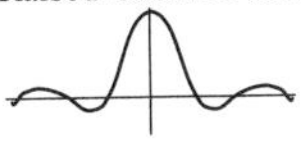	$\frac{2f_0\mathrm{sn}(2f_0t)\cos(2\pi\kappa f_0t)}{1-(4\kappa f_0t)^2}$	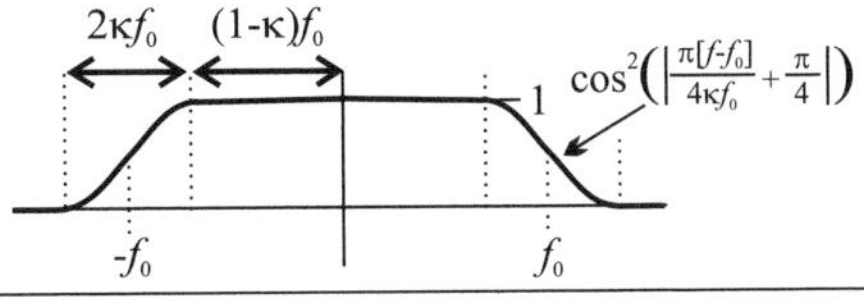	

Table B.3 Properties of the Fourier Transform

PROPERTY	TIME DOMAIN	FREQUENCY DOMAIN
Basic Pair Notation	lowercase function, $s(t)$	uppercase function, $S(f)$
Transformation Definition	$\int_{-\infty}^{+\infty} S(f)\exp(j2\pi ft)\,df$	$\int_{-\infty}^{+\infty} s(t)\exp(-j2\pi ft)\,dt$
Duality	$S(t)$	$s(-f)$
Linearity	$Ax(t)+By(t)$	$AX(f)+BY(f)$
Scaling	$s(at)$	$\frac{1}{a}S(\frac{f}{a})$
Conjugation	$s^*(t)$	$S^*(-f)$
Time Reversal	$s(-t)$	$S^*(f)$
Time Shift	$s(t-t_0)$	$S(f)\exp(-j2\pi ft_0)$
Frequency Shift	$s(t)\exp(j2\pi f_0t)$	$S(f-f_0)$
Cosine Modulation	$s(t)\cos(2\pi f_0t)$	$\frac{1}{2}[S(f-f_0)+S(f+f_0)]$
Sine Modulation	$s(t)\sin(2\pi f_0t)$	$\frac{1}{2j}[S(f-f_0)-S(f+f_0)]$
Convolution	$\int_{-\infty}^{+\infty} x(\zeta)y(t-\zeta)\,d\zeta$	$X(f)Y(f)$
Multiplication	$x(t)y(t)$	$\int_{-\infty}^{+\infty} X(\zeta)Y(f-\zeta)\,d\zeta$
Differentiation	$\frac{d^n s(t)}{dt^n}$	$(j2\pi f)^n S(f)$
Integration	$\int_{-\infty}^{t} s(\tau)\,d\tau$	$\frac{1}{j2\pi f}S(f)$
Time Replication	$\sum_{n=-\infty}^{+\infty} s(t-nT)$	$\frac{1}{T}\sum_{n=-\infty}^{+\infty} S(\frac{n}{T})\delta(f-\frac{n}{T})$
Time Sampling	$\sum_{n=-\infty}^{+\infty} s(nT)\delta(t-nT)$	$\frac{1}{T}\sum_{n=-\infty}^{+\infty} S(f-\frac{n}{T})$

Table B.4 Steps for Calculating a Time–Doppler Fourier Transform

	PRETRANSFORM	TABLE LOOKUP	POSTTRANSFORM
Forward	$s(t)$	$s(t) \rightarrow S(f)$	$S(\omega) = S(f)\Big\vert_{f=\frac{\omega}{2\pi}}$
Inverse	$S(f) = S(\omega)\Big\vert_{\omega=2\pi f}$	$s(t) \leftarrow S(f)$	$s(t)$

Forward Transform: Calculate the forward transform of $s(t)$ from the tables to get $S(f)$. Then make the substitution $f = \frac{\omega}{2\pi}$.

Inverse Transform: Starting with $S(\omega)$, make the substitution $\omega = 2\pi f$. Now use the tables to calculate the time-domain function, $s(t)$.

that frequency–delay channel analysis is *backwards* from the conventional time–frequency definitions. Therefore, we have to apply the rules in Table B.5 before using the conventional Fourier transform tables.

Table B.5 Steps for Calculating a Frequency–Delay Fourier Transform

	PRETRANSFORM	TABLE LOOKUP	POSTTRANSFORM
Forward	$S(f) = s(f)$	$s(t) \leftarrow S(f)$	$S(\tau) = s(t)\Big\vert_{t=\tau}$
Inverse	$s(t) = S(\tau)\Big\vert_{\tau=t}$	$s(t) \rightarrow S(f)$	$s(f) = S(f)$

Forward Transform: Because the polarity of the frequency-delay transform is reversed, the frequency domain is the base domain. Starting with the frequency–domain function, $s(f)$, find this function form on the right-hand side of the transform tables. Its corresponding time-domain function, $S(t)$, is the forward transform. Once you find it, substitute τ for t.

Inverse Transform: Starting with a function, $S(\tau)$, substitute t for τ. Then transform this to the frequency domain using the function form in the left-hand side of the transform tables.

Example B.3: Frequency Domain to Delay Domain

Problem: Calculate the delay-domain transform of the cosine pulse function, $s(f) = -10\cos\left(\frac{\pi f}{W}\right) \mathrm{u}\left(\frac{W}{2} - |f|\right)$.

Solution: From Table B.5, we should prepare for the forward transform by treating $s(f)$ as if it were in the frequency domain:

$$S(f) = -10\cos\left(\frac{\pi f}{W}\right) \mathrm{u}\left(\frac{W}{2} - |f|\right)$$

In this way, the forward transform from frequency to delay is actually an *inverse* transform from frequency to time in Table B.1. Applying duality, we can use the cosine pulse/offset sinc pair:

$$s(t) = -5W\left[\mathrm{sn}\left(Wt + \frac{1}{2}\right) + \mathrm{sn}\left(Wt - \frac{1}{2}\right)\right] \longleftrightarrow S(f)$$

After the transform is complete, Table B.5 instructs us to substitute τ for t and assume that this is the delay transform, $S(\tau)$:

$$S(\tau) = -5W\left[\mathrm{sn}\left(W\tau + \frac{1}{2}\right) + \mathrm{sn}\left(W\tau - \frac{1}{2}\right)\right]$$

B.4 Space–Wavenumber Transforms

The space–wavenumber Fourier pairs are very similar to the time–Doppler pairs, since both are radian frequency transforms. The procedure for performing the transform is given in Table B.6.

Table B.6 Steps for Calculating a Space–Wavenumber Fourier Transform

	PRETRANSFORM	TABLE LOOKUP	POSTTRANSFORM
Forward	$s(t) = s(r)\Big\vert_{r=t}$	$s(t) \rightarrow S(f)$	$S(k) = S(f)\Big\vert_{f=\frac{k}{2\pi}}$
Inverse	$S(f) = S(k)\Big\vert_{k=2\pi f}$	$s(t) \leftarrow S(f)$	$s(r) = s(t)\Big\vert_{t=r}$

Forward Transform: Starting with a space-varying function, $s(r)$, replace r with the variable t. Calculate the forward transform of $s(t)$ from the transform tables to get $S(f)$. Then make the substitution $f = \frac{k}{2\pi}$.

Inverse Transform: Starting with $S(k)$, make the substitution $k = 2\pi f$. Now use the transform tables to calculate the time-domain function, $s(t)$.

Example B.4: Space Domain to Wavenumber Domain

Problem: Find the Fourier transform in the wavenumber domain for the function of linear space, $s(r) = \mathrm{J}_0(2\pi r)$.

Solution: According to Table B.6, start with a time-domain function, $s(t)$, by performing a simple $r = t$ substitution: $s(t) = \mathrm{J}_0(2\pi t)$. Now we can begin the transformation to the frequency domain using the tables. We will use the Bessel/U-shape pairs:

$$\mathrm{J}_0(2\pi t) \longleftrightarrow \frac{\mathrm{u}(1 - |f|)}{\pi\sqrt{1 - f^2}}$$

The posttransform operation is a substitution of $f = \frac{k}{2\pi}$ to produce

$$S(k) = \frac{\mathrm{u}(1 - |\frac{k}{2\pi}|)}{\sqrt{\pi^2 - (\frac{k}{2})^2}}$$

B.5 Trigonometric Relationships

Trigonometric relationships are often used in the calculation of Fourier transform pairs. Table B.7 has been included as a general summary of useful trigonometric relations.

Table B.7 Useful Trigonometric Identities

Geometrical Definitions $\cos(\theta) = \frac{x}{r}$ $\sin(\theta) = \frac{y}{r}$ $\tan(\theta) = \frac{y}{x}$ 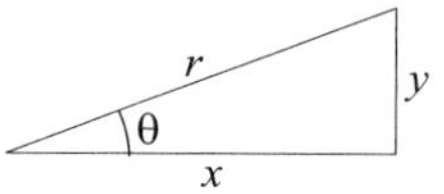 $\cos^2(\theta) + \sin^2(\theta) = 1$	**Complex Exponents** $\exp(j\theta) = \cos(\theta) + j\sin(\theta)$ $\cos(\theta) = \frac{\exp(j\theta)+\exp(-j\theta)}{2}$ $\sin(\theta) = \frac{\exp(j\theta)-\exp(-j\theta)}{j2}$
Double-Angle Formulas $\cos(2\theta) = \begin{cases} 2\cos^2(\theta) - 1 \text{ or} \\ 1 - 2\sin^2(\theta) \text{ or} \\ \cos^2(\theta) - \sin^2(\theta) \end{cases}$ $\sin(2\theta) = 2\cos(\theta)\sin(\theta)$	**Series Expansion** $\exp(j\theta) = \sum_{n=0}^{\infty} \frac{(j\theta)^n}{n!}$ $\cos(\theta) = \sum_{n=0}^{\infty} \frac{(-1)^n \theta^{2n}}{(2n)!}$ $\sin(\theta) = \sum_{n=0}^{\infty} \frac{(-1)^n \theta^{2n+1}}{(2n+1)!}$
Addition Formulas $\cos(\theta_1+\theta_2) = \cos(\theta_1)\cos(\theta_2) - \sin(\theta_1)\sin(\theta_2)$ $\sin(\theta_1+\theta_2) = \cos(\theta_1)\sin(\theta_2) + \sin(\theta_1)\cos(\theta_2)$	**Half-Angle Formulas** $\cos\left(\frac{\theta}{2}\right) = \sqrt{\frac{1+\cos(\theta)}{2}}$ $\sin\left(\frac{\theta}{2}\right) = \sqrt{\frac{1-\cos(\theta)}{2}}$
Inverse Derivatives $\frac{d\sin^{-1}(x)}{dx} = \frac{1}{\sqrt{1-x^2}}$ for $\|x\| \leq 1$ $\frac{d\cos^{-1}(x)}{dx} = \frac{-1}{\sqrt{1-x^2}}$ for $\|x\| \leq 1$ $\frac{d\tan^{-1}(x)}{dx} = \frac{1}{1+x^2}$	**Derivatives** $\frac{d\sin(\theta)}{d\theta} = \cos(\theta)$ $\frac{d\cos(\theta)}{d\theta} = -\sin(\theta)$ $\frac{d\tan(\theta)}{d\theta} = 1 + \tan^2(\theta)$

Appendix C

RANDOM PROCESS THEORY

This appendix reviews the basic terminology and definitions of random process theory. Our discussion is by no means comprehensive, so a reader without prior exposure to probability or random process theory should consult a fundamental text in the field.

C.1 Definitions

Most probability theory is based on the concept of a random variable. This section presents the definitions and terminology associated with random variables and random processes.

C.1.1 Random Variables

An *experiment*, in terms of probability theory, is any set of repeatable procedures that produces an unpredictable *outcome*. Examples of simple experiments are flipping a coin, rolling a die, or driving to work through traffic. For the coin, the outcome is either heads or tails. For a standard die, the outcome is one of the six sides facing upwards when the die comes to rest. For a commute to work, the outcome may be defined as the time required to reach the job site.

With the terminology of *experiment* and *outcome*, it is possible to give the following formal definition for a random variable:

> A **random variable** maps the **outcome** of an **experiment** to a number.

For the case of the coin, the two possible outcomes are abstract concepts: *heads* or *tails*. We cannot apply analytical mathematics to this situation unless we assign a numerical value to these results. To remedy this, we define a random variable, C, that takes on the value of 0 if the outcome of the coin experiment is heads and 1 if the outcome of the coin experiment is tails.

The mapping of a random variable is completely arbitrary. We could just as easily have assigned the value of π for the outcome of heads and the value of -200 for tails, but convenience usually dictates a logical mapping. For example, the die experiment produces an upward face with 1 through 6 dots, so it is natural to construct a random variable that takes on values of 1, 2, 3, 4, 5, and 6 for each corresponding case. Likewise, for experiments with a continuum of outcomes (as opposed to the discrete outcomes of the coin and die experiments), the mapping is usually related to the physical units of the outcome. Random fading channels, for example, are mapped by a random variable to a numerical value that represents either the *voltage* or *current* induced at the terminals of an antenna.

We are interested in doing more than listing the outcomes of experiments; we want to assign *probabilities* to *events*. An event is simply a specified set of outcomes. For the die experiment, an event could be simply a roll of 1; or an event could be a roll of 3 and 6; or any number besides 2. For a random channel experiment, an event may be the received voltage envelope dropping below a certain threshold voltage.

Although there are many ways to define the term *probability*, the most straightforward is in terms of frequency of occurrence. A *probability* is a number between 0 and 1 that describes how often an event occurs. The typical notation relating the probability of an event is given below:

$$\Pr[\text{Event}] = \text{Numerical Probability}$$

So, to describe the probability of an event in the die experiment, let D be the random variable mapping the outcome to the corresponding number, 1 through 6. (The convention is to use a capitalized letter to denote real random variables.) Let a roll of 3 be the event we characterize. This event is expressed mathematically as $D = 3$. We know that if the die is fair, a roll of any side has equal probability. We would then write the probability of this event as

$$\Pr[D = 3] = \frac{1}{6}$$

You can read the above notation as "the probability of the random variable D being equal to 3 is one-sixth."

C.1.2 Random Processes

A continuous complex random process, $\tilde{x}(t;\zeta)$, may be thought of as a continuum of random variables strung together in time. Each point in time of the function $\tilde{x}(t;\zeta)$ represents a single random variable. The properties of each of these random variables along with their interdependent characteristics define the behavior of the random process.

Note that the variable ζ was appended to the argument of the random process $\tilde{x}(t;\zeta)$; this variable is the *ensemble index*. A random process is like a random variable in that both represent experimental output. Since a random process is either a sequence or continuum of random variables, each point on a single random process is an *outcome*. The collection of outcomes for a single random process experiment is called a *realization*. (Table C.1 provides a helpful summary of the definitions used to describe random variables and random processes.)

The set of all possible realizations of a random process is called an *ensemble*. The value ζ provides an index into which a single realization may be specified. For example, $\tilde{x}(t;\zeta_1)$ and $\tilde{x}(t;\zeta_2)$ are two different time-varying functions of the random process $\tilde{x}(t;\zeta)$. Figure C.1 illustrates this concept of the ensemble of a complex random process.

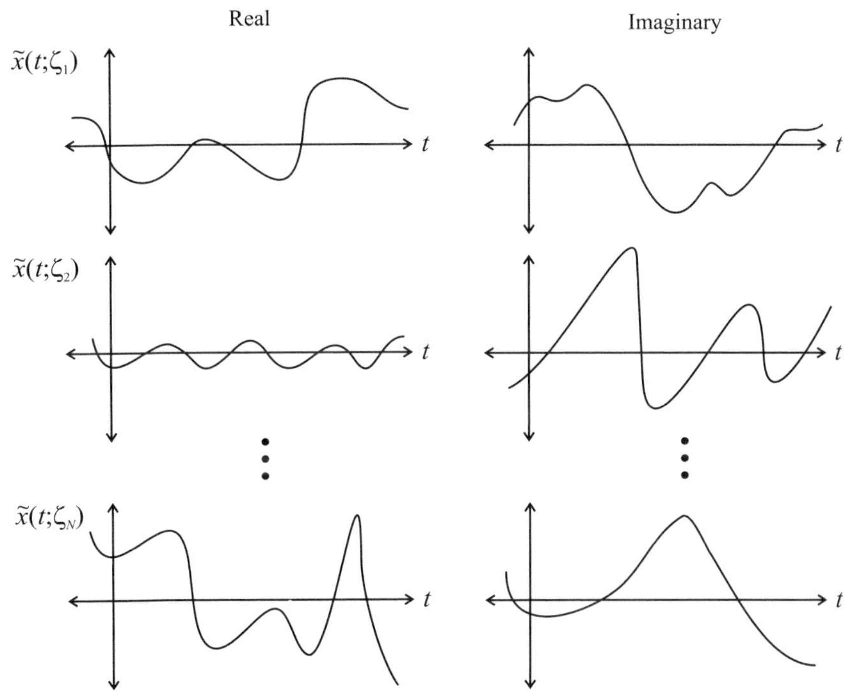

Figure C.1 The time-varying random process, $\tilde{x}(t;\zeta)$, has a stochastic ensemble consisting of many individual realizations.

> *Note: Lose the* ζ
>
> As is the convention in random process theory, the ensemble index ζ will be dropped from the representation of most random processes after this section. The presence of ζ is implicit if we simply write $\tilde{x}(t)$. Otherwise, the notation becomes cumbersome when we begin discussing multivariable random processes.

Table C.1 Glossary of Random Process Terminology

ensemble	set of all possible realizations of a random process
event	set of experimental outcomes
experiment	repeatable set of procedures resulting in an outcome
outcome	distinct result of an experiment
random process	mapping of experimental outcomes to functions
random variable	mapping of experimental outcomes to numbers
realization	outcome of a process experiment
trial	single performance of an experiment

C.2 Probability Density Functions

This section describes the class of functions used to describe first-order statistics of random variables.

C.2.1 Definitions

A random variable may be characterized by either a *cumulative distribution function* (CDF) or a *probability density function* (PDF). The CDF represents the probability that a random variable lies below a threshold. Formally, we define the CDF of a random variable X as

$$F_X(x) = \Pr[X \leq x]$$

For this case, the event is an outcome, X, that falls beneath the threshold x. Based on this definition, we can state the following properties about a CDF:

1. $F_X(-\infty) = 0$
2. $F_X(+\infty) = 1$
3. $0 \leq F_X(x) \leq 1$ for all x
4. $F_X(x_2) \geq F_X(x_1)$ for all $x_2 \geq x_1$

All CDFs must have these properties.

> *Note: Conventions of CDFs and PDFs*
>
> In this book, CDFs are always denoted as F, and PDFs are always denoted as f, with their subscript denoting the random variable they describe. The lowercase version of the random variable is used in the argument of each. Thus, the function $F_U(u)$ is recognized as the CDF of the random variable U - and the PDF for this same random variable would logically be $f_U(u)$.

The PDF is defined as the derivative of the CDF. The PDF $f_X(x)$ may be calculated from the CDF using the following equation:

$$f_X(x) = \frac{dF_X(x)}{dx} \tag{C.2.1}$$

It is important to remember that while the CDF has units of probability, a PDF has units of *marginal probability* (probability per unit of X). As a result, the value of $f_X(x)$ does not have to be less than 1, although it must always be nonnegative.

Both PDF and CDF may be used to calculate the probability of a *range event*, such as the probability of X lying on the interval $(x_1, x_2]$. If we desire to use the CDF for this calculation, we can write

$$\Pr[x_1 < X \leq x_2] = F_X(x_2) - F_X(x_1) \tag{C.2.2}$$

If we desire to use the PDF for this calculation, we can write

$$\Pr[x_1 < X \leq x_2] = \int_{x_1}^{x_2} f_X(x)\, dx \tag{C.2.3}$$

The calculation is simpler if we have the CDF (no integration involved). However, there are some very useful features of a PDF that make it the primary function for characterizing random variables.

A graph of the PDF helps us visualize the frequency of occurrence of a continuous random variable. Let us construct a *histogram* of the random variable X. A histogram is a plot of the numerical outcomes of many trials of the same experiment. The resulting values are grouped by "bins" - a set of ranges that span all possible values of X - and are tabulated in a bar graph.

Examples of histograms are shown in Figure C.2. These examples compare the PDF and the histogram of an exponentially distributed random variable, X. The mathematical definition for the exponential PDF is

$$f_X(x) = \alpha \exp(-\alpha x) \mathrm{u}(x) \tag{C.2.4}$$

where α is a positive constant. As we perform more trials of this experiment, the shape of the histogram becomes smoother. For numerous trials, the shape of the histogram approaches that of the PDF.

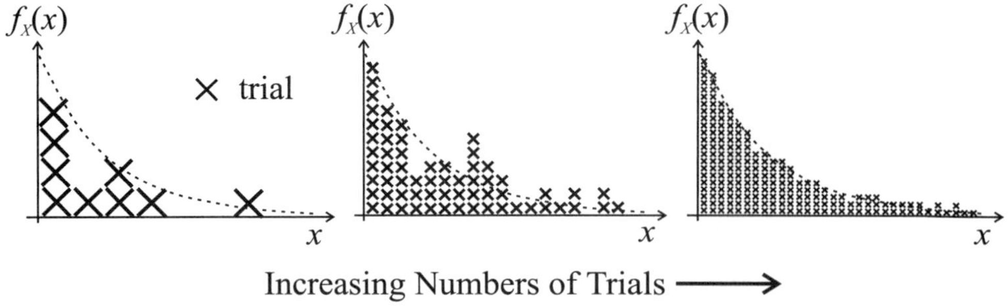

Figure C.2 A histogram of trials begins to resemble the PDF as the number of trials increases.

C.2.2 Joint Distributions

We also use joint CDFs and PDFs to describe the distributions of two or more random variables. Joint CDFs are defined in a similar manner to their single-variable counterparts. For a CDF of random variables X and Y, the probabilistic definition is

$$F_{XY}(x, y) = \Pr[X \leq x \text{ and } Y \leq y]$$

Again, the joint PDF is the derivative of the joint CDF - this time with respect to both random variables:

$$f_{XY}(x, y) = \frac{\partial^2 F_{XY}(x, y)}{\partial y \, \partial x}$$

We could easily extend these definitions to joint PDFs and joint CDFs of N random variables.

In the study of baseband channel models, the joint PDF has a very useful application: characterizing a *complex* random variable. Since a complex random variable may be represented as a pair of real-valued random variables, the behavior of a complex random variable is best characterized by a joint PDF. We may choose to use real and imaginary component random variables (X,Y) or magnitude and phase random variables (R,Φ) to represent the complex random variable:

$$Z = X + jY = R \exp(j\Phi)$$

The real and imaginary components correspond to Cartesian coordinates in the complex plane. The magnitude and phase components correspond to the polar coordinates in the complex plane. The choice of representation is usually a matter of convenience.

Regardless of whether $F_{XY}(x, y)$ or $F_{R\Phi}(\rho, \phi)$ characterizes a complex random variable, the interpretation of the joint PDF is a two-dimensional extension of Figure C.2. This concept is illustrated in Figure C.3. If the outcome of an experiment maps to a complex random variable, then we may run many trials of the experiment to construct a histogram of outputs in the complex plane. Numerous trials will yield a histogram that takes the shape of the PDF.

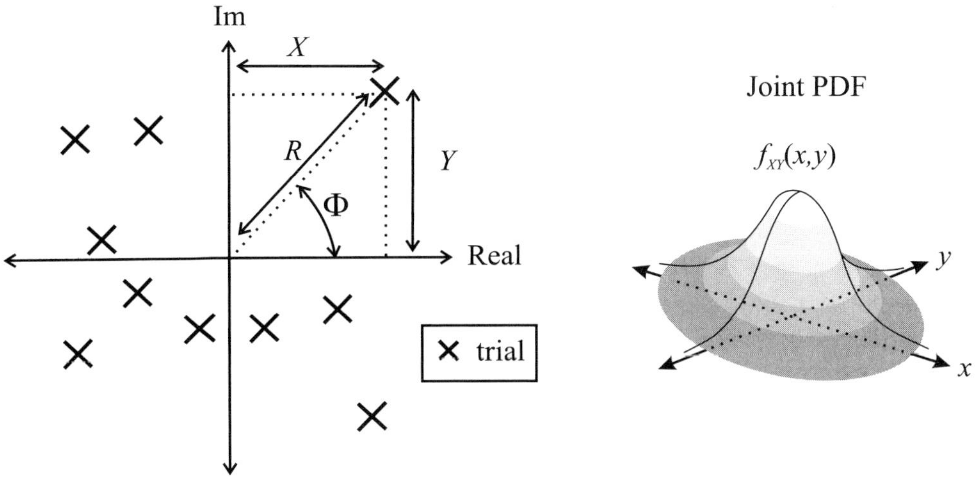

Figure C.3 A joint PDF is used to represent complex random variables whose outcomes are represented in Cartesian (X,Y) or polar (R,Φ) coordinates on the complex plane.

C.2.3 Computing Statistics

A *statistic* is a quantifiable property of a random variable. The most common statistics are calculated directly from the PDF of a random variable. To calculate a statistic, we introduce the concept of an *ensemble average.* The operation of ensemble average, $\mathrm{E}\{(\cdot)\}$, is defined to be

$$\mathrm{E}\{(\cdot)\} = \int_{-\infty}^{+\infty} (\cdot)\, f_X(x)\, dx \qquad \text{(C.2.5)}$$

Equation (C.2.5) averages an arbitrary quantity, $(\cdot)$, over the PDF of a random variable.

The arbitrary quantity in Equation (C.2.5) is usually a function that results in a certain statistic. For the general case, a statistic based on the function $G(x)$ is calculated by theoretically (using ensemble average) or experimentally (using a collection of data drawn without bias from an ensemble). We would write these two statistical operations as

$$\begin{array}{cc} \text{Theoretical Calculation} & \text{Experimental Calculation} \\ \mathrm{E}\{G(X)\} = \int_{-\infty}^{+\infty} G(x)\, f_X(x)\, dx & \left\langle G(X) \right\rangle = \lim_{N\to\infty} \frac{1}{N} \sum_{i=1}^{N} G(X_i) \end{array} \qquad \text{(C.2.6)}$$

where $< \cdot >$ represents the operation of averaging over a set of unbiased data points, $\{X_i\}$. It is a fundamental theorem of probability that, given an infinite set of unbiased data points, the same statistic calculated using either theoretical PDFs or experimental data are identical.

Table C.2 lists the common names, functions, and symbols for calculating statistics from PDFs or data points. Although analytically defined statistics, such as those in Equation (C.2), are the most common in analysis, not all statistics may be defined by a simple $G(x)$. For example, the peak or median values of a random variable are statistics that cannot be defined using Equation (C.2.6).

Table C.2 Common Statistics for a Random Variable, X

Common Symbols	$G(X)$	Description
μ, $\bar{x}$	X	mean, first moment
$\overline{x^2}$	X^2	mean-squared, second moment
$\overline{x^n}$	X^n	nth moment
σ^2, $\overline{(x-\bar{x})^2}$	$(X-\mu)^2$	variance, standard deviation squared
H_X	$X \ln(X)$	entropy

Note: Linearity of Averaging

Ensemble averaging is a linear operator - an important and convenient characteristic. The operation $\mathrm{E}\{(\cdot)\}$ may be taken through integrals, summations, and nonrandom quantities.

C.3 Functions of Random Variables

Often, it is necessary to calculate the PDF of a random variable after a function has acted upon it. For the general case, a function $g(x)$ acts upon a random variable X to produce a new random variable Y with a different PDF. Mathematically, we write this operation as $Y = g(X)$. The complexity of mapping $f_X(x)$ to $f_Y(y)$ depends on the nature of $g(x)$.

C.3.1 Functions With Inverses

The simplest mapping occurs when the function $g(x)$ has a clearly defined inverse, $g^{-1}(x)$, that is also a valid function. Examples of real functions with simple inverses are $g(x) = 6x$, $g(x) = x^3$, or $g(x) = \exp(x)$. Examples of functions that do not have simple inverses are $g(x) = x^2$, $g(x) = |x|$, or $g(x) = \cos(x)$.

Using the simple relationship $x = g^{-1}(y)$, we may equate the probabilities of the following events for monotonically increasing $g(x)$:

$$\Pr[Y \leq y] = \Pr[X \leq g^{-1}(y)]$$

This relationship may be written in terms of the CDFs of X and Y:

$$F_Y(y) = \int_{-\infty}^{g^{-1}(y)} f_X(x)\, dx \tag{C.3.1}$$

Taking the derivative with respect to y on both sides produces a general expression that maps $f_X(x)$ to $f_Y(y)$ for any invertible function $g(x)$:

$$\begin{aligned} f_Y(y) &= \left|\frac{d[g^{-1}(y)]}{dy}\right| f_X(g^{-1}(y)) \\ &= \left|\frac{dy}{dx}\right|^{-1} f_X(g^{-1}(y)) \end{aligned} \tag{C.3.2}$$

Example C.1 shows how to use Equation (C.3.2) to map PDFs through a function $g(x)$ with a simple inverse.

Example C.1: Linear Mapping

Problem: Derive an expression for calculating $f_Y(y)$ from $f_X(x)$ if the two random variables are related by the linear function: $Y = mX + b$.

Solution: First, calculate the following expressions that describe our PDF mapping

$$\underbrace{g(x)}_{y} = mx + b \qquad \underbrace{g^{-1}(y)}_{x} = \frac{y}{m} - \frac{b}{m} \qquad \frac{dy}{dx} = m$$

After this step, the quantities are mechanically inserted into Equation (C.3.2) to produce the final answer:

$$f_Y(y) = \frac{1}{|m|} f_X\left(\frac{y}{m} - \frac{b}{m}\right)$$

C.3.2 Multiroot Functions

If the function $g(x)$ has no simple inverse, we must resort to describing its inverse in terms of multiple roots. If the equation $y = g(x)$ has n roots, $\{x_1, x_2, \cdots x_n\}$, then we may write Equation (C.3.1) as

$$F_Y(y) = \sum_{i=1}^{n} \int_{-\infty}^{x_i} f_X(x)\, dx \quad \text{where } y = g(x_1) = g(x_2) = \cdots = g(x_n) \tag{C.3.3}$$

Taking the derivative of this equation produces the general expression for mapping $f_X(x)$ to $f_Y(y)$:

$$\begin{aligned} f_Y(y) &= \sum_{i=1}^{n} \left|\frac{dx_i}{dy}\right| f_X(x_i) \\ &= \sum_{i=1}^{n} \frac{f_X(x_i)}{|g'(x_i)|} \quad \text{where } g'(x) = \frac{dg(x)}{dx} \end{aligned} \tag{C.3.4}$$

Example C.2 shows how to use Equation (C.3.4) to map PDFs through a function $g(x)$.

Example C.2: Sinusoidal Mapping

Problem: If X is a random variable that follows a uniform distribution over the interval $[0,\pi]$, find the PDF of Y if $Y = \sin(X)$.

Solution: Descriptions of the function, $g(x)$, and the PDF, $f_X(x)$, are shown below:

$$g(x) = \sin(x)$$

$$g'(x) = \cos(x)$$

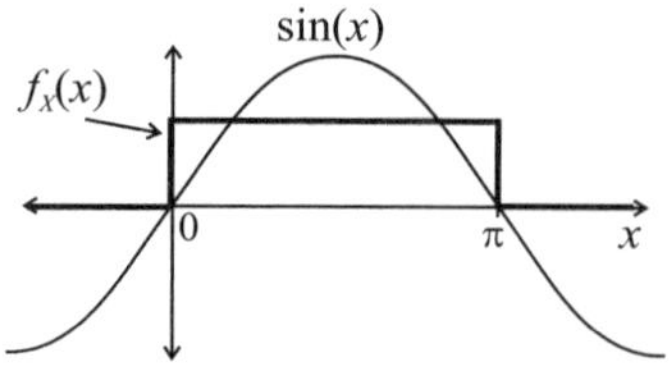

There are an *infinite* number of roots for $-1 \leq y \leq 1$ and no roots outside this range. However, since the PDF for X is nonzero only over the range of $[0,\pi]$, we may summarize the roots of $y = g(x)$ for our analysis as

$$\begin{aligned} y < 0 &\ : \ \text{no roots} \\ 0 \leq y \leq 1 &\ : \ x_1 = \sin^{-1}(y), \quad x_2 = \pi - \sin^{-1}(y) \\ y > 1 &\ : \ \text{no roots} \end{aligned}$$

There are useful roots only over the range $0 \leq y \leq 1$. For this range, we may use Equation (C.3.4) to calculate the PDF for Y:

$$f_Y(y) = \frac{1/\pi}{|\cos(\sin^{-1}(y))|} + \frac{1/\pi}{|\cos(\pi - \sin^{-1}(y))|}$$

which may be simplified using trigonometric identities so that the following full solution results:

$$f_Y(y) = \begin{cases} y < 0 & : \ 0 \\ 0 \leq y \leq 1 & : \ \frac{2}{\pi\sqrt{1-y^2}} \\ y > 1 & : \ 0 \end{cases}$$

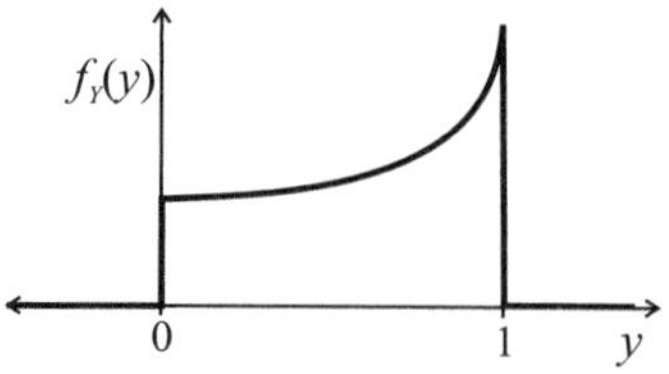

Appendix D

GLOSSARY

D.1 Mathematical Symbols and Conventions

x	real (passband)
X	Fourier transform
x^{-1}	inverse function
dx, ∂x	partial
$\|\|x\|\|$	nearest integer
$\lceil x \rceil$	ceiling (closest integer$\geq x$)
$\lfloor x \rfloor$	floor (closest integer$\leq x$)
$\frac{dx(r)}{dr}$	first derivative
$\frac{d^n x(r)}{dr^n}$	nth derivative
$\dot{x}$	first time derivative
$\tilde{x}$	complex (baseband)
$\arg\{\tilde{x}\}$	argument (phase) of $\tilde{x}$
$\|\tilde{x}\|$	magnitude (envelope)
$\vec{\mathrm{x}}$	vector
$\hat{\mathrm{x}}$	unit vector
$\vec{\mathrm{x}} \cdot \vec{\mathrm{y}}$	scalar product
$\|\vec{\mathrm{x}}\|$	Cartesian length
$\tilde{\vec{\mathrm{x}}}$	complex vector
∇	$= \frac{\partial}{\partial x}\hat{\mathrm{x}} + \frac{\partial}{\partial y}\hat{\mathrm{y}} + \frac{\partial}{\partial z}\hat{\mathrm{z}}$
$\nabla\tilde{x}$	gradient of $\tilde{x}$
$\nabla^2\tilde{x}$	Laplacian $(\nabla \cdot \nabla\tilde{x})$
$\nabla \cdot \tilde{\vec{\mathrm{x}}}$	divergence of $\tilde{\vec{\mathrm{x}}}$
$\nabla \times \tilde{\vec{\mathrm{x}}}$	curl of $\tilde{\vec{\mathrm{x}}}$
$\mathbf{X}$	matrix
$\tilde{\mathbf{X}}$	complex matrix
$\tilde{x}^*$, $\tilde{\mathbf{X}}^*$	complex conjugate
$\tilde{\mathbf{X}}^T$, $\tilde{\vec{\mathrm{x}}}^T$	transpose
$\tilde{\mathbf{X}}^\dagger$	Hermitian
$\tilde{\mathbf{X}}^{-1}$	matrix inverse
$\{x_i\}$	series collection
x_i	series element
x_{ij}	row i, col j element
$F_X(x)$	CDF of X
$f_X(x)$	PDF of X
C_X	autocorrelation of X
$\bar{C}_X$	autocovariance of X
ϱ_X	unit autocovariance of X
S_X	PSD of X
$\bar{x}$, μ_X	mean of X
$\overline{x^n}$	nth moment of X
σ_X^2	variance of X
$x(t) \otimes y(t)$	convolution

D.2 Acronym List

3WDP	Three-Wave with Diffuse Power
ADP	Angle-Delay Profile
AGC	Automatic Gain Control
AM	Amplitude Modulation
ASK	Amplitude Shift Keying
AWGN	Additive White Gaussian Noise
BER	Bit-Error Rate
BPSK	Binary Phase Shift Keying
CDF	Cumulative Density Function
CDMA	Code Division Multiple Access
CIR	Channel Impulse Response
C-SLAC	Correlated phase Stochastic Local Area Channel
DPSK	Differential Phase Shift Keying
DQPSK	Differential Quadrature Phase Shift Keying
EGC	Equal Gain Combining
FM	Frequency Modulation
FSK	Frequency Shift Keying
GMSK	Gaussian Minimum Shift Keying
GSM	Global System for Mobile communication
IC	Integrated Circuit
IQ	In-phase and Quadrature
I-SLAC	Independent phase Stochastic Local Area Channel
ITU	International Telecommunications Union
LAN	Local Area Network
LMDS	Local Multipoint Distribution Service
MIMO	Multiple Input, Multiple Output
MISO	Multiple Input, Single Output
MRC	Maximum Ratio Combining
MSK	Minimum Shift Keying
MTMR	Multiple Transmitter, Multiple Receiver
NCC	Normalized Channel Convention
NII	National Information Infrastructure

PCC	Physical Channel Convention
PCS	Personal Communications Service
PDF	Probability Density Function
PAP	Power Angle Profile
PDP	Power Delay Profile
PLA	Perfect Linear Array
QAM	Quadrature Amplitude Modulation
QPSK	Quadrature Phase Shift Keying
RAM	Random Access Memory
RF	Radio Frequency
RMS	Root Mean Square
SINR	Signal-to-Interference+Noise Ratio
SIMO	Single Input, Multiple Output
SISO	Single Input, Single Output
SLAC	Stochastic Local Area Channel
SNR	Signal-to-Noise Ratio
SS	Spread Spectrum
SSS	Strict-Sense Stationary
SVD	Singular Value Decomposition
TDMA	Time Division Multiple Access
TWDP	Two-Wave with Diffuse Power
UHF	Ultra-High Frequency
U-SLAC	Uncorrelated phase Stochastic Local Area Channel
UWB	Ultra-Wideband
VHF	Very High Frequency
WLAN	Wireless Local Area Network
WSS	Wide-Sense Stationary
WSS-US	Wide-Sense Stationary, Uncorrelated Scattering

BIBLIOGRAPHY

[Abd00] A. Abdi, H. Hashemi, and S. Nader-Esfahani, "On the PDF of the Sum of Random Vectors," *IEEE Transactions on Communications*, vol. **48**, no. 1, pp. 7–12, Jan 2000.

[Abr70] M. Abramowitz and I. A. Stegun, Eds., *Handbook of Mathematical Functions*, Dover Publications, New York, 9th edition, 1970.

[Ala98] S. Alamouti, "Space Block Coding: A Simple Transmitter Diversity Technique for Wireless Communications," *IEEE Journal on Selected Areas of Communications*, vol. **16**, pp. 1451–1458, Oct 1998.

[And00] J. B. Andersen, "Antenna Arrays in Mobile Communications: Gain, Diversity, and Channel Capacity," *IEEE Antennas and Propagation Magazine*, vol. **42**, no. 2, pp. 12–16, April 2000.

[And01] M. R. Andrews, P. P. Mitra, and R. deCarvalho, "Tripling the Capacity of Wireless Communications Using Electromagnetic Polarization," *Nature*, vol. **409**, pp. 316–318, 18 Jan 2001.

[Bel63] P. A. Bello, "Characterization of Random Time-Variant Linear Channels," *IEEE Transactions on Communications Systems*, pp. 360–393, Dec 1963.

[Bel64] P. A. Bello, "Time–Frequency Duality," *IEEE Transactions on Information Theory*, vol. **IT-10**, pp. 18–33, Jan 1964.

[Ben48] W. R. Bennett, "Distribution of the Sum of Randomly Phased Components," *Quarterly Journal of Applied Mathematics*, vol. **5**, pp. 385–393, April 1948.

[Bor80] M. Born and E. Wolf, *Principles of Optics: Electromagnetic Theory of Propagation, Interference, and Diffraction of Light*, Pergamon Press, New York, 6th edition, 1980.

[Bra91] W. R. Braun and U. Dersch, "A Physical Mobile Radio Channel Model," *IEEE Transactions on Vehicular Technology*, vol. **40**, no. 2, pp. 472–482, May 1991.

[Bro97] G. S. Brown, "Rough Surface Scattering (EE 6104): Course Notes," Tech. Rep., Virginia Tech, Fall 1997.

[Bro98] G. S. Brown, "Analytical Electromagnetics (EE 6114): Course Notes," Tech. Rep., Virginia Tech, Fall 1998.

[Chi00] D. Chizhik, M. J. Gans, and R. A. Valenzuela, "Capacities of Multi-Element Transmit and Receive Antennas: Correlations and Keyholes," *IEEE Electronics Letters*, pp. 1099–1100, 22 June 2000.

[Cla68] R. H. Clarke, "A Statistical Theory of Mobile-Radio Reception," *Bell System Technical Journal*, vol. **47**, pp. 957–1000, 1968.

[Cou93] L. W. Couch, *Digital and Analog Communication Systems*, Macmillan, New York, 4th edition, 1993.

[Cou98a] A. J. Coulson, *Characterization of the Mobile Radio Multipath Fading Channel*, Ph.D. thesis, University of Auckland, New Zealand, August 1998.

[Cou98b] A. J. Coulson, A. G. Williamson, and R. G. Vaughan, "A Statistical Basis for Lognormal Shadowing Effects in Multipath Fading Channels," *IEEE Transactions on Communications*, vol. **46**, no. 4, pp. 494–502, April 1998.

[Cov91] T. M. Cover and J. A. Thomas, *The Elements of Information Theory*, John Wiley & Sons, New York, 1991.

[Dev87] D. M. J. Devasirvatham, "A Comparison of Time Delay Spread and Signal Level Measurements Within Two Dissimilar Office Buildings," *IEEE Transactions on Antennas and Propagation*, vol. **AP-35**, no. 3, March 1987.

[Dur98a] G. D. Durgin and T. S. Rappaport, "A Basic Relationship Between Multipath Angular Spread and Narrowband Fading in a Wireless Channel," *IEE Electronics Letters*, vol. **34**, no. 25, pp. 2431–2432, 10 Dec 1998.

[Dur98b] G. D. Durgin, T. S. Rappaport, and H. Xu, "Radio Path Loss and Penetration Loss Measurements in and around Homes and Trees at 5.85 GHz," *IEEE Transactions on Communications*, vol. **46**, no. 11, pp. 1484–1496, Nov 1998.

[Dur99a] G. D. Durgin and T. S. Rappaport, "Effects of Multipath Angular Spread on the Spatial Cross-Correlation of Received Voltage Envelopes," in *49th IEEE Vehicular Technology Conference*, Houston, TX, May 1999, vol. 2, pp. 996–1000.

[Dur99b] G. D. Durgin and T. S. Rappaport, "Three Parameters for Relating Small-Scale Temporal Fading to Multipath Angles-of-Arrival," in *PIMRC '99*, Osaka, Japan, Sep 1999, pp. 1077–1081.

[Dur99c] G. D. Durgin, T. S. Rappaport, and D. A. de Wolf, "More Complete Probability Density Functions for Fading in Mobile Communications," *49th IEEE Vehicular Technology Conference*, vol. **2**, pp. 985–989, May 1999.

[Dur00a] G. D. Durgin and T. S. Rappaport, "Spatial Channel Modeling for Wireless Communications," in *Wireless Communications for the New Millenium*, N. Morinaga, R. Kohno, and S. Sampei, Eds. Kluwer, Boston, 2000.

[Dur00b] G. D. Durgin and T. S. Rappaport, "Theory of Multipath Shape Factors for Small-Scale Fading Wireless Channels," *IEEE Transactions on Antennas and Propagation*, vol. **48**, no. 5, pp. 682–693, May 2000.

[Dur01a] G. D. Durgin, V. Kukshya, and T. S. Rappaport, "Joint Angle and Delay Spread Statistics for 1920 MHz Peer-to-Peer Wireless Channels," in *Antennas and Propagation Symposium*, Boston, July 2001.

[Dur01b] G. D. Durgin, S. Sampei, and N. Morinaga, "Computer Simulation of Multiple Transmitter, Multiple Receiver Wireless Channels," in *YRP Wireless Summit*, Yokosuka, Japan, July 2001.

[Dur01c] G. D. Durgin, S. Sampei, and N. Morinaga, "Design of Multi-Antenna Wireless Systems in Multipath Environments," in *Wireless and Personal Multi-media Communications '01*, Aalborg, Denmark, Sep 2001.

[Dur02] G. D. Durgin, T. S. Rappaport, and D. A. de Wolf, "New Analytical Models and Probability Density Functions for Fading in Wireless Communications," *IEEE Transactions on Communications*, vol. **50**, no. 6, pp. 1005–1015, June 2002.

[Dur03] G. D. Durgin, V. Kukshya, and T. S. Rappaport, "Wideband Measurements of Angle and Delay Dispersion for Outdoor and Indoor Peer-to-Peer Radio Channels at 1920 MHz," *IEEE Transactions on Antennas and Propagation*, May 2003.

[Ebi91] Y. Ebine, T. Takahashi, and Y. Yamada, "A Study of Vertical Space Diversity for a Land Mobile Radio," *Electronics and Communications in Japan*, vol. **74**, no. 10, pp. 68–76, 1991.

[Egg95] P. C. F. Eggers, "Angular Dispersive Mobile Radio Environments Sensed by Highly Directive Base Station Antennas," in *PIMRC '95*, Toronto, Canada, Sep 1995, pp. 522–526.

[Eli93] A. C. Elitzur and L. Vaidman, "Quantum Mechanical Interaction-Free Measurements," *Foundations of Physics*, vol. **23**, no. 7, pp. 987–997, July 1993.

[Ert98] R. B. Ertel, P. Cardieri, K. W. Sowerby, T. S. Rappaport, and J. H. Reed, "Overview of Spatial Channel Models for Antenna Array Communication Systems," *IEEE Personal Communications Magazine*, vol. **5**, no. 1, pp. 10–22, Feb 1998.

[Esp73] R. Esposito and L. R. Wilson, "Statistical Properties of Two Sine Waves in Gaussian Noise," *IEEE Transactions on Information Theory*, vol. **19**, pp. 176–183, March 1973.

[Fle96] B. H. Fleury, "An Uncertainty Principle for WSS Processes and Its Application to WSSUS Systems," *IEEE Transactions on Communications*, vol. **44**, no. 12, pp. 1632–1634, Dec 1996.

[Fle00] B. H. Fleury, "First- and Second-Order Characterization of Direction Dispersion and Space Selectivity in the Radio Channel," *IEEE Transactions on Information Theory*, vol. **46**, no. 6, pp. 2027–2044, Sep 2000.

[Fos96] G. J. Foschini, "Layered Space–Time Architecture for Wireless Communication in a Fading Environment When Using Multi-Element Antennas," *Bell Labs Technical Journal*, pp. 41–59, Autumn 1996.

[Fos98] G. J. Foschini and M. J. Gans, "On Limits of Wireless Communication in a Fading Environment When Using Multiple Antennas," *Wireless Personal Communications*, vol. **6**, no. 3, pp. 311–335, March 1998.

[Gan72] M. J. Gans, "A Power-Spectral Theory of Propagation in the Mobile Radio Environment," *IEEE Transactions on Vehicular Technology*, vol. **VT-21**, no. 1, pp. 27–38, Feb 1972.

[God97] L. C. Godara, "Applications of Antenna Arrays to Mobile Communications, Part I: Performance Improvement, Feasibility, and System Considerations," *Proceedings of the IEEE*, vol. **85**, no. 7, pp. 1031–1060, July 1997.

[Gra94] I. S. Gradshteyn and I. W. Ryzhik, *Table of Integrals, Series, and Products*, Academic Press, New York, 5th edition, 1994.

[Gre62] P. Green Jr., "Radar Astronomy Measurement Techniques," Tech. Rep., MIT Lincoln Lab., Dec 1962.

[Has79] H. Hashemi, "Simulation of the Urban Radio Propagation Channel," *IEEE Transactions on Vehicular Technology*, vol. **VT-28**, no. 3, Aug 1979.

[Has93] H. Hashemi, "The Indoor Radio Propagation Channel," *Proceedings of the IEEE*, vol. **81**, no. 7, pp. 943–968, July 1993.

[Jak71] W. C. Jakes, "A Comparison of Specific Space Diversity Techniques for Reduction of Fast Fading in UHF Mobile Radio Systems," *IEEE Transactions on Vehicular Technology*, vol. **VT-20**, no. 4, pp. 81–91, Nov 1971.

[Jak74] W. C. Jakes, Ed., *Microwave Mobile Communications*, IEEE Press, New York, 1974.

[Lib99] J. C. Liberti and T. S. Rappaport, *Smart Antennas for Wireless CDMA Communications*, Prentice Hall, Upper Saddle River, NJ, 1999.

[Nag00] A. F. Naguib and A. R. Calderbank, "Space-Time Coding for High Data Rate Wireless Communications," in *Wireless Communications for the New Millenium*, N. Morinaga, R. Kohno, and S. Sampei, Eds. Kluwer, Boston, 2000.

[Nak60] M. Nakagami, "The m-Distribution – A General Formula of Intensity Distribution of Rapid Fading," in *Statistical Methods in Radio Wave Propagation*, W. C. Hoffman, Ed. Pergamon, Elmsford, NY, 1960.

[Nic20] J. W. Nicholson, "Generalisation of a Theorem Due to Sonine," *Quarterly Journal of Pure and Applied Mathematics*, vol. **48**, pp. 321–329, 1920.

[Ogi91] J. A. Ogilvy, *Theory of Wave Scattering from Random Rough Surfaces*, Adam Hilger, Philadelphia, 1991.

[Oss64] J. F. Ossanna, "A Model For Mobile Radio Fading Due to Building Reflections: Theoretical and Experimental Fading Waveform Power Spectra," *Bell System Technical Journal*, pp. 2935–2971, Nov 1964.

[Pap91] A. Papoulis, *Probability, Random Variables, and Stochastic Processes*, McGraw Hill, New York, 3rd edition, 1991.

[Par92] J. D. Parsons, *The Mobile Radio Propagation Channel*, John Wiley & Sons, New York, 1992.

[Pat99] N. Patwari, G. D. Durgin, T. S. Rappaport, and R. J. Boyle, "Peer-to-Peer Low Antenna Outdoor Radio Wave Propagation at 1.8 GHz," in *49th IEEE Vehicular Technology Conference*, Houston, TX, May 1999, vol. 1, pp. 371–375.

[Pro89] J. G. Proakis, *Digital Communications*, McGraw Hill, New York, 2nd edition, 1989.

[Rap02a] T. S. Rappaport, *Wireless Communications: Principles and Practice*, Prentice-Hall Inc., Upper Saddle River, NJ, 2nd edition, 2002.

[Rap02b] T. S. Rappaport, A. Annamalai, R. M. Buehrer, and W. H. Tranter, "Wireless Communications: Past Events and a Future Perspective," *IEEE Communications Magazine*, pp. 148–161, May 2002.

[Reu74] D. O. Reudink, "Properties of Mobile Radio Propagation Above 400 MHz," *IEEE Transactions on Vehicular Technology*, Nov 1974.

[Ric44] S. O. Rice, "Mathematical Analysis of Random Noise," *Bell System Technical Journal*, vol. **23**, pp. 282–332, July 1944.

[Ric45] S. O. Rice, "Mathematical Analysis of Random Noise (continued)," *Bell System Technical Journal*, vol. **24**, pp. 46–156, Jan 1945.

[Ric48] S. O. Rice, "Statistical Properties of a Sine Wave Plus Random Noise," *Bell System Technical Journal*, vol. **27**, no. 1, pp. 109–157, Jan 1948.

[Rum86] W. D. Rummler, R. P. Coutts, and M. Liniger, "Multipath Fading Channel Models for Microwave Digital Radio," *IEEE Communications Magazine*, vol. **24**, no. 11, pp. 30–42, Nov 1986.

[Sha48] C. E. Shannon, "A Mathematical Theory of Communications," *Bell Systems Technical Journal*, vol. **27**, pp. 379–423 and 623–656, 1948.

[Siw01] S. Siwamogsatham and M. P. Fitz, "Improved High-Rate Space-Time Trellis Codes via Orthogonality and Set Partitioning," in *Proceedings of WPMC '01*, Aalborg, Denmark, 2001.

[Skl01] B. Sklar, *Digital Communications: Fundamentals and Applications*, Prentice Hall, 2nd edition, 2001.

[Spe00] Q. H. Spencer, B. D. Jeffs, M. A. Jensen, and A. L. Swindlehurst, "Modeling the Statistical Time and Angle of Arrival Characteristics of an Indoor Multipath Channel," *IEEE Journal on Selected Areas of Communications*, vol. **18**, no. 3, pp. 347–360, March 2000.

[Sta94] H. Stark and J. W. Woods, *Probability, Random Processes, and Estimation Theory for Engineers*, Prentice Hall, Upper Saddle River, NJ, 2nd edition, 1994.

[Ste94] R. Steele, *Mobile Radio Communications*, IEEE Press, New York, 1994.

[Tar98] V. Tarokh, N. Seshadri, and A. R. Calderbank, "Space-Time Codes for High Data Rate Wireless Communication: Performance Criterion and Code Construction," *IEEE Transactions on Information Theory*, pp. 744–765, Mar 1998.

[Tur95] A. M. D. Turkmani, A. A. Arowojolu, P. A. Jefford, and C. J. Kellett, "An Experimental Evaluation of the Performance of Two-Branch Space and Polarization Diversity Schemes at 1800 MHz," *IEEE Transactions on Vehicular Technology*, vol. **44**, no. 2, pp. 318–326, May 1995.

[Ung82] G. Ungerboeck, "Channel Coding with Multilevel/Phase Signals," *IEEE Transactions on Information Theory*, vol. **28**, pp. 56–67, Jan 1982.

[Vau00] R. G. Vaughan, "Spaced Directive Antennas for Mobile Communications by the Fourier Transform Method," *IEEE Transactions on Antennas and Propagation*, vol. **48**, no. 7, pp. 1025–1032, July 2000.

[Win98] J. H. Winters, "Smart Antennas for Wireless Systems," *IEEE Personal Communications*, vol. **1**, pp. 23–27, Feb 1998.

[Wu95] Y. Wu and W. Y. Zou, "Orthogonal Frequency Division Multiplexing: A Multicarrier Modulation Scheme," *IEEE Transactions on Consumer Electronics*, vol. **41**, no. 3, pp. 392–399, Aug 1995.

[Xu99] H. Xu, T. S. Rappaport, R. J. Boyle, and J. H. Schaffner, "38 GHz Wideband Point-to-Multipoint Radio Wave Propagation Study for a Campus Environment," in *IEEE 49th Vehicular Technology Conference*, Houston, TX, May 1999.

[Yac00] M. D. Yacoub, "Fading Distributions and Co-Channel Inteference in Wireless Systems," *Radio Science*, vol. **42**, no. 1, pp. 150–158, Feb 2000.

INDEX

B

C

D

S

T

U

V

W